A DELTA-MAN IN YEBU

Occasional Volume of the Egyptologists' Electronic Forum

No. 1

Edited by
A. K. Eyma and **C. J. Bennett**

A Delta-man in Yebu: Occasional Volume of the Egyptologists' Electronic Forum No. 1

Universal Publishers/uPUBLISH.com
USA • 2003

ISBN: 1-58112-564-X

www.uPUBLISH.com/books/eyma_bennett.htm

Table of Contents

Preface

It is with great pleasure that I present to you this volume, produced by the Egyptologists' Electronic Forum (EEF). The EEF is a scholarly e-mail discussion forum for mainstream Egyptology, hosted by Yale University. Since its establishment in 1998, the EEF has become a leading on-line forum for discussion of Egyptological issues.[1] Its subscribers include many well-known Egyptologists, as well as graduate students and interested amateurs. The current volume arose from a suggestion by one of our members, Federico Rocchi, that this new electronic community could also prove to be a vehicle for disseminating ideas and research in a more permanent way. In line with the digital nature of the EEF, the current volume is being printed via new, Internet-based processes (print-on-demand) and is also being released in electronic format (e-book).

The title of this volume comes from a saying found in the *Tale of Sinuhe*, an ancient Egyptian literary text, with which the exiled Egyptian Sinuhe comments on his stay abroad: "As if a Delta-man saw himself in Yebu, a marsh-man in Ta-Seti."[2] In the case of the present title it points at the two media that this volume tries to bridge: the paper and digital worlds. For many contributing Egyptologists this is the first time that they have published a paper in a volume that is produced by an e-community, and that will not appear via traditional printing. In that sense, Egyptology (still very much a traditional paper discipline) is the Delta-man travelling to a distant southern city. But it is also unusual for an e-group to publish something on paper, and in that sense it is EEF that is the stranger in Elephantine (Yebu). Of course we hope and expect that you will not experience Sinuhe's feelings of discomfort while reading this volume, and that you will conclude that Delta and Yebu are after all part of the same Egypt, and that the new medium (the Internet) can make a useful contribution to the old discipline of Egyptology.

The papers in the volume cover a wide spectrum of Egyptological topics and have been grouped according to five broad themes that may be discerned: royalty in ancient Egypt, scarabs and funerary items, archaeology and early Egypt, Egyptology – past, present and future, and ancient Egyptian language, science and religion.

Sincere thanks go to the contributing authors, to my fellow editor Chris Bennett, to David Lorton (for proofreading and editorial advice), and to the following persons for providing suggestions and assistance in a variety of forms: Federico Rocchi, Leslie Bailey, Michael Tilgner, Tamara Siuda, Sarah Parcak, Troy Sagrillo and Michael Schreiber.

Aayko Eyma
ayma@tip.nl
EEF moderator

---

[1] For information about EEF, please see one of the following URLs: http://welcome.to/EEF, http://egyptology.tk, or http://www.netins.net/showcase/ankh/eefmain.html

[2] Sinuhe B43, R65-66. Cf. A. M. Blackman, *Middle-Egyptian Stories*, Bibliotheca Aegyptiaca II, Bruxelles 1972, pp. 17-18; M. Lichtheim, *Ancient Egyptian Literature*, Vol. 1, University of California Press 1975, p. 225. Ta-Seti (*t3-sty*) is rendered "Nubia" by Lichtheim and most other translators. But the parallelism between the two lines would suggest that the term more likely refers to the 1st Upper Egyptian nome, in which Elephantine (Yebu, *3bw*) was located. That is, in both lines the southern-most region of Egypt is set against the most northern part of the country (the marshy lagoons [*ẖ3t*] of the Delta [*idḥw*]). Compare the Turin Coronation Inscription of Horemhab, line 22, which states that this king restored the temples "from the *ẖ3t* of *idḥ*(*w*) to *t3-sty*", i.e. in the whole land; see *BAR* III §31 and Alan Gardiner's "Notes on the Story of Sinuhe", Chapter II, "Comments on the Text Part IV", p. 87, available online at the following URL: http://www.cwru.edu/UL/preserve/Etana/notes_story_sinuhe/notes_story_sinuhe.htm

The Institution of Kingship in Ancient Egypt*

David Lorton

In studying the civilization of ancient Egypt, we are at the mercy of the surviving evidence, which is uneven in its temporal and spatial distributions, and highly uneven in its content. With regard to architecture, to cite only one example, our evidence is dominated by tombs and temples. With the special exception of Amarna, our evidence for domestic architecture and such texts as we have reflecting the lives of the non-elite stem from atypical sites, namely the villages of the workmen who labored on the royal funerary monuments, at Kahun, Deir el-Medina, and now at Giza as well.

When it comes to texts, the preserved record is also highly skewed. Temples and tombs yield texts we call "historical" and "biographical," though these labels are really misnomers.[1] Biographical texts can contain information about events that occurred in the lives of individuals, but these events are always selected so as to portray people achieving success in the service of their community, the state, or their temple, and they often include—and sometimes just boil down to—stock phrases to the effect that the deceased was one "who clothed the naked," "gave bread to the hungry," and so forth. In short, these texts can easily be compared to the genre of eulogies in our own culture: with their assurances of the goodly life the deceased had led, they serve to encode the values that constituted the accepted norms of society and to transmit them to the next generation. Indeed, these texts were displayed in the more "public" parts of tombs; sometimes they were even carved on the façades.

With regard to the historical texts composed on behalf of kings, there are some containing a fair amount of detail, but they constitute exceptions to the rule. For the most part, such allusions as are made to historical events are brief and lacking in the kind of details that provide grist for the mill of the political historian. Fondness for the inclusion of royal epithets increased over time; by the Ramesside period, they constituted nearly all the content of many a text, with the result that allusions to historical events can seem to have been included as hardly more than an afterthought. By way of a loose analogy, the content of royal inscriptions can be compared to that of an American president's State of the Union address: little of the personality of the incumbent emerges from this material, stress rather being placed on his presidential—or ancient Egyptian kingly—qualities and on events that exemplify his successful execution of his office. In his dissertation on the kingship, Nicolas-Christophe Grimal took note of the rather impersonal monarch who seems to emerge from these texts.[2] In the writer's review of Grimal's book, it was countered that a number of kings—particularly of the Middle and New Kingdoms—have vivid enough personalities in the surviving textual record or in their royal portraiture, and that the phenomenon described by Grimal was not that of an impersonal monarch, but rather that of the royal office itself, devoid of its occupant.[3] With the passage of some time and the opportunity for further reflection, the writer would now modify that observation and suggest that we think in terms of two personas that inhabited a king. On the one hand, there was his individual human persona, which was usually not stressed, but whose expression must have been allowable given that it

---

* This is the publication of a paper given at the annual meeting of the Society of Biblical Literature on November 18, 1995. The remarks are intended to complement the excellent treatment of the topic by R. J. Leprohon, "Royal Ideology and State Administration in Pharaonic Egypt," in *Civilizations of the Ancient Near East*, ed. J. M. Sasson, vol. 1 (New York: Charles Scribner's Sons, 1995), 273–287.

[1] On the points made in this and the following paragraph, see also the writer's remarks in "Legal and Social Institutions of Pharaonic Egypt," in *Civilizations of the Ancient Near East*, vol. 1, 346.

[2] N. C. Grimal, *Les Termes de la propagande royale égyptienne de la XIXe dynastie à la conquête d'Alexandre*, Mémoires de l'Académie des Inscriptions de Belles-Lettres 6 (Paris, 1986), 719-721.

[3] D. Lorton, *Chronique d'Égypte* 68 (1993), 100-102.

sometimes does occur, as just noted. And on the other hand, there was what could be called the divine royal persona, a single, immortal, self-regenerating persona that transcended the individual mortal monarch but occupied each individual reigning king. As a result of this belief, while individual kings could and did come and go, there was a sense in which, through the three thousand years of ancient Egypt's history, there was only one king: this divine royal persona who was expressed as "Horus" or as the "living royal *ka*." And it is suggested here that the qualities expressed in the royal epithets studied by Grimal, which he refers to as describing an impersonal monarch and which the writer ascribed to the royal office, are in fact the ideal qualities of that divine persona.

The preceding was a digression that anticipated some of the discussion to follow; it is now time to return to the matter of the nature of the preserved textual evidence. Some topics, such as medicine and astronomy, are represented by relatively small text corpora, and at any given time only a few scholars are to be found at work on them. Law is an important area of potential relevance to those with comparative interest, but again, the documentation is relatively small in quantity; we have a small corpus of royal decrees from the Old Kingdom through the New Kingdom, and while there is a significant number of procès-verbaux of cases at law before the Late Period, nearly all the preserved documentation stems from a single site, the village of Deir el-Medina mentioned earlier, and it is confined to a period of about two and one-half centuries. Most scholars who concentrate on the textual documentation principally study, not surprisingly, the genres represented by the majority of the surviving texts: history, literature, and religion. But it is religion that accounts for the vast majority of all the surviving documentation from ancient Egypt, whether archaeological, architectural, artistic, or textual, and the textual sources cover a wide range of concerns: one is religion in the (for us) more ordinary sense of concern with the divine; another is the funerary literature, of which major corpora have survived; a third is represented by magic; and finally, there is the ideologically-weighted—that is to say, the religiously-weighted—material related to the institution of kingship.

It should be noted that this seemingly odd circumstance need not be attributed to special religious obsessions on the part of the ancient Egyptians, though of course their beliefs and practices were important to them, but stems rather from the topographical peculiarities of the land. The domestic architecture of the town sites in the Nile floodplain was of mud brick, a circumstance not conducive to its survival in any case. Moreover, for the most part, ancient town sites lie under the present water-table, under cultivated land, or under modern urban sites. Many mounds that actually survived in the Delta down to modern times have been attacked by *sebakhin* or even bulldozed away during the last century, as the expanding population of the country led to a need for more land for cultivation. Indeed, had it not been for its unusual and exceptional desert location, Akhenaten's capital city at Amarna would have been lost to us, and with it the Amarna letters, which have proved such an invaluable source of information regarding history, international law, and language in the second millennium. Our principal textual sources, then, stem from the walls of stone temples, erected principally in urban sites, or from the stelae and statues set up in their forecourts, and from those sites which are located in the dry climate of the desert. Amarna and the workmen's villages have already been mentioned, and there have been recent excavations at Ain-Asil, the site of a late Old Kingdom town located in the oasis of Dakhla in the Western Desert. And, of course, there are the royal mortuary temples, and the royal and private tombs, which were either buildings constructed of stone or excavated into the rock cliffs that frame the Nile valley. These tombs are not only sources of scenes and texts carved or painted on their walls. Much of the textual material that has survived to us on wood, leather, and papyrus, including literary, magical, and medical texts, owes its preservation to the fact that it was placed in tombs, or was left behind in the workmen's villages or the royal mortuary temples.

Considerable space has been devoted here to the nature of our sources, because it explains how it is that certain areas receive a great deal of attention in the scholarly literature while others remain relatively ignored. It is religion that accounts for the vast majority of the evidence we have from Egypt. The office of kingship has had religious aspects in many cultures, Egypt's being no exception, and because Egypt's king was divine, there is much talk in the evidence about his divinity and his relationship to the gods and goddesses of the land.

And yet, the king both reigned and ruled. He was the head of state, and in certain periods, he was the head of an empire in Africa and Asia. There must have been political intrigues, and the king must have had elite constituencies to satisfy, but such matters are mostly absent from the histories of Egypt, for the simple reason that they are mostly absent from the surviving evidence. They are not, however, entirely absent. Manetho, who was of course not contemporary with the event, mentioned that king Teti of Dynasty VI was assassinated by his bodyguards.[4] From the same dynasty, there is some evidence to the effect that the rule of Pepi I began with a struggle against a counter-king named Userkare.[5] From Dynasty XII, we know of an assassination attempt against Amenemhet I.[6] A fragmentary stela of his coregent and successor, Sesostris I, mentions civil unrest in Elephantine, though it has to remain uncertain whether this was really some sort of rebellion, in the sense of an attempt to institute a new political order or at least a new regime, or just a riot of major proportions.[7] From a rare source, a fragment of a palace day-book of king Sobekhotep III of Dynasty XIII, we have a similar report of the quelling of unrest in a city.[8] There was an attempt on the life of Ramesses III of Dynasty XX.[9] And finally, we have the report that in Dynasty XXVI Amasis usurped the throne from Apries and allowed him to be killed, though he then had him buried with full royal honors.[10]

There is another telling incident, one that has gone unappreciated in the secondary literature, and the writer must thank Dr. Scott Morschauser, who is planning to write on it, for permitting mention of it here. Ramesses II of Dynasty XIX ruled for sixty-seven years,[11] and if there were any kings with long reigns who seem to have defined what it meant to be an absolute monarch, it would be Amenophis III of the preceding Dynasty XVIII and Ramesses II. But in the first year of his sole reign, Ramesses visited the temple of Karnak to appoint a new high priest. The priests staged oracles, and his choices were rejected, as it was believed, by the god himself. Finally, the god—that is to say, obviously the priests—accepted a man named Nebwenenef.[12] Unfortunately, we shall probably never know the details behind this incident. But Amun of Karnak was the chief god of the country, his priesthood was the richest and most powerful in the land, and we seem to have here, in a report of an incident dating to the outset of this king's reign, an account of a test of wills between the king and the priesthood of Amun, one of the most important of his constituencies.

Ramesses' beginning would thus appear to have been somewhat shaky, and one escapade nearly cost him his life at Kadesh. In his fifth year, Ramesses led his army into Syria-Palestine, and when he

---

[4] W. G. Waddell, trans., *Manetho* (Cambridge, Mass.: Harvard University Press, 1940), 51-53. Waddell follows E. Meyer in identifying the Othoês or Othius of the epitomes, which ascribe the king inconsistently to Dynasty V and Dynasty VI, with Teti; see ibid., 53, n. 2.

[5] H. Goedicke, "Userkare," in *Lexikon der Ägyptologie*, vol. 6 (Wiesbaden: Otto Harrassowitz, 1992), col. 901.

[6] See A. Volten, *Zwei altägyptische politische Schriften,* Analecta Aegyptiaca 4 (Copenhagen, 1945), 110-111. For an English translation, see M. Lichtheim, *Ancient Egyptian Literature,* vol. 1: *The Old and Middle Kingdoms* (Berkeley: University of California Press, 1973), 137.

[7] The text is published and discussed by W. Helck, "Die Weihinschrift Sesostris' I. am Satet-Tempel von Elephantine," *Mitteilungen des Deutschen Archäologischen Instituts Kairo* 34 (1978), 69–78.

[8] See D. B. Redford, *Pharaonic King-Lists, Annals and Day-Books,* SSEA Publication 4 (Mississauga, Ont.: Benben Publications 1986), 108.

[9] Sir Alan Gardiner, *Egypt of the Pharaohs* (New York: Oxford University Press, 1966), 289–292.

[10] Ibid., 361–362.

[11] The figure for the length of the reign of Ramesses II is supplied by the Abydos stela of Ramesses IV; see J. H. Breasted, *Ancient Records of Egypt,* vol. 4 (reprint ed., New York: Russell and Russell Inc., 1962), 228.

[12] K. A. Kitchen, *Ramesside Inscriptions, Historical and Biographical,* vol. 3 (Oxford: B. H. Blackwell, 1980), 283. For a translation and discussion, see K. Sethe, "Die Berufung eines Hohenpriesters des Amon unter Ramses II.," *Zeitschrift für aegyptische Sprache und Altertumskunde* 44 (1907), 30–35. The understanding of the text here follows that of Sethe and of Kitchen, *Pharaoh Triumphant: The Life and Times of Ramesses II, King of Egypt* (Warminster: Aris & Phillips, 1982), 46–47.

was encamped with his vanguard, an attack by Hittite chariotry caused a panic in his army; in the fighting that ensued, Ramesses nearly lost his life. His personal prowess saved him, and he officially claimed a victory . . . but so did the Hittites. There seems to have been a stalemate in this contest between the two great powers, and while Ramesses was able to vaunt his own prowess in his inscriptions, this incident—at least, as reported—was scarcely a credit to his army. And yet, he boasted of it in texts of which we have a number of copies.[13] Perhaps we can say this incident marked the onset of Egypt's decline as a major power in Asia, but we would say it with the advantage of hindsight. What must be noted here is that Ramesses' army failed him, and that in some way we cannot fully fathom today, his commissioning of inscriptions commemorating this fact marks yet another rift between him and an important constituency, the armed forces.

Perhaps there are further cases of this sort in the surviving evidence which have somehow gone unnoticed, but in all likelihood they are few in number. But to the small number of references to specific historical incidents bearing on the practice of kingship we can add two important texts that speak in a sense to the theory of kingship, though not so much to theological concerns as to the practical aspects of the exercise of the office.

One of these texts is the poetic report regarding the battle of Kadesh. There, speeches of Ramesses II define his relationship to his army, as well as his relationship to the god Amun. The speeches have been analyzed by Scott Morschauser.[14] From this analysis, it is clear that each of the two relationships is characterized by a set of mutual obligations, and that in both cases, the obligations are the same. Further, interestingly enough, the mutual obligations that bind the king and his military and his vassals on the one hand, and the king and the god on the other hand, are essentially the same set of obligations as are to be found in the legal and covenantal stipulations of the ancient Near Eastern vassal treaties and the Hebrew Bible. Morschauser's analysis helps us to see the king's status as an intermediary figure between other mortals and the divine, and it also shows us that the Egyptian king was not a despot who ruled arbitrarily, but rather a monarch whose claim to the loyalty and support of his followers was inextricably related to the obligations to them which he, for his own part, was expected to fulfill.

The second source is a didactic text called the Instruction for Merikare.[15] The instruction is supposedly the product of a king of the Heracleopolitan Period for his successor, though there has been some disagreement as to whether that was actually the case, or whether it was a work of the Middle Kingdom, fictitiously set in the past.[16] Our earliest copies of the text, however, can be dated paleographically to Dynasty XVIII, and one of them has been dated by Georges Posener to the reign of Amenophis II or Thutmosis IV.[17] Comparing the text to the concerns with the king's divinity to be found in many a royal inscription, Ramses Moftah has noted the very much this-worldly focus of the text.[18] Its badly broken opening section has been studied by Philippe Derchain,[19] who has been able to show that it advises the king to be a teacher to his followers, and this is followed by practical advice to advance his officials. Next, the king is admonished to do justice and to not oppress the widow or

[13] For English translations of the texts, see M. Lichtheim, *Ancient Egyptian Literature,* vol. 2: *The New Kingdom* (Berkeley: University of California Press, 1976), 60–72. See also the important study of the texts by A. H. Gardiner, *The Kadesh Inscriptions of Ramesses II* (Oxford: Oxford University Press, 1960).

[14] S. Morschauser, "Observations on the Speeches of Ramesses II in the Literary Record of the Battle of Kadesh," in *Perspectives on the Battle of Kadesh,* ed. H. Goedicke, (Baltimore: Halgo, 1985), 123–206.

[15] For an English translation, see Lichtheim, *Ancient Egyptian Literature,* vol. 1, 99–108.

[16] Perhaps the first to argue for a Dynasty XII date was G. Björkman, "Egyptology and Historical Method," *Orientalia Suecana* 13 (1964), 32. The earlier dating has been defended by E. Blumenthal, "Die Lehre für König Merikare," *Zeitschrift für aegyptische Sprache und Altertumskunde* 107 (1980), 40–41.

[17] See G. Posener, "Philologie et archéologie égyptiennes," *Annuaire du Collège de France* 62 (1962), 291–293.

[18] See R. Moftah, *Studien zum ägyptischen Königsdogma im Neuen Reich,* Deutsches Archäologisches Institut Kairo Sonderschrift 20 (Mainz: Philipp von Zabern, 1985), 78-79.

[19] P. Derchain, "Éloquence et politique: L'Opinion d'Akhtoy," *Revue d'Égyptologie* 40 (1989), 37–47.

punish wrongfully, and the like, and he is reminded of the Judgment of the Dead, to which even he will be subject. Then, the king is advised to promote individuals from a social group called the *djamu,* who are a source of soldiers, and to reward them with goods and fields, and to promote capable people regardless of their social class. What follows seems somewhat mixed in content: religious piety is advised, along with good relations with the south of the country in this period of political division. Advice not to usurp the monuments of previous kings is followed by a description of conditions in the Delta and a famous passage disparaging Asiatics. Good conditions in the north of the country are mentioned again, along with a praise of the office of kingship, and then an admission of failure when the desecration of a cemetery was followed by a military defeat. The omniscience of the divine is invoked, along with yet another enjoinder to piety. Finally, there is a section with religious content, including the following important passage regarding the creator god's beneficence towards humankind: "Having constructed a shrine behind them, when they weep, he hears. It is for them that he made rulers in the egg, a lifter to lift from the back of the weak of arm. It is for them that he made Magic (*hekau*) as weapons, to ward off what might happen."[20] As the writer has suggested elsewhere,[21] this list is more meaningful to our modern sensibilities if we read it in reverse order. First, the god has made it possible for humans to help themselves in the face of adversity with Magic. Second, the institution of kingship, and the bureaucratic organization and maintenance of social order implied by its mention, makes life easier even for the relatively helpless, as was noted earlier in the text. And finally, in cases of suffering that governmental institutions cannot address—life-threatening illness is an example that comes easily to mind—the god will hear, and presumably respond to, people's weeping. It should be clear enough that although religious matters are duly noted in this text, it is nevertheless, as Moftah maintains, overwhelmingly this-worldly in its orientation. And in this last section, it is particularly striking that although the kingship is presented as a remedy for human suffering, it is noted as *only one* of three such remedies.

Thus far, mention has been made of two single sources, the poem on the battle of Kadesh and the Instructions for Merikare. In addition to these individual texts, we have a group of sources—namely, the inscriptions and representations from the tombs of the royal officials at Amarna—created within the space of no more than a dozen years and containing a wealth of material that can be studied synchronically. In a paper that appeared in 1993,[22] the writer pointed to a remarkable coincidence that certainly seems to be more than *just* a coincidence, given that our earliest copies of the Instruction for Merikare date to Dynasty XVIII. In Merikare, as just outlined, there is a stress on the king as teacher, on the loyalty the king can promote by rewarding his followers,[23] and on the promotion of commoners, including the social group called *djamu*; this combination of themes is unique to this text in the hieratic literature. But these same themes are stressed in the tombs at Amarna, and their combination there is also unique in the private tombs of Egypt. There is clearly a relationship between the two, though whether it is a direct or an indirect one could be a subject of disagreement. But more important in the present context is the fact that, while considerable stress is placed on the divinity of the king in the Amarna tombs, there is also a stress on these practical aspects of kingship.

In 1979, in the context of a review article,[24] the writer suggested an approach to the kingship employing a model derived from the work of Fritz Kern on early medieval kingship.[25] This interest in

[20] Merikare, P 134–138.

[21] D. Lorton, "God's Beneficent Creation: Coffin Texts Spell 1130, the Instructions for Merikare, and the Great Hymn to the Aten," *Studien zur Altägyptischen Kultur* 20 (1993), 142.

[22] D. Lorton, "The Instruction for Merikare and Amarna Ideology," *Göttinger Miszellen* 134 (1993), 69–83.

[23] The theme of loyalism in Middle Kingdom sources and at Amarna has been studied by J. Assmann, "Die 'Loyalistische Lehre' Echnatons," *Studien zur Altägyptischen Kultur* 8 (1980), 1–32. On the phenomenon of loyalism earlier in Dynasty XVIII, see also the writer's remarks in *Göttinger Miszellen* 134 (1993), 69–70.

[24] D. Lorton, "Towards a Constitutional Approach to Ancient Egyptian Kingship," *JAOS* 99 (1979), 460–465.

[25] S. B. Chrimes, trans., *Kingship and Law in the Middle Ages* (New York and Evanston, 1970).

Kern's work did not derive from the specifics of his study—one cannot, of course, read medieval Europe back into ancient Egypt—but rather from the categories into which he divided his study, for these clearly represent productive questions that one could ask of any kingship. Kern's categories are (1) the divine sanction of the monarchical principle, (2) the individual's right to the throne, (3) the consecration of the king, (4) the king and the law, and (5) the right to resistance. It should be clear enough that this list certainly takes religious matters into account, but I tried to show how it can be used to take a more holistic approach to the institution, one that integrates both the theory and the practice of kingship in ancient Egypt.

Curiously enough, in his dissertation on the kingship, Grimal objected to this approach, though he conceded the point that we sometimes do catch a glimpse of a "humanity of the king far removed from dogma."[26] But it must be insisted that the study of the kingship is more than the study of its dogma. An effort such as Grimal's dissertation, while it is both necessary and invaluable, is nevertheless inevitably one-dimensional. To focus exclusively on royal epithets and descriptive phrases, and their theological information regarding the king's divinity, without taking advantage of information such as what we are told in the Instruction for Merikare, or what we find in the tombs at Amarna, or the incidents from the reign of Ramesses II noted above, results in a one-sided and potentially misleading picture of what the ancient Egyptians thought about the institution of kingship and its holders. In this connection, it can also be noted that while the much later Demotic Chronicle might bear traces of foreign influence in its negative evaluations of the reigns of certain monarchs, its approach has early antecedents in the Instruction for Merikare, as noted, and in the Restoration Stela of Tutankhamun.[27]

Turning to the issue of the king's divinity, the notion of an individual who was a mortal, and yet somehow also divine, seems sufficiently removed from Western experience that it has proved to be a dilemma for Egyptologists. Perhaps the most extreme formulation of what the divinity of the king might have meant to the ancient Egyptians was made by Henri Frankfort in his book *Kingship and the Gods,*[28] and in the wake of Posener's critique of Frankfort's approach as long ago as 1960,[29] we have remained without a comprehensive, magisterial treatment of the topic. There was, however, an important advance in 1985, when Lanny Bell published a study of what the Egyptians called the "living royal *ka*" at the temple of Luxor in the time of Amenophis III, where, in his opinion, the king experienced an annual ritual regeneration that entailed this "living royal *ka.*"[30] The earliest evidence for just what this entity might be stems from the Late Period, when a hymn from the reign of the Persian emperor Darius informs us that the "living royal *ka*" was a hypostasis (the Egyptian word is *ba*) of Amun-Re, the solar creator god,[31] but the earlier material studied by Bell seems consistent with this information.

In the course of recent (and as yet unpublished) work on the texts from the tombs at Amarna, the present writer came upon another clue to what constituted the divine aspect of the king, one that is expressed by the term *hau,* which is typically translated, according to context, as "flesh," "limbs," or "body." At Amarna, apparently in the course of a ritual performed at sunrise every morning, when the king offered a Maat symbol and prayed, the king was said to be "embraced" by the rays of the sun, and

---

[26] Grimal, *Termes de la Propagande,* 288, n. 911.

[27] On the Demotic Chronicle, see J. H. Johnson, "The Demotic Chronicle as an Historical Source," *Enchoria* 4 (1974), 1–17. For an English translation of the text of Tutankhamun, see J. A. Wilson, in *Ancient Near Eastern Texts Relating to the Old Testament*, ed. J. B. Pritchard, 2d ed. (Princeton: Princeton University Press, 1955), 251–252. The texts express the theme of reversals of fortune as stemming from offenses against the gods.

[28] H. Frankfort, *Kingship and the Gods: A Study of Ancient Near Eastern Religion as the Integration of Society and Nature* (Chicago: University of Chicago Press, 1948).

[29] G. Posener, *De la divinité du pharaon,* Cahiers de la Société Asiatique 15 (Paris: Imprimerie Nationale, 1960).

[30] L. Bell, "Luxor Temple and the Cult of the Royal Ka," *Journal of Near Eastern Studies* 44 (1985), 251–294.

[31] See the writer's treatment in "The Invocation Hymn at the Temple of Hibis," *Studien zur altägyptische Kultur* 21 (1994), 168 and 184–185.

as a result of this embrace, he "emerged" (*peri*) from the god Aten. This emergence from Aten, which is also described as the god "bearing" (*mesi*) the king, is further explained as an "emergence" from the "rays" (*setut*) of the sun god, from his "body" (*khet*), and from his "limbs" or "flesh" (*hau*). But this term *hau,* which in the Middle Kingdom is written as *ha,* can be traced back to the reign of Sesostris I of Dynasty XII, when it occurs in the White Chapel at Karnak in statements regarding the king's relationship to the sun god Re and to the Ennead. A single, important example can suffice here: the god Atum says to the king, "It is by his making you into a single *ha* with himself that your father Re made your great rank of King of Upper and Lower Egypt."[32] A rendering along the lines of "It is by his making you into a single flesh with himself. . ." seems at first sight to yield no real sense, and the writer would suggest that, if only by way of a heuristic bridge, we render *ha* as "substance": "It is by making you into a single substance with himself that your father Re made your great rank of King of Upper and Lower Egypt." We are not told just how this is done, but the significance lies in *what* was done: in some manner (at Amarna, this was done when the rays of the sun "embraced" the king), the sun god Re effected a change in the human individual that made him into a "single substance" with himself, thereby conferring on him his "rank" of "King of Upper and Lower Egypt." The significance of this information becomes clearer when we realize that it helps us to understand an otherwise puzzling statement at the beginning of the Story of Sinuhe, a work probably authored in the reign of the same king, Sesostris I. There, the date of the death of Amenemhet I, Sesostris' father and coregent, is reported, and we are then told, "The god entered his horizon, the King of Upper and Lower Egypt, Sehetepibre, he ascended to the sky, he being joined with the sun disk (*itn,* "Aten"), the divine *ha* being merged with the one who made him (or possibly, "it")."[33] This passage cannot refer to the funeral, which would have taken place later, after a lengthy mummification period, nor can the "divine *ha*" that ascends to the sky be the king's corpse, which of course remained on earth and was buried. The passage makes real sense only when understood in the light of the information in the passage from the White Chapel cited earlier: it is a special "substance" created by the sun or the sun god, one that made Amenemhet divine (a "god") and "King of Upper and Lower Egypt," that left him at his death and returned to his maker in the sky. The Story of Sinuhe is a literary text, but what we are told here clearly reflects or became official royal ideology, for we find it paraphrased in nonliterary sources in Dynasty XVIII in reporting the death of Amenophis I,[34] and repeated almost verbatim in reporting the death of Thutmosis III.[35]

Returning to the texts at Amarna, there is material there which serves to provide a link between the religious wording of royal ideology and the more practical side of kingship as we find it in the Instruction for Merikare. The texts employ a vocabulary of creation, one used in reference to both the god and the king. Three terms employed in describing creation (including the god's creation of the world) are used of the king as creator of his followers: *iri* "to make," *qed* "to build," and *sekheper* "to create," and the names of what were divinities in the traditional religion were applied to the king: Hapy ("Inundation"), Shay ("Fate"), and Renenet ("Bounty"). Most often, such matters are mentioned in passing in the hymns at Amarna, but certain passages supply contexts that serve to explain what these terms meant to the ancients. Thus, for example, we find Akhenaten addressed as "my lord, who builds like Aten, plentiful of goods, Inundation surging daily, who sustains Egypt <with> silver and gold like the sands of the shores."[36] Or again, the king is called, "my god, who built me, the one who ordained good things for me, the one who created me, who gave me income and provided my needs with his *ka,*

[32] P. Lacau and H. Chevrier, *Une Chapelle de Sésostris Ier à Karnak* (Cairo: Institut Français d'Archéologie Orientale, 1956), 124, no. 344.

[33] R. Koch, *Die Erzählung des Sinuhe*, Bibliotheca Aegyptiaca 17 (Brussels, 1990), 4.1–5.7.

[34] *Urk.* IV, 54.15–17.

[35] Ibid., 896.1–3.

[36] *p3y.i nb ḳd mi itn ꜥš3 ḫt ḥꜥpy ḥr ḥw rꜥ nb sꜥnḫ Kmt <m> ḥḏ nbw mi šꜥw n wḏbw*, M. Sandman, *Texts from the Time of Akhenaten,* Bibliotheca Aegyptiaca 8 (Brussels, 1938), 91.12–13 (tomb of Ay).

the ruler, who made me a person."[37] A third example will suffice here: in yet another hymn, reference is made to "every official whom pharaoh, l. p. h., builds in the entire land, to whom he gives silver, gold, clothing, goods, and cattle every year."[38] From these passages, it should be clear enough that in calling Akhenaten a creator and an Inundation these texts identify the king as the promoter and rewarder of his loyal officials. To us today, such expressions might seem cynical in the extreme, but we must recall that they are part of a larger, rich theological expression in which the king's ability to do these things stems from a *homoousia,* a consubstantiality, of him and the solar creator god, as noted earlier.

As the study of the material from the tombs at Amarna progresses, it is becoming increasingly clear that there was little new under the sun there.[39] The theme of the king as creator of his followers can be traced back to the Middle Kingdom, when the verb *sekheper* could be used of the king appointing his officials, as when one of these is called "(the one) whom the Horus, lord of the palace, created,"[40] while another official is said to have been "created as sole companion."[41] It can be added that the king as rewarder and promoter is closely tied to another theme, which we call "loyalism," and in an important study, Jan Assmann has shown a close connection between expressions of this theme in the Middle Kingdom and at Amarna.[42]

Turning to the question of potential directions for future research, the following can be suggested. Fortunately, our need for compendia of relevant information in the form of royal titles, epithets, and descriptive phrases is mostly satisfied. Hans Goedicke's collection of Old Kingdom material from outside the Pyramid Texts bearing on the kingship has long been available,[43] as has been that of Elke Blumenthal on the First Intermediate Period and the Middle Kingdom.[44] More recently, Nicolas-Christoph Grimal's monumental publication of epithets and descriptive phrases from the Ramesside and later periods has nearly completed this indispensable first step.[45] Curiously enough, this leaves us without a reference work of this sort for the Second Intermediate Period and Dynasty XVIII,[46] and it can only be hoped that this gap will be filled in the relatively near future.

An understanding of the ideology of kingship does not automatically flow, however, from the act of preparing such compendia. Material regarding the divine figure Kamutef ("Bull-of-his-mother," a phrase that means what it looks like it means), a probably related text dealing with the ongoing self-reengendering of the divine royal persona (see above) as Horus, attested in four copies from Dynasty XIII to the Ptolemaic period,[47] Osiris' posthumous engendering of Horus (these last two themes have

---

[37] *p3y.i nṯr ḳd (w)i p3 š3 n.i nfrw p3 sḫpr (w)i di n.i ʿḳw ir ẖrt.i m k3.f p3 ḥk3 ir wi m rmṯ* , ibid., 24.1–2 (tomb of Panehesy).

[38] *sr nb ḳd pr-ʿ3 ʿ.w.s. m t3 r-ḏr.f di n.f ḥḏ nbw ḥbs ḥnw ng3w(?) tnw rnpt*, ibid., 79.4–6 (tomb of Tutu). For the reading of the first word of the passage as *sr,* cf. ibid., 55.17 (tomb of Rames).

[39] See D. B. Redford, "The Sun-disc in Akhenaten's Program: Its Worship and Antecedents, I," *Journal of the American Research Center in Egypt* 13 (1976), 47–53; J. Assmann, *Studien zur altägyptischen Kultur* 8 (1980), 1–32; and the present writer's remarks in *Studien zur altägyptischen Kultur* 20 (1993), 135–137 and *Göttinger Miszellen* 134 (1993), 69–83.

[40] E. Blumenthal, *Untersuchungen zum ägyptischen Königtum des Mittleren Reiches*, vol. 1: *Die Phraseologie*, Abhandlungen der Sächsishen Akademie der Wissenschaften zu Leipzig, Philosophisch-historische Klasse 61/1 (Berlin, 1970), 286.

[41] Ibid., 371. For further occurrences, see ibid., 285–286 and 322.

[42] See the study cited in note 23 above. Expressions of loyalism can in fact be traced back to the Old Kingdom; see *Urk.* I 195.6–9.

[43] H. Goedicke, *Die Stellung des Königs im Alten Reich,* Ägyptologische Abhandlungen 2 (Wiesbaden: Otto Harrassowitz, 1960).

[44] See note 40 above.

[45] See note 2 above.

[46] While a fair amount of representative material is to be found in R. Moftah, *Studien zum ägyptischen Königsdogma,* this work was not intended to be a compendium along the lines of those cited just above.

[47] The text has been studied by H. O. Lange, *Ein liturgisches Lied an Min,* Sitzungsberichte der Preussischen Akademie der Wissenschaften 28 (1927), 331–338, citing examples from a private stela of Dynasty XIII date, the temple of Seti I at Abydos (Dynasty XIX), and the temple of Horus at Edfu. To these examples should be added the one in N. de G. Davies,

to do with the "divine royal persona" mentioned above), as well as the divine birth legend making the reigning monarch the child of the sun god, all represent controlling concepts that must be taken into account in a full explication of Egyptian thought regarding the kingship. A first attempt along these lines was in fact made by Helmuth Jacobsohn, in his monograph *Die dogmatische Stellung des Königs in der Theologie der alten Ägypter.*[48] Unfortunately, his treatment was too brief and his arguments insufficiently cogent, and his work has been essentially ignored. The writer believes that Jacobsohn's basic insights were correct, though the material he reviewed must be dealt with in greater depth, as it has been in part by Lanny Bell in his work on the "living royal *ka*" with respect to the temple of Luxor[49] and by Dieter Müller in his investigation of the sexual aspect of the concept *ib* "heart" in ancient Egyptian.[50]

Past attempts at syntheses on the topic of the kingship have tended to draw on evidence from various periods. Conservatism in Egyptian religious belief can be cited to justify such an approach, though the obvious need to take into account the likelihood of developments over time can also be cited in objection to it. The publication of compendia of material from specific periods of Egyptian history reflects an awareness of this issue. But approaching the matter from a slightly different perspective, given the highly uneven temporal distribution of the evidence, a good case could be made for the study of the ideology of kingship from an extremely synchronic point of view, that is, in specific reigns from which we have a significant amount of evidence. Good candidates for such study would be the reign of Sesostris I, including the information from his White Chapel at Karnak, that of Hatshepsut and her funerary monument at Deir el-Bahri, that of Amenophis III and the information from the temple of Luxor, that of Akhenaten and the information from the boundary stelae and the private tombs at Amarna, that of Ramesses II, including the cults of the royal colossi, and that of Ramesses III, with the information from his monument at Medinet Habu.

And finally, it cannot be forgotten that the institution of kingship was a practical one in ancient Egypt, for as noted, the king both reigned and ruled. While work on the religious aspects of the kingship represents a front on which there is still much progress to be made, it has proceeded at the expense of the near total neglect of the exercise of the office, with the result that there can even seem to be an unbridgeable gap between the theory and the practice of kingship. As indicated, however, there is evidence bearing on the practical side of the royal office, as well as evidence bearing on how this evident gap was bridged in the minds of the ancient Egyptians. The quantity of religiously weighted information on the kingship makes this aspect of it an understandable preoccupation, but in the longer run, we must strive to make this institution an object not just of intellectual history, but of social history as well.[51]

Dr. David Lorton
Baltimore, MD, U.S.A.

The Temple of Hibis in el Khargeh Oasis, Metropolitan Museum of Art Egyptian Expedition 17 (New York, 1953), vol. 3, pl. 32, upper left; most of this last version has been translated by E. Cruz-Uribe, *Hibis Temple Project,* vol. 1 (San Antonio: Van Siclen Books, 1991), 124–125. For a partial translation, see also Frankfort, *Kingship and the Gods,* 189.

[48] H. Jacobsohn, *Die dogmatische Stellung des Königs in der Theologie der alten Ägypter*, Ägyptologische Forschungen 8 (Glückstadt, 1955). Jacobsohn's approach was in part anticipated by Frankfort; see *Kingship and the Gods*, 180.

[49] See note 30 above.

[50] D. Müller, "Die Zeugung durch das Herz in Religion und Medizin der Ägypter," *Orientalia*, N.S. 35 (1967), 247–274.

[51] The above remark is not intended to suggest that intellectual and social-historical approaches are inherently unrelated, as should be clear enough from the theological explication of Akhenaten's relationship to his followers at Amarna. The writer's earlier contribution, cited in note 21 above, was an initial attempt at a more holistic approach; in this vein, see also A. J. Spalinger, "The Concept of the Monarch During the Saite Epoch: An Essay of Synthesis," *Orientalia* N.S. 47 (1978), 12–36. On the potential for integrating intellectual and social approaches, see also the writer's remarks in *Studien zur Altägyptische Kultur* 20 (1993), 149–154.

Aspects of the Deification of Some Old Kingdom Kings

Amr Aly Aly Gaber

The present article catalogues a corpus of different documents dealing with the deification of some Old Kingdom kings. The term "deification" is here used for posthumous attestations of these kings in a divine role. The documents not only testify to the persistence (or re-emergence) of a popular cult for a particular deceased Old Kingdom king, but their analysis may help to throw light on the different ways that were employed to reflect the deification of these kings. It also shows the need to reconsider some of those ways. The study covers seven kings.

I) King Sneferu:

1) Monuments found at *Dahshur* showing king Sneferu invoked in the *hotepdinesu* formula:

a) Three hieroglyphic lines on the front part of a limestone statuette of a person called *Khenty-ka* read[1]:

ḥtp-di-nsw Snfr-w(i) (Ḥr) nbw di.f prt-ḫrw t ḥnḳt k3(w) 3pd(w) šs mnḫt (n k3 n) ḥry-tp ḫry-ḥbt Ḫnty-k3

"An offering which the king gives [to] Sneferu, the golden (Horus), that he may give invocation-offerings [consisting of] bread, beer, oxen, fowl, alabaster, and clothing (to the *ka* of) the chief lector priest, *Khenty-ka*."

b) On the right side of a limestone group-statue, probably representing a husband and wife, an inscription reads[2]:

[Snfr-w(i)] ḥtp-di-(nsw) di.(f) prt-ḫrw t ḥnḳt k3(w) 3pd(w) šs mnḫt (n k3) n Sp-n-mwt

"An offering which (Sneferu) gives, that he may give invocation-offerings [consisting of] bread, beer, oxen, fowl, alabaster, and clothing (to the *ka*) of *Sepenmut*."

c) A limestone cylindrical altar, dating to the Middle Kingdom, was found in the funerary temple of the Bent Pyramid of king Sneferu. It is 64 cm. high, the diameter of its top is 22.5 cm., and the upper part has a cylindrical depression with a diameter of 9 cm. and a depth of 4 cm. An inscription which goes around the top of the altar reads[3]:

Ḥr Nb-m3ʿt ḥtp-di-nsw Ptḥ Snfr-w(i)

The eight columns covering the body of the altar read from left to right[4]:

(di.sn) n k3 n Snfr-w(i)-snb ḫ3 t ḥnḳt n k3 n mty s3w Snfr-w(i) ḥtp-ḏf3w n k3 n wʿb Snfr-w(i)-šri ṯ3w nḏm n ʿnḫ n k3 n wʿb ʿ3 Snfr-(wi) m3ʿ-ḫrw ḫ3 k3(w) 3pd(w)

---

[1] A. Fakhry, *The Monuments of Sneferu at Dahshur*, vol. II, *The Valley Temple,* part II - *The Finds,* Cairo (1961), p. 38, no. 70, fig. 337.
[2] *Ibid.*, p. 20, no. 7, fig. 296.
[3] A. Fakhry, *The Monuments of Sneferu at Dahshur*, vol. I, *The Bent Pyramid,* Cairo (1959), p. 86, fig. 52.
[4] *Ibid.*, p. 87.

"Horus Nebmaat, an offering which the king gives [to] Ptah and [to] *Sneferu,* (that they may give) to the *ka* of *Sneferu-seneb* a thousand of bread and beer, and to the *ka* of the controller of the phyles *Sneferu* an offering of food, and to the *wab*-priest *Sneferu-sheri* the sweet breath of life, and to the *ka* of the chief *wab*-priest *Sneferu,* justified, a thousand of oxen and fowl."

d) On the front side of a limestone torch-shrine of the chief-sculptor *Seshenu,* which dates to the Middle Kingdom, an inscription reads[5]:

ḥtp-di-(nsw) Ḥr Nb-mꜣꜥt nswt-bity Snfr-w(i)

"(An offering which the king gives) [to] Horus Nebmaat, the King of Upper and Lower Egypt, Sneferu."

e) An inscription on the lap of a fragment of a Middle Kingdom limestone statuette of a person called *Ameny* reads[6]:

ḥtp-di-nsw snfr-(wi) Ḥr nbw di.f prt-ḫrw t ḥnḳt ... n ...Ỉmny-Snfr-w(i)

"An offering which the king gives [to] Snefer(u), the golden Horus, that he may give invocation-offerings [consisting of] bread and beer [to the *ka*] of [...] *Ameny-Sneferu.*"

f) The upper surface of the kilt of a fragment of a Middle Kingdom limestone statue bears the following text[7]:

ḥtp-di-nsw Ptḥ-Skr-Wsir nb ꜥnḫ S[nfr]-w(i) Ḥr Nb-mꜣꜥt prt-ḫrw t ḥnḳt kꜣ(w) ꜣpd(w) šs mnḫt ... n kꜣ n

"An offering which the king gives [to] Ptah-Sokar-Osiris, the lord of life, and [to] S(nefer)u, Horus Nebmaat, [that they may give] invocation-offerings [consisting of] bread, beer, oxen, fowl, alabaster, and clothing [...] to the *ka* of [...]."

2) Monuments found in the *Sinai* showing different aspects of the deification of king Sneferu:

a) A tablet (40 cm. x 21 cm.) found at *Wadi-Maghara* has the following inscription, dated to the reign of Amenemhet III[8]:

ḥꜣt-sp 2 ḫr ḥm n (nswt-bity) N(y)-mꜣꜥt-Rꜥ di ꜥnḫ ḏt ḥtp-di-nsw Ḏhwty nb ... t ḥnḳt wꜥbt n Ỉti rpꜥt n nswt-bity Snfr-w(i) ... nṯrw n kꜣ n šd(i)-wḥꜥt Ḫnty-ḫty

"Year 2 under the majesty of the King of Upper and Lower Egypt, Nimaatre, given life eternally. An offering which the king gives [to] Thot, the lord of the [...], [...] pure bread and beer to *Iti*, the

[5] On the same surface a cartouche with only a *ka* sign appears. A. Fakhry was inclined to believe that the cartouche belonged to king Amenemhet II and not to king Senusert I, as the person who owned the shrine lived during the reign of the first king; see A. Fakhry, *The Monuments of Sneferu at Dahshur*, vol. II, part II, pp. 63-65, pls. LXVIII, LXIX.
[6] *Ibid.*, p. 31, no. 43.
[7] *Ibid.*, p. 28, no. 32.
[8] JE 38572; *PM* VII, p. 342; A. H. Gardiner and T. E. Peet, *The Inscriptions of Sinai*, I, EES, London (1952), pl. XI 24.

hereditary noble of the King of Upper and Lower Egypt, Sneferu, […] gods, to the *ka* of the remover of the scorpions[9], *Khenty-khety.*"

b) A rock inscription dated to king Amenemhet IV's reign which once existed at *Wadi Maghara*, now destroyed, had the following text[10]:

ḥ3t-sp 6 ḫr ḥm n nswt-bity n N(y)-m3ʿt-(ḫrw)-Rʿ di ʿnḫ ḏt mry Spd nb i3btt mry Ḥwt-Ḥr nbt Mfk3t mry Snfr-w(i)

"Year six under the Majesty of the King of Upper and Lower Egypt, Nimaat(kheru)re, beloved of Soped, lord of the east, beloved of Hathor, lady of the Turquoise-country, beloved of Sneferu."[11].

c) On a round-topped stele (45 cm. x 27 cm.) made of red sandstone, king Amenemhet III is mentioned as[12]:

mry nswt-bity Snfr-(wi) m3ʿ-ḫrw

"Beloved of the King of Upper and Lower Egypt, Snefer(u), justified."[13]

d) A part of a graffito found at *Serabit el-Khadim*, which dates to king Amenemhet III's reign, reads[14]:

nswt-bity Snfr-w(i) m3ʿ-ḫrw di.f ʿnḫ ḏd w3s n nswt-bity N(y)-(m3ʿt-Rʿ)

"The King of Upper and Lower Egypt, Sneferu, justified. May he give life, stability, and dominion to the King of Upper and Lower Egypt, Ni(maatre)."

e) A part of an inscription dating to the Middle Kingdom, at *Rôd el-ʿAîr*, reads[15]:

mrr ḥsw sw Snfr-w(i) Ḥr Nb-m3ʿt ḏdt.f ḳbḥw mw m3ḫ snṯr ḫ3 m t ḥnḳt k3(w) 3pdw(w) šs mnḫt ḫt nbt nfrt wʿbt ʿnḫt nṯr im n k3 n šd(i)-wḥʿt ʾInpw-nḫt

"He who wishes Sneferu, Horus Nebmaat, to praise him should say: cool water, burning incense, thousand of bread, beer, oxen, fowl, alabaster, clothing, all good and pure things on which a god lives to the *ka* of the remover of the scorpions, *Anubis-nekht.*"[16]

---

[9] This title appeared during the Middle Kingdom. For other variants, see *Wb.* I, p. 350, 9; W. A. Ward, *Index of Egyptian Administrative and Religious Titles of the Middle Kingdom*, Beirut (1982), p. 178, no. 1538. Another but similar title, *ḫrp Srḳt,* which might be a magician who plays an intermediary role between a private individual and the god, appears from the Old Kingdom till the Ptolemaic period, see A. H. Gardiner, "Professional Magicians in Ancient Egypt", *PSBA* 39 (1917), pp. 34-44. S. Sauneron points out that it means a magician rather than a doctor; cf. S. Sauneron, "Le «chancelier du dieu» ( ) dans son double rôle d'embaumeur et de prêtre d'Abydos", *BIFAO* 51 (1952), p. 147. H. Gauthier believed that this title might mean "he who fights the scorpions" or "he who heals the bites", see H. Gauthier, "Découvertes récentes dans la nécropole Saïte d'Héliopolis", *ASAE* 33 (1933), p. 33.

[10] A. H. Gardiner, T. E. Peet, *The Inscriptions of Sinai*, I, pl. XI 35; *PM* VII, p. 343.

[11] A. H. Gardiner, T. E. Peet and J. Černý, *The Inscriptions of Sinai*, II, p. 72.

[12] This piece, of unknown provenance, is now in Manchester University Museum (no. 980), see *ibid.*, I, pl. XXXVI 104.

[13] A. H. Gardiner, T. E. Peet and J. Černý, *The Inscriptions of Sinai*, II, p. 108.

[14] A. H. Gardiner, T. E. Peet, *The Inscriptions of Sinai*, I, pl. XLVII 124a; *PM* VII, 349; J. Vandier, "Un nouvel Antef de la XIe Dynastie", *BIFAO* 36 (1936-1937), p. 109, fig. 2.

[15] A. H. Gardiner, T. E. Peet, *The Inscriptions of Sinai*, I, pl. XCIII 502.

[16] A. H. Gardiner, T. E. Peet, and J. Černý, *The Inscriptions of Sinai*, II, p. 518.

f) A part of a graffito at *Rôd el-ʿAîr* dating to the Middle Kingdom reads[17]:

mrr ḥsw sw Snfr-w(i) Ḥr Nb-mꜣʿt pḥt.f pr.f m ḥtp ḏdt.f ḳbḥw mꜣḫ sntr ʿw S-n-Wsrt nb imꜣḫ

"He who wishes Sneferu, Horus Nebmaat, to praise him is he who will reach his home in peace, he should say: cool water and burning incense, for the interpreter[18], *Senusret*, lord of honor."[19]

g) The gods featuring in document (2b) are also mentioned in an inscription in *Wadi Maghara*, dated to year 42 of king Nimaatre (Amenemhet III), about an expedition led by the treasurer Senusret-seneb-Sebekkhi, who is called[20]:

ḥsy n Ḥwt-Ḥr nbt ḫꜣst Mfkꜣt Spdw nb iꜣbtt Snfr-w(i) Ḥr nb ḫꜣswt

"praised of Hathor, lady of the Turquoise-country, (of) Sopdu, lord of the east, (of) Sneferu, (and) (of) Horus, lord of the foreign lands."

3) In the northern necropolis of *Abydos*, A. Mariette found a rectangular limestone base, dating to the Middle Kingdom. A part of the hieroglyphic inscription on the base reads[21]:

"An offering which the king gives [to] Sokar, [to] Osiris, and [to] the King of Upper and Lower Egypt, Sneferu, [that they may give] invocation-offerings [consisting of] bread and beer, to the *ka* of the overseer of the cattle, *Djedu.*"

4) On a relief (Cairo JE 41469) in *Deir el-Medineh*, Ramesside period, the owner is represented offering a pot of perfume to Amenhotep I, Ahmose-Nefertary, Meretseger, and a fourth person. The fourth person's figure is accompanied with the following cartouche[22]: [*S*]*nfr*.

---

[17] A. H. Gardiner, T. E. Peet, *The Inscriptions of Sinai*, I, pl. XCIV 510.

[18] It seems that this title, when given to Egyptians, means "those who know the Nubian language", as leaders engaged in Nubia, but when given to Nubians it means "the Egyptianized Nubians", see L. Habachi, *The Sanctuary of Heqaib*, AV 33 (1985), p. 28, footnote (a). W. A. Ward believes that it means "overseer of the foreign mercenaries", see W. A. Ward, *op. cit. [note 9]*, p. 13, no. 59. R. O. Faulkner translated it as "caravan-leaders" (lit. "overseer of dragomans"), who were specialists in desert travel and warfare, most of them living near the Nubian frontier, see R. O. Faulkner, "Egyptian Military Organisation", *JEA* 39 (1953), p. 43. As for A. H. Gardiner's translation "overseer of the dragomans", see A. H. Gardiner, "The Reading of the Geographical Term [hieroglyphs]", *JEA* 43 (1957), p. 7. H. Goedicke accepted the translation of "foreigner" as he points to the Asiatics who appeared in the funerary temple of Sahure of the 5th Dynasty and who bore this title. But he refused the translation of "interpreter" or "dragoman", see H. Goedicke, "The Title [hieroglyphs] in the Old Kingdom", *JEA* 46 (1960), p. 64. It could also mean "scholar", see F. W. Read, "The Sense of the Word [hieroglyphs]", *BIFAO* 13 (1917), pp. 141-142. A. H. Gardiner points out that in the Old Kingdom it means "interpreter" or "dragoman", while in the Middle Kingdom it means "foreigner", see A. H. Gardiner, "The Egyptian Word for 'Dragoman'", *PSBA* 37 (1915), p. 125. T. E. Peet called attention to the fact that a certain Khay, buried at Saqqara, bore the title of "interpreter of a difficult science", see T. E. Peet, "A Further Note on the Egyptian word for 'Dragoman'", *PSBA* 37 (1915), p. 224. G. Jéquier translated it as "interpreter", cf. G. Jéquier, "The Sign [hieroglyph]", *PSBA* 37 (1915), p. 252.

[19] A. H. Gardiner, T. E. Peet, *The Inscriptions of Sinai*, I, p. 219.

[20] *BAR* I §721-723; *LD* II, 137g.

[21] A. Mariette, *Catalogue Général des monuments d'Abydos découverts pendant les fouilles de cette ville*, Paris (1880), p. 588, no. 1496; A. Erman, "Geschichtliche Inschriften aus dem Berliner Museum", *ZÄS* 38 (1900), p. 121. This object is mentioned in the *CGC* as of unknown origin, see H. O. Lange and H. Schäfer, *Grab und Denksteine des Mittleren Reiches*, II, *CGC* 20400-20780, Berlin (1908), p. 375, no. 20742.

[22] G. Legrain, "Notes d'inspection", *ASAE* 9 (1908), pp. 57-59.

5) Priesthood of Sneferu:

According to a decree dating to the reign of king Pepi I[23], the priests of the pyramids of king Sneferu were exempted from taxes. This privilege indicates that the funerary cult of Sneferu flourished during the 6th Dynasty. However, no priests for the king in question are known from the end of the Middle Kingdom until the Late Period, when the cult of the king regained its popularity. A stele in the British Museum (no. 380) mentions a priest of Sneferu[24]. A sarcophagus, now in the Louvre Museum (no. D.13) and dating to either the Saite period or the Ptolemaic period, mentions a priest of Ptah called *Ankhhep*, who was also a priest of Sneferu[25].

Evaluation

The above documents dealing with the deification of king Sneferu were found mainly in two sites: *Dahshur* and the *Sinai.*

According to documents (1a)-(1f) and (3), many dead persons invoked Sneferu in the *hotepdinesu* formula. He was invoked alone in documents (1a), (1b), (1d), and (1e), but in documents (1c), (1f), and (3), he was invoked together with other deities: Ptah, Sokar, and Osiris.

In regard to document (1c), A. Fakhry concluded that this altar was presented to the temple of Sneferu for the benefit of the *kas* of the four named persons, who seem to have been connected with the cult of Sneferu. He also pointed out that either these men were named after Sneferu or their names contained the name of Sneferu in their construction, something which was very common among the families of the priests connected with the temples of the king during the 12th Dynasty[26].

Documents (1a)-(1f), dating to the Middle Kingdom, were all found at *Dahshur* in the funerary and valley temples of the Bent Pyramid. It seems that the cult of the deified king Sneferu gained popularity at this site. Only document (3), also Middle Kingdom, was found at *Abydos.* In it, Sneferu is invoked together with Sokar and Osiris, and A. Erman agreed that the king was invoked as a god[27].

In documents (2b) and (2c), found in the *Sinai,* kings Amenemhet III and Amenemhet IV are mentioned as the "beloved of Sneferu", as if beloved of a god. This may call to our attention the case of another deified, but in this case non-royal, person: Imhotep. King Nectanebo II (30th Dynasty) was called "beloved of Imhotep"[28].

In documents (2b) and (2g), Sneferu is mentioned among the gods of Sinai; in (2b) his name is written without a cartouche, but in (2g) inside a cartouche.

Other artifacts that were found in the *Sinai,* namely documents (2e) and (2f), show that dead persons urged those who wish to be praised by Sneferu to recite many offerings such as bread, beer, and cool water to them.

In document (2d) from the *Sinai*, king Sneferu was not invoked to give ordinary offerings, but to give king Amenemhet III life, prosperity, and health. J. Vandier drew attention to the fact that king Sneferu's cult was very popular during the 12th Dynasty, especially in the *Sinai.* He and L. Habachi believed that the reason for this was that the ancient Egyptians during the 12th Dynasty regarded

---

[23] L. Borchardt, "Ein Königserlaß aus Dahschur", *ZÄS* 42 (1905), pp. 1-11; cf. also I. E. S. Edwards, *The Pyramids of Egypt*, London (1949), p. 91.

[24] A. Erman, *ZÄS* 38, p. 122; H. Gauthier, *LdR*, MIFAO XVII/I (1907), p. 67.

[25] A. Erman, *ZÄS* 38, p. 122; H. Gauthier, *LdR*, MIFAO XVII/I (1907), p. 67.

[26] A. Fakhry, *The Monuments of Sneferu at Dahshur,* vol. I, p. 87.

[27] A. Erman, *ZÄS* 38, p. 121.

[28] This monument was wrongly attributed to Nectanebo I in M. L'abbé Paul Tresson, "Sur deux monuments Égyptiens inédits de l'époque d'Amasis et de Nectanebo 1er", *Kêmi* 4 (1931), p. 149. It was correctly assigned to Nectanebo II in: D. Wildung, *Egyptian Saints. Deification in Pharaonic Egypt*, New York (1977), p. 46; *id.*, *Imhotep und Amenhotep Gottwerdung im alten Ägypten*, MÄS 36, p. 46.

Sneferu as the first king to send campaigns to the quarries and mines of the *Sinai*, for he undertook great mining activities there[29]. Moreover, the names of his predecessor kings were not familiar to the ancient Egyptians during the 12th Dynasty[30]. F. Petrie further pointed out that Sneferu was the earliest king who was adored in the temple of *Serabit el-Khadem* in the *Sinai*[31].

As for document (4), it seems that Sneferu received offerings together with the deified king Amenhotep I and the deified queen Ahmose-Nefertary. G. Legrain expressed the opinion that the name in the cartouche in question should be read as *Tanefer*, who he identified as a royal person, but it is now generally believed the name is to be read as Sneferu[32].

All the above evidence clearly shows that king Sneferu was deified after his death during the Middle Kingdom in *Dahshur* and in the *Sinai*. There are two isolated references from *Abydos* (Middle Kingdom) and *Deir el-Medineh* (New Kingdom) that may need a special explanation.

II) King Khufu:

1) A limestone statue preserved in the Pushkin Museum (no. 5575) represents a man called *Memi* and his wife *Aki*, seated with their legs crossed[33]. An inscription on the left side of a dorsal pillar of the statue reads[34]:

ḥtp-di͗-nsw Ḫw(i͗).f-w(i͗) di͗.f prt-ḫrw t ḥnḳt k3(w) 3pd(w) n k3 n Mmi͗

"An offering which the king gives [to] Khufu that he may give invocation-offerings [consisting of] bread, beer, oxen, and fowl to the *ka* of *Memi*."

A part of another inscription on the right side of the dorsal pillar reads[35]:

ḥtp-di͗-nsw Ḫw(i͗).f-w(i͗) di͗.f prt-ḫrw t ḥnḳt k3(w) 3pd(w) n k3 n nbt pr Di͗kw ḫ3 m t ḥnḳt n k3 n ʿki͗

"An offering which the king gives [to] Khufu that he may give invocation-offerings [consisting of] bread and beer to the *ka* of the lady of the house, *Diku*, a thousand of bread and beer to the *ka* of *Aki*."

In this document, king Khufu was invoked by a couple in the *hotepdinesu* formula as a god. J. Vandier dated the stele to the end of the Middle Kingdom / the beginning of the Second Intermediate Period, and more precisely to king Amenemhet IV's reign[36]. W. Barta dated the stele to the 13th Dynasty[37].

---

[29] J. Vandier, *op. cit. [note 14]*, pp. 109-111; L. Habachi, *Features of the Deification of Ramesses II,* ADAIK 5 (1969), p. 46.

[30] J. Vandier, *op. cit. [note 14]*, pp. 109-111.

[31] W. M. F. Petrie, C. T. Currelly, *Researches in Sinai*, London (1906), pp. 123-124.

[32] G. Legrain, *ASAE* 9, pp. 57-59; B. Bruyère, *Meret Seger à Deir el Médineh*, MIFAO 58, Cairo (1930), p. 210, fig. 109; D. B. Redford, *Pharaonic King-lists, Annals and Day-Books,* SSEA Publication IV, Mississauga (1986), p. 49; A. McDowell, "Awareness of the Past in Deir el-Medina", in R. J. Demarée and A. Egberts, *Village Voices. Proceedings of the Symposium 'Texts from Deir el-Medina and their Interpretation', Leiden, May 31- June 1, 1991,* Leiden (1992), p. 100.

[33] This statue was formerly in the collection of W. Golénscheff, no. 3114, see J. Vandier, "Le groupe de Mémi et d'Akhou", in *ДРЕВНИЙ ЕГИПЕТ*, Moscow (1960), p. 105. See also *id.*, *Manuel d'Archéologie Égyptien, tome III: les grandes époques. La Statuaire*, Paris (1958), p. 241. The statue has also been published by S. Hodjash and O. Berlev, *The Egyptian Reliefs and Stelae in the Pushkin Museum of Fine Arts*, Leningrad (1982), no. 42.

[34] J. Vandier, *op. cit.*, p. 105.

[35] *Ibid.*, p. 105.

[36] *Ibid.*, pp. 106-109.

[37] W. Barta, *Aufbau und Bedeutung der altägyptischen Opferformel,* ÄgFo 24 (1968), p. 232.

Evaluation

In light of the available information, it seems that king Khufu was deified after his death during the 13th Dynasty. His name was invoked a long time after his death. Z. Hawass proposes that his son, Khafre, worshipped king Khufu as Re. He believes that one of the niches in the upper temple of Khafre's complex was dedicated to Khufu's statue. He also suggests that the two boats found in pits located to the south of Khufu's pyramid functioned as solar boats for Khufu as Re[38]. But we have to take into consideration that there is no textual or archaeological evidence to confirm his hypothesis.

III) King Sahure:

1) A scene on the causeway of king Sahure shows a bark returning to Egypt after a journey. On its deck, Egyptian sailors accompanied by Asiatics are represented bowing and raising their hands in the attitude of adoration. The text above the scene reads[39]:

ỉ3w n.k S3ḥ-w(ỉ)-Rʿ nṯr ʿnḫw m33.n nfr.k

"Adoration to you, Sahure, god of the living, [when] we see your beauty."

2) On several blocks and steles found in the funerary complex of king Sahure, the goddess Sekhmet of Sahure is mentioned.

a) An inscription on a stele, showing a man bringing an offering to the goddess Sekhmet, reads[40]:

ḥm-nṯr Sḫmt S3ḥ-w(ỉ)-Rʿ

"The priest of Sekhmet [of] Sahure."

b) On the upper register of a limestone funerary stele, the goddess Sekhmet is represented standing on the left side with a lioness head surmounted by a solar disc with cobra. On the right side, a man is shown standing and raising his hands in the attitude of adoration in front of an altar, two *nemset*-vases, and lotus flowers. The text accompanying Sekhmet reads[41]:

Sḫmt S3ḥ-(wỉ)-Rʿ ỉrt Rʿ ḥry-tp ʾItn nbt pt ḥnwt nṯrw

"Sekhmet [of] Sahure, the Eye of Re, which is upon the sun disc, lady of the sky, mistress of the gods."

c) Another inscription on a limestone stele reads[42]:

---

[38] Z. Hawass, *The Funerary Establishments of Khufu, Khafra, and Menkaura during the Old Kingdom*, Ph. D. diss., University of Pennsylvania (1987), p. 629.

[39] C. Boreux, *Études de nautique Égyptienne, l'art de navigation en Egypt jusqu'à la fin de l'ancien Empire*, MIFAO 50 (1925), p. 475, fig. 187; D. B. Redford, *Egypt Canaan, and Israel in Ancient Times*, Cairo (1993), pp. 51-52, fig. 4; see also K. Sethe, *Urk.* I, p. 169, 14. The causeway of the pyramid complex of king Sahure is 200 m. long. It was built from limestone. For the pyramid complex of king Sahure, see A. Fakhry, *The Pyramids*, Chicago (1969), pp. 171-175.

[40] L. Borchardt, *Das Grabdenkmal des Königs SA⁴ḤU-REʿ, Band I: der Bau*, Leipzig (1910), p. 127; for other priests dating to the 19th Dynasty, see H. Gauthier, *LdR*, MIFAO 17/I, p. 113, XXIII, XXIV.

[41] L. Borchardt, *op. cit.*, pp. 122-123, pl. 167.

[42] *Ibid.*, p. 129, pl. 175.

Sḫmt S3ḥ-w(i)-Rᶜ

"Sekhmet [of] Sahure"

3) A priest called *Horib* who lived during the Ptolemaic period possessed the title of[43]:

ḥm-nṯr n Sḫmt n(t) pr Sḫmt n S3ḥ-(wi)-Rᶜ

"Priest of Sekhmet of the temple of Sekhmet of Sahure."

4) A part of an inscription on a votive statue, found at *Karnak*, reads[44]:

nṯr nfr Ḫpr-k3-Rᶜ di ᶜnḫ ḏt ir.n.f m mnw.f n it.f S3ḥ-w(i)-Rᶜ

"The good god, Kheperkare, given life eternally, he has made it as a monument for his father Sahure."

Evaluation

The first document should be considered with much caution since it is not certain that the Egyptian sailors, accompanied by foreigners (Asiatics), adored king Sahure himself, as his figure is missing. On the other hand, the accompanying text proves that they adored the king when they saw him. The peculiar title *nṯr ᶜnḫw*, "the god of the living", may call to our attention a similar title that was used by another deified, albeit non-royal, person, Isi of the Sixth Dynasty[45].

Documents (2a)-(2c) were discovered in the funerary complex of king Sahure at *Abusir*, and are connected with a phenomenon that occurred frequently in the New Kingdom: several deities were connected to kings and were invoked as "deity of PN".

A. I. Sadek points out that the cult of Sekhmet of Sahure was located in the southern part of the funerary temple of king Sahure at *Abusir* and flourished there from at least the reign of Thutmosis III until the fifth regnal year of Darius II. For example, a graffito of a priest of Sekhmet of Sahure, a man called Pasmin, attests to the existence of the cult in the fifth year of Amasis[46]. Sadek could however not provide an explanation of why the cult of Sekhmet of Sahure came about in the New Kingdom.

Sekhmet of Sahure had two festivals that were celebrated in the village of *Deir el-Medineh* on the 16th day of the first month of *Peret*, and on the 11th day of the fourth month of *Peret*[47]. Borchardt confirms that in the New Kingdom there were festivals that were celebrated for the lioness-headed goddess Sekhmet of Sahure[48].

[43] E. Otto, "Eine Memphitische Priester Familie des 2. Jh. v. chr.", *ZÄS* 81 (1956), p. 123; for another example on the sarcophagus of a certain Ahmose, now in the Museum of Berlin, no. 38, see H. Gauthier, *LdR*, MIFAO 17/I, p. 113, XXV.

[44] This statue was found by G. Legrain at the Karnak cachette (no. 421), see H. Gauthier, *LdR*, MIFAO 17/I, p. 112, XVII, p. 273, XXX.

[45] See M. Alliot, "Un nouvel example de vizir divinisé dans l'Égypt ancienne", *BIFAO* 37 (1937-1938), p. 136; *id., Rapport sur les fouilles de Tell Edfou (1933)*, FIFAO 10/2 (1935), pp. 28, 30, 38; H. de Meulenaere, "La statue d'un contemporain de Sébekhotep IV", *BIFAO* 69 (1971), pp. 61-62; G. Daressy, "Monuments d'Edfou datant du Moyen Empire", *ASAE* 17 (1917), p. 240; R. Engelbach, "Report on the Inspectorate of Upper Egypt from April 1920 to March 1921", *ASAE* 21 (1921), pp. 65-66; *id.*, "Steles and Tables of Offerings of the Late Middle Kingdom from Tell Edfû", *ASAE* 22 (1922), p. 114.

[46] A. I. Sadek, *Popular Religion in Egypt during the New Kingdom*, HÄB 27 (1987), pp. 29-34.

[47] *Ibid.*, pp. 172-173.

[48] L. Borchardt, "Die Ausgrabung des Totentempels Königs Sahure bei Abusir 1907/8", *MDOG* 37 (1908), p. 29.

The New Kingdom combinations of deities with kings[49] (and probably once with a queen[50]) have not received much attention, except for the deities of king Ramesses II. Although it has been suggested that such "Deity of Ramesses II" combinations merely suggest some kind of special relation of the king to the god or his statue (personal cult or royal temple building), many scholars believe that they point at an identification of the king with the god[51]. If the latter applies, then "Sekhmet of Sahure" could equally be a form of the deified Sahure.

Document (2a) attests to a priest of the goddess Sekhmet of Sahure. In documents (2b) and (2c), Sahure's name does not have a cartouche when appearing in this goddess-king combination. Document (3) indicates that the cult of Sekhmet of Sahure may have survived till the Ptolemaic period.

According to document (4), king Senusret I set up a votive statue to king Sahure in the Karnak temple. It is odd to find a king in the Middle Kingdom making a statue for the king in question, an event that seems to have been the result of Sahure gaining a special reputation. However, this evidence should be taken with much caution as Senusret I also dedicated a statue to king Niuserre[52].

IV) King Menkauhor:

1) On the lower register of a New Kingdom limestone stele preserved in the Louvre Museum (B50), a person called *Thouthou*[53] and his wife *Mau* are represented standing and raising their hands in the attitude of adoration in front of *Duamutef* and the deified king Menkauhor[54]. The king in question wears the *nemes*-headdress, a short kilt, and holds in his right hand the *heqa*-sign and in his left hand the *ankh*-sign. The inscription accompanying Menkauhor reads[55]:

Wsỉr nb tꜣwy Mn-kꜣw-Ḥr mꜣꜥ-ḫrw

"Osiris, the lord of the two lands, Menkauhor, justified."

2) On another limestone stele preserved in the Louvre Museum (B 48 = E 3028), which belongs to the tomb of a person called *Amuneminet*, king Menkauhor is represented standing before an altar with a *nemset*-vase and lotus flowers[56]. He wears the *nemes*-headdress, a short kilt, and holds the *was*-scepter and the *hedj*-mace in his left hand and the *ankh*-sign in his right hand. The vulture goddess *Nekhbet* is depicted hovering above the king to protect him[57].

---

[49] For a survey of the deities of Thutmosis I, Thutmosis III, Ramesses II, Merenptah, and Ramesses III, see the Appendix of this article.

[50] Namely queen Tiy of the 18th dynasty, see the Appendix.

[51] The different opinions, with references, are listed in the Appendix.

[52] H. Gauthier, *LdR,* MIFAO 17/I, p. 129, XVIIIB.

[53] M. Ibrahim discovered another block of limestone from the tomb of Thouthou in the course of the excavations in Saqqara, and on it Thouthou and his wife Mau are represented. He bore the titles of "physician" and "scribe", while his wife had the title of "chantress of Amun", see M. I. Aly, "New Kingdom Scattered Blocks from Saqqara", *MDAIK* 56 (2000), p. 229.

[54] A. Mariette found this stele in the Serapeum of Saqqara, see H. Gauthier, *LdR*, MIFAO 17/I, p. 132; J. Berlandini-Gernier, "Varia Memphitica I (I)", *BIFAO* 76 (1976), p. 315, footnote 5.

[55] The stele was executed in low relief, see F. Jonckheere, *Les médecins de l'Égypte pharaonique, essai de prosopographie*, Brussels (1958), p. 72, no. 78, fig. 24, p. 162; see also J. Berlandini-Gernier, "La pyramide «ruinée» de Sakkara-Nord et Menkaouhor", *BSFÉ* 83 (1978), p. 25, fig. 2; *id.*, "La pyramide «ruinée» de Sakkara-Nord et le roi Ikaouhor-Menkaouhor", *RdÉ* 31 (1979), p. 19, pl. 2b; *PM* III[2], p. 820.

[56] J. Berlandini-Gernier, *op. cit. [note 54]*, pp. 303-309; see also *id.*, *BSFÉ* 83, fig 1; *id.*, *RdÉ* 31, p. 19, pl. 2a; E. Graefe, "Das Grab des Vorstehers der Kunsthandwerker und Vorstehers der Goldschmiede, Ameneminet, in Saqqara", in *Memphis et ses nécropoles au Nouvel Empire, nouvelles données, nouvelles questions, Actes du colloque international CRNS Paris, 9 au 11 Octobre 1986*, ed. A. Zivie, Paris (1988), p. 52.

[57] J. Berlandini-Gernier, *op. cit. [note 54]*, pls. LIII, LIV.

<u>*Evaluation*</u>

It has long been suggested that document (1) is a stele from the funerary temple of king Menkauhor. But J. Berlandini-Gernier proposed that the stele is a part of a tomb in Saqqara which belongs to a certain *Thouthou* who was a doctor[58], and that it dates to the reign of Ay or Horemheb[59].

Document (2) was originally also dated to the Old Kingdom. However, J. Berlandini-Gernier believes that it should be dated to the New Kingdom, to the reign of Tutankhamun or Horemheb[60], and also believes that this fragment supplies evidence for the existence of a cult of the dead king in *Saqqara* during the New Kingdom[61].

As both these two documents were found in *Saqqara* and (if we follow the interpretations of Berlandini-Gernier) date to the New Kingdom, it seems that king Menkauhor was deified during that period at that site.

V) King Unas:

1) At the top of a limestone rectangular stele found to the south of the causeway of Unas at *Saqqara*, there are two hieroglyphic lines which read[62]:

ḥtp-di-nsw Ptḥ nswt-bity Wnis mꜣʿ-ḫrw di.f prt-ḫrw t ḥnḳt kꜣ(w) ꜣpd(w) šs mnḫt snṯr mrḥt ḫt nbt nfrt wʿbt ʿnḫ nṯr im n kꜣ n innw nṯr r šbw.f ḏwi psḏt iit.s ḫry-ḥbt Wnis-m-sꜣ.f

"An offering which the king gives [to] Ptah, and [to] the King of Upper and Lower Egypt, Unas, justified, that he may give invocation-offerings [consisting of] bread, beer, oxen, fowl, alabaster, clothing, incense, ointment, every good and pure thing on which the god lives to the *ka* of the one who brings the god to his food, and who summons the Ennead[63] that she comes, the lector priest, *Unasemsaf*."

---

[58] *Ibid.*, p. 315.

[59] *Ibid.*, p. 315, footnote 5.

[60] *Ibid.*, pp. 309-312.

[61] *Ibid.*, p. 313.

[62] Ahmed M. Moussa discovered this stele on December 2, 1965. It is preserved in the magazine of Saqqara, no. 16110 Saqqara 2-12-1965, see A. M. Moussa, "A Stele from Saqqara of a Family Devoted to the Cult of King Unas", *MDAIK* 27 (1971), p. 81, pls. XIII, XIV.

[63] The Egyptian word indicating the Ennead is *psḏt*. The great Ennead is that of Heliopolis. It consists of Atum, Shu, Tefnut, Geb, Nut, Osiris, Nephtys, Seth, and Isis, see A. H. Gardiner, *Egyptian Grammar*, p. 291, footnote 8; H. Brunner, "Neunheit", *LÄ* IV (1982), cols. 473-474. But it is not a rule that every Ennead should consist of nine deities. The Ennead of Abydos consists of seven gods: Khnum of Herwer, Khnum of the cataract, Thot, Horus of Letopolis, Harendotes, Wepwawet of Lower Egypt, and Wepwawet of Upper Egypt, see *ibid.*, col. 475. J. G. Griffiths believed that the two Enneads of the Pyramid Texts, "The mighty and the great Enneads", represent Upper and Lower Egypt. He also added that the little Ennead of the gods is probably a re-interpretation, perhaps based on a misreading of the Pyramid Texts, see J. G. Griffiths, "Some Remarks on the Enneads of the Gods", *Orientalia* 28 (1959), pp. 34-56. Also W. Barta points out that during the Old Kingdom the *psḏt wrt* belongs to Upper Egypt, and the *psḏt ʿꜣt* belongs to Lower Egypt, see W. Barta, *Untersuchung zur Götterkreis der Neunheit*, MÄS 28 (1973), pp. 50ff. The lesser and the greater Enneads could be represented as the Nile gods uniting the two lands, as on the statues of king Senusert III from Lisht. They could also be sister, mother, and daughter of a god, see L. Troy, "The Ennead: The collective as goddess, a commentary on textual personification", in *The Religion of the Ancient Egyptians: Congnitive Structures and Popular Expressions, Proceedings of Symposia in Uppsala and Bergen 1987 and 1988*, ed. G. Englund, Boreas 20 (1989), pp. 59-69.

2) On the back of a group statue of a certain *Sermaat* and his wife *Khenmet* found in the valley temple of king Unas[64], there are nine hieroglyphic lines. A part of the text reads[65]:

ḥtp-di-nsw Ptḥ-Skr nswt-bity Wnis mꜣʿ-ḫrw di.f prt-ḫrw t ḥnḳt kꜣ(w) ꜣpd(w) šs mnḫt sntr mrḥt ḥtpt ḏfꜣw rnpy nbt ḥnḳt nbt nfrt wʿbt ʿnḫt ntr im n kꜣ n imꜣḫy ḫr nṯr ʿꜣ nb pt imy-r pr-ḥsb it mḥ m ʿnḫ-tꜣwy šmꜣʿw n Ptḥ rsy inb.f nb tꜣwy Sr-mꜣʿt

"An offering which the king gives [to] Ptah-Sokar, and [to] the King of Upper and Lower Egypt, Unas, justified, that he may give invocation-offerings [consisting of] bread, beer, oxen, fowl, alabaster, clothing, incense, ointment, all food offerings, all good and pure things on which the god lives to the *ka* of the one honored by the great god, the lord of the sky, the steward of the reckoning of the Lower-Egyptian barley in *Ankhtawy*, the singer of Ptah south-of-his-wall, the lord of the two lands, *Sermaat.*"[66]

Evaluation

The two documents in which king Unas is invoked in the *hotepdinesu* formula were both found in the vicinity of the funerary complex of this king. H. Altenmüller believes that king Unas was venerated during the Middle Kingdom in *Saqqara*, and that his cult was a local one[67]. He also gathered a number of examples in which the deceased persons are called "honored by the King of Upper and Lower Egypt Unas"[68].

VI) King Teti:

1) A broken statue belonging to a certain *Amenwahsu* is preserved in the Borély Museum in Marseilles[69]. It dates to the 19th Dynasty[70]. The left side bears a relief that represents *Amenwahsu* kneeling and raising his hands in the attitude of adoration before king Teti, whose name occurs behind him without a cartouche[71]:

Tti Mr(y)-n-Ptḥ

"Teti Mery-en-Ptah"

King Teti is shown standing inside his pyramid. He holds a stick in his right hand and the *ankh*-sign in the other one. A part of the legend in front of *Amenwahsu* reads[72]:

rdit iꜣw n Wsir sn tꜣ Tti Mr(y)-n-Ptḥ ir n sš wḏḥw n nb tꜣwy Imn-wꜣḥ-sw

---

[64] This statue was discovered on December 2, 1972, see A. Moussa and H. Altenmüller, "Ein Denkmal zum Kult des Königs Unas am Ende der 12. Dynastie", *MDAIK* 31 (1975), p. 93.
[65] *Ibid.*, p. 92, fig.1.
[66] *Ibid.*, pp. 94-95.
[67] H. Altenmüller, "Zur Vergöttlichung des Königs Unas im Alten Reich", *SAK* I (1974), p. 1; A. Moussa and H. Altenmüller, *MDAIK* 31, p. 97; A. Labrousse and A. Moussa, *Le temple d'accueil du complexe funéraire du roi Ounas*, BdÉ 111 (1996), p. 11.
[68] For these examples, see H. Altenmüller, *op. cit.*, pp. 4-7.
[69] E. Naville, "Le roi Teta Merenptah", *ZÄS* 16 (1878), p. 69; *PM* III², p. 729.
[70] J. Malek, "A Meeting of the Old and New. Saqqâra During the New Kingdom", *Studies in Pharaonic Religion and Society in Honor of J. Gwyn Griffiths*, ed. A. B. Lloyd, EES, Occasional Publications 8 (1992), p. 68.
[71] On the right side of the statue, Amunwahsu's wife is depicted kneeling in the same attitude as her husband before a statue inside a pyramid, see E. Naville, *ZÄS* 16, pl. IV; J. Capart, *Recueil de monuments Égyptiens*, Brussels (1902), pl. XLV.
[72] *Ibid.*

"Giving adoration to Osiris, and kissing the earth [for] Teti Mery-en-Ptah. Made by the scribe of the offering-table of the lord of the Two Lands, *Amenwahsu*."

2) The lower register of a limestone stele found at *Saqqara* shows two standing men[73]. The first one raises his hands in adoration, while the other one offers birds. A part of the accompanying text reads[74]:

rdit ỉ3w n Ttỉ sn t3 n nṯr ˁ3 di.f ˁnḫ wḏ3 snb nḏm-ib ḥr šms k3.f imy-r mrt Mn-nfr-mry-Ptḥ

"Giving adoration to Teti, and kissing the earth for the great god, that he may give life, prosperity, health and happiness to the follower of his *ka*, the overseer of the pastureland [of the pyramid complex], *Men-nefer-mery-Ptah*."

3) A stele of a certain *Nebnekhet* was found in the mortuary temple of king Teti at *Saqqara*. On the stela, which could be attributed to the 19th Dynasty, the owner offers to Ptah, Re-Horakhty, and Teti[75].

4) Two other New Kingdom steles were also found in the mortuary temple of the king in question. On the first one, king Teti is represented seated on a throne in front of an offering table. The king wears a long kilt, the double crown, and holds the *heqa* and *ankh*-sign[76]. On the second stele, a person is depicted adoring Teti[77].

5) A stele, probably from the 19th Dynasty, shows on its recto a figure of a man called *Amenemhet* offering flowers to a seated figure of Osiris. Behind the god, king Teti is depicted, standing and holding a stick in his left hand and an *ankh*-sign in his right hand[78]. The verso bears the following legend[79]:

s3 Rˁ Ttỉ di.f pr(t)-(ḫrw)

"Son of Re, Teti, that he may give invocation-[offerings]."

6) An inscription on a limestone stele of a certain *Monthuhotep* attests to the invocation of king Teti in the *hotepdinesu* formula. It was found in *Saqqara* but its exact provenance is not recorded. The first line of the inscription reads[80]:

ḥtp-di-nsw s3 Rˁ Ttỉ Inpw (or: *ḥry sšt3*) *Mnṯw-ḥtp m3ˁ-ḫrw*

"An offering which the king gives to the son of Re, Teti, and Anubis [that they may give invocation-offerings to] *Monthuhotep*, justified" (or: "An offering which the king gives to the Son of Re, Teti [that he may give invocation-offerings to] the overseer of secrets[81], *Monthuhotep*, justified.")

---

[73] JE 36852; P. Lacau, *Stèles de la XVIIIe dynastie*, CGC 34065-34189, Cairo (1957), pp. 234 -235, no. 34188, pl. LXXI; *PM* III², p. 395; A. Barasanti, "Un monument du roi Teti", *ASAE* 13 (1913), p. 255.
[74] P. Lacau, *op. cit.*, p. 235; A. Barasanti, *op. cit.*, p. 255.
[75] JE 36855; J. Malek, *op. cit. [note 70]*, p. 68; *PM* III², p. 395.
[76] J. E. Quibell, *Excavations at Saqqarah (1907-1908), vol. III*, Cairo (1909), p. 114, pl. 4 right.
[77] J. Malek, *op. cit.*, p. 68; J. E. Quibell, *op. cit.*, pl. 4 right; *PM* III², p. 395.
[78] J. Malek, *op. cit.*, p. 68, pl. VII 2, 1,2; *id.*, *JEA* 75, p. 73; *PM* III², p. 572.
[79] J. Malek, *op. cit.*, pl. VII 2.
[80] H. G. Fischer, *Egyptian Studies I, Varia*, New York (1976), p. 59, fig. 1.

7) There is a limestone stele preserved in the Boston Museum of Fine Arts (Inventory no. 25.635), which belongs to a certain *Ptah-Sety*. On the upper register, king Teti is represented standing before Osiris[82] and offering two vases to him. The king's cartouche is followed by *di ʿnḫ* "given life"[83].

8) On a limestone fragment stele that was found by Quibell in the Monastery of St. Jeremias at *Saqqara*, king Teti is shown standing before a person's hands raised in adoration. The king's head is surmounted by a solar disc flanked by two cobras[84].

9) On the upper register of a round-topped stele which belongs to a person called *Ay*, king Teti is shown offering lotus flowers to Osiris followed by Horus and Isis. This stele was discovered in one of the isolated tombs of the *Memphite Serapeum*[85].

10) On the left side of the lower register of a limestone block that was discovered at *Saqqara*, a king or a royal statue is represented standing. He wears the *khat*-headdress and an elaborate kilt. Before the head of the king is a cartouche, which reads: *Tti Mry-Ptḥ*. On the right side, a standing man is shown raising his hands in adoration[86].

Evaluation

Concerning the first document, J. Berlandini believes that the adorer worships a statue of Teti Mery-en-Ptah that was placed before his pyramid, as the text refers to it with the word *twt*[87]. Despite the fact that the provenance of the statue is unknown, J. Malek suggests that it may have come from the vicinity of Teti's pyramid[88].

E. Naville pointed out that during the 19th Dynasty the cult of king Teti was still practiced by the priests, a tradition that was continued under the kings of the Saitic and Ptolemaic periods[89].

The epithet "Beloved of Ptah" in the 1st (and 10th) document was attributed by J. Malek to the late 18th Dynasty[90]. J. Yoyotte suggested that it may indicate that Teti was chosen by Ptah to be a king[91].

---

[81] H. G. Fischer, *ibid.*, pointed out that the sign could not be taken as the name of Wepwawet or Anubis. Dealing with the different forms of the signs that can be read as Anubis, E. Mahler did not attest the form in question and identified the object between the forelegs of the jackal as a staff, see E. Mahler, "The Jackal-gods on Ancient Egyptian Monuments", *PSBA* 36 (1914), p. 148-149. H. G. Fischer pointed out that the name of the god would normally come before the name of the king, but he gives many examples of exceptions to this rule, examples in which the name of the god follows that of the king, see *ibid.*, p. 59, footnote 6. He also contradicts the idea that this could be a title of Monthuhotep, to be read *ḥry-sštꜣ* "overseer of secrets", see H. G. Fischer, *op. cit.*, pp. 59, 61. For this title, see *Wb.* IV, p. 299 (2,3). Senenmut, who served during the reign of Hatshepsut, held the title of "overseer of secrets of the 'house of the morning'", see W. C. Hayes, "The Sarcophagus of Senenmut", *JEA* 36 (1950), pl. 7, l. 50. He proposes that it could be read as *stꜣt* "Aroura", which would mean arouras are given to Monthuhotep, see H. G. Fischer, *op. cit.*, p. 61. For this word, see *Wb.* IV, p. 356 (1, 4).

[82] C. Firth found this stele at Saqqara. It dates back to the end of the 18th Dynasty or early 19th Dynasty, see D. Dunham, "Four New Kingdom Monuments in the Museum of Fine Arts, Boston", *JEA* 21 (1935), p. 148; *PM* III², p. 572. Its dimensions are 1.25 m. x 51 cm. x 18 cm., see also J. Malek, *op. cit. [note 70]*, p. 68, pl. VII 2, and fig. 3.

[83] D. Dunham, *op. cit.*, pl. XVII 2; S. D'Auria, P. Lacovara, and C. H. Roehrig, *Mummies and Magic. The Funerary Arts of Ancient Egypt*, Boston (1988), p. 158, fig. 109.

[84] JE 40693; *PM* III², p. 667; J. Malek, *op. cit. [note 70]*, p. 68; B. Grdseloff, "Le roi Iti divinisé", *ASAE* 39 (1939), p. 394, fig. 17; D. Wildung, *Die Rolle ägyptischer Könige im Bewußtsein ihrer Nachwelt I*, MÄS 17 (1969), pp. 97-99, pl. VIII 2.

[85] For this stele, see J. Malek, *op. cit. [note 70]*, p. 68; *PM* III², 781.

[86] M. I. Aly, *op. cit. [note 52]*, pp. 226-228.

[87] J. Berlandini-Gernier, *op. cit. [note 54]*, p. 313.

[88] *Ibid.*

[89] For the existence of one of these priests, see E. Naville, *ZÄS* 16, p. 72.

As no other god is mentioned in document no. (2), it seems that king Teti is referred to as the great god. Here, Teti grants the adorer life, prosperity and health, and not the usual offerings for which the deceased invoked the deities.

According to documents (1), (2), (3), (4, 2nd stele), (8) and (10), king Teti was worshiped by the owners of these steles, in one case (document no. (3)) among the deities.

In document no. (6), king Teti is invoked in the *hotepdinesu* formula. J. Malek believes that also in document no. (5), the deified king Teti was invoked in a *hotepdinesu* formula as a god[92].

Concerning document no. (7), P. Gunn commented that the formula *di ꜥnḫ* would indicate real worship of the dead king as a god, still involved in human life, rather than a funerary cult[93]. However, P. C. Smither is of the opinion that it is an epithet often used by the kings in their divine aspects[94].

Although document (9) does not show a deification of king Teti, it at least shows that he was held in some kind of veneration so many centuries after his death.

M. I. Aly assumes that Teti was venerated during the New Kingdom as a saint or a *sheikh* rather than a god[95]. According to J. Berlandini-Gernier, Teti was venerated as the founder of the Sixth Dynasty, and his funerary temple was used as a chapel for his cult, as may be deduced from the ex-votos discovered there[96].

VII) King Pepi II:

1) There is a red sandstone group-statuette preserved in the Cairo Museum which represents a seated couple. It bears an inscription on its back, which reads[97]:

ḥtp-di-nsw Nfr-kꜣ-Rꜥ mꜣꜥ-ḫrw prt-ḫrw t ḥnḳt kꜣ(w) ꜣpd(w) n šs mnḫt n imꜣḫwt sꜣt Ḥwt-Ḥr prt-ḫrw t ḥnḳt kꜣ(w) ꜣpd(w) n kꜣ n Ḏꜣw-m-nḫt rn sꜣt n Kꜣw-Ptḥ

"An offering which the king gives [to] Neferkare, justified, [that he may give] invocation-offerings [consisting of] bread, beer, oxen, fowl, alabaster, and clothing to the honored Sat-Hathor, [and] invocation-offerings [consisting] of bread, beer, oxen, fowl, alabaster, and clothing to the *ka* of *Djauemnekht*, daughter of *Kauptah*."

2) Another statuette preserved in the Cairo Museum represents a striding man. An inscription engraved on the short dorsal pillar of the statuette reads[98]:

imꜣḫw ḫr Nfr-kꜣ-Rꜥ mry-n-Rꜥ

"Honored by Neferkare, beloved of Re."

---

[90] J. Malek, *op. cit. [note 70]*, p. 68.

[91] J. Yoyotte, "À propos de la parentée féminine du roi Téti (VIe Dynastie)", *BIFAO* 57 (1958), p. 96, footnote 5.

[92] J. Malek, *op. cit. [note 70]*, p. 58.

[93] D. Dunham, *op. cit. [note 82]*, p. 147, footnote 1.

[94] P. C. Smither, "The Writing of *ḤTP-DI-NSW* in the Middle and New Kingdoms", *JEA* 25 (1939), p. 37.

[95] M. I. Aly, *op. cit. [note 52]*, p. 228.

[96] J. Berlandini-Gernier, *op. cit. [note 54]*, p. 314.

[97] JE 51480. This statue was found in front of the door of the store in the vestibule of the antechamber, see G. Jéquier, *Le monument funéraire de Pepi II*, part III, *Les approches du temple*, Cairo (1940), p. 31, pl. 50, fig. 11. See also W. Barta, *op. cit. [note 36]*, p. 56, footnote 5.

[98] JE 51170; G. Jéquier, *op. cit.*, p. 31, pl. 50, fig. 12. For this formula used by dead persons to be honored by kings such as Khafre, Menkaure, Shepseskaf, Userkaf, and Sahure, see S. Hassan, *The Great Pyramid of Khufu and Its Mortuary Chapel, Excavations at Giza, Season 1938-39*, vol. X, Cairo (1960), p. 55.

Evaluation

In the first document, king Pepi II is invoked alone in the *hotepdinesu* formula, so as to grant the deceased offerings. The epithet *mꜣꜥ-ḫrw* after his name strongly suggests that Pepi II was already dead at the time when the statuette was placed in his funerary complex. This evidence implies that the deification of the king took place after his death and that his funerary complex was the cult center of the deified king.

Concerning the second document, it is worthy to mention that the *imꜣḫw ḫr* formula could be used for both deities and kings. However, caution is required, because the formula was also used for ordinary persons in the Old Kingdom. The following examples may illustrate the latter:

A lady called *Reput-ka* (?), the wife of *Abdu*, was entitled in their tomb in *Giza* as[99]:

ʾImꜣḫwt ḫr hy.s

"Honored by her husband."

And another person, *Sekhem-ka-Re*, was called[100]:

ʾImꜣḫw ḫr it.f

"Honored by his father."

Yet another woman called *Thetwet*, wife of a certain *Seshemu*, was mentioned as[101]:

ʾImꜣḫwt ḫr hy.s rꜥ nb

"Honored by her husband every day."

Finally, a very rare formula reads[102]:

ʾImꜣḫw ḫr rmṯ

"Honored by the people"

These examples show that the *imꜣḫw ḫr* formula was an ordinary formula that sometimes could be used by the deceased wishing to be honored by one of his relatives.

---

[99] Adel-Moneim Abu-Bakr, *Excavations at Giza 1949-1950*, Cairo (1953), pp. 69, 77.

[100] S. Hassan, *Excavations at Giza*, vol. IV, 1932-1933, Cairo (1943), p. 103; *id.*, *The Great Pyramid of Khufu and Its Mortuary Chapel, Excavations at Giza*, vol. X; p. 54. For another example of a person called Iwen-Re, son of king Khafre, see *id., Excavations at Giza*, vol. VI, part III, 1934-1935, Cairo (1950), pp. 31, 33.

[101] S. Hassan, *The Great Pyramid of Khufu and Its Mortuary Chapel, Excavations at Giza*, vol. X, p. 57; *id.*, *Excavations at Giza*, vol. III, 1931-1932, Cairo (1941), p. 82, fig. 70, p. 83.

[102] *Ibid.*, p. 88.

General Comments and Conclusions:

Six general points may be considered first when going through all the listed documents.

The study encountered only seven Old Kingdom kings for whom there are documents available that deal with their deification (i.e. posthumous attestations of these kings in a divine role). Other Old Kingdom kings for whom there is not enough material to prove or to deny their deification were not included[103].

The common factor among these kings is that they were all deified after their death. The aspects dealing with their deification ranged mainly between invocation in the *hotepdinesu* formula and adoring the king so as to grant the deceased offerings or life, prosperity and health. The deified kings were invoked alone or with other deities.

In regard to deified kings invoked in the *hotepdinesu* formula, two kings can be added to W. Barta's list of deified kings[104]: king Unas and king Teti.

If the "deities of the kings" (e.g. Sekhmet of Sahure) were deified forms of the kings, as some authors believe, then they should be considered as one of the aspects of the deification of the kings, for these forms were invoked in the *hotepdinesu* formula, and were adored and presented with offerings. At any rate, such deities deserve a closer (and broader) study, which is not to be confined to the deities of Ramesses II in isolation, as seems to be the case presently. It is hoped that the appendix to this article may be an incentive to such a study.

In the *Lexikon der Ägyptologie*[105], several persons are mentioned as being deified by the *im3ḫw ḫr* formula: Pepi II (see document (VII.2) above), Djedefhor[106], Ptahhotep II[107], and Kagemni[108]. However, the idea of deification by this *im3ḫw ḫr* ("honored by") formula must be reconsidered, since this formula was used in regard to deities, kings and common (non-deified) persons alike.

The cults of the deified kings seem to have been limited to specific sites, and did not prevail all over the Nile Valley. These sites were mainly in the vicinity of the funerary complexes of the king in question, but there are a few exceptions.

In light of the above six points, it may be asked: why were these kings deified and why at the specific sites that attest to their deification?

Several decades ago, G. Jéquier suggested that those funerary cults that were still going on in later times were due to priests who found an economic advantage in exploiting the piety of the private people, people seeking to obtain a share in the offerings for their own dead or the intercession of the deified king[109]. This explanation suggests that the priests were responsible for the deification of the dead kings – kings who had nothing to do with their own deification. In line with this interpretation it could be pointed out that the funerary priests of the king must have benefited from the offerings personally. Finally, this theory also ascribes to those priests the revival of the importance of the dead kings thousands of years after these kings had died.

A hypothesis which could be added is that the cult of those deified kings might have appeared during times when the ancient Egyptians faced many difficulties and problems, especially during the Intermediate Periods, times in which people might have had an easy access to the funerary complexes

---

103 Such as the case of king Userkaf, see J. Malek, *op. cit. [note 70]*, p. 71.

104 W. Barta, *op. cit. [note 36]*, p. 232.

105 H. Goedicke, "Vergöttlichung", *LÄ* VI (1985), col. 990.

106 H. Junker, *Giza VII, Der Ostabschnitt des Westfriedhofs*, vol. I, Vienna and Leipzig (1944), p. 24. See also H. Goedicke, "Ein Verehrer des Weisen *ḎDFḤR* aus dem Späten Alten Reich", *ASAE* 55 (1958), p. 49; see also C. M. Zivie, *Giza au Deuxième Millénaire*, BdÉ 70 (1976), p. 31. See H. Goedicke, *ASAE* 55, p. 35, 45.

107 S. Hassan, *Mastabas of Ny-ʿankh-Pepy and Others, Excavations at Saqqara, 1937-1938*, vol. II, Cairo (1975), p. 70, pl. 64.

108 C. Firth and B. Gunn, *Teti Pyramid Cemeteries,* vol. I, IFAO, Excavations at Saqqara, Cairo (1926), p. 223.

109 *Ibid.*; see also G. Jéquier, *Le Mastabat Faraoun,* IFAO, Fouilles à Saqqara, Cairo (1928), p. 32.

of some kings where they left their offerings and placed their steles. The cults survived and found their way among the people during the following dynasties. But it seems that, with very few exceptions (e.g. the case of Sneferu), such cults were usually not strong enough to continue until the Late Period.

This model, in which such popular local cults grew out of the royal funerary cults and in which the dead king assumed the function of patron of the necropolis in the vicinity of his funerary complex (and as such functioned locally in the *hotepdinesu* formula), seems to fully explain the documents pertaining to Sneferu in Dahshur, and to Unas, Teti and Pepi II in Saqqara.

The one occurrence (Second Intermediate Period) of Khufu in a *hotepdinesu* formula is less clear, apart from showing that the king was deified, and the reasons behind two New Kingdom documents refering to a deified Menkaure, both from Saqqara, are unknown.

Sahure is the only one of our seven kings whose deification is not irrefutable, seeing the uncertainties that surround the interpretation of the documents pertaining to him. The popular "Sekhmet of Sahure" cult was established in the king's funerary complex at Abusir, and may refer to a deified form of the king – but a definitive conclusion would require more research on the "deities of the kings" concept. The cult seems to have gained a more than local significance (cf. the mention at Deir el-Medineh), likely because a well-known goddess was involved. Why the cult was established in the New Kingdom remains unexplained[110]. Also the other document, about sailors adoring king Sahure, is not easy to interpret unequivocally (what was adored – the living king or a statue of the dead king?).

The case of Sneferu requires special consideration. The documents from Dahshur fit the model, but the two exceptional documents from Abydos and Deir el-Medineh do not. A. McDowell convincingly suggests that the occurrence in Deir el-Medineh likely was the result of the fact that in this village the *Prophecy of Neferty* was widely read, a work in which Sneferu features prominently[111]. In the Middle and New Kingdoms, Sneferu had become the prototype of a "good king" and was renowned as pyramid builder[112]; perhaps this generic sentiment is also behind the isolated occurrence in Abydos, which otherwise must remain unexplained.

Sneferu's remarkable position in the Sinai, as deified patron of the region, also defies the standard model outlined above. R. Ventura suggests that the cult was transfered to Sinai by workmen who had previously been engaged in the building of the pyramids of the 12th Dynasty kings at Dahshur, a place where Sneferu's cult did thrive[113]. Via this hypothesis, the standard model indirectly plays a role. However, this solution alone does not suffice: not only is Sneferu's position in the Sinai documents very remarkable, featuring among the gods, but this is also the only instance among our documents in which a deified king is linked as patron to a living king (documents (I.2b) and (I.2d)). This indicates that what may have started as just a transferred popular cult became a state cult. The 12th Dynasty kings not only built their tombs where Sneferu had done so (Dahshur), but also explicitly compared their mining activities in Sinai with those of the Old Kingdom king[114]. In short, Sneferu's elevation to patron of the Sinai, based on the fact that he was remembered for his pioneering mining activities in

---

110 Sahure seems to have been remembered for the fact that he probably was the first to launch a major expedition to Punt. The Palermo Stone prominently mentions this expedition (*BAR* I §161, II §247), and a later Punt expedition under king Isesi refers to it (*Urk.* I, 131). Just before(?) the establishment of the "Sakhmet of Sahure" cult in Sahure's funerary complex there was another major Punt-expedition, namely queen Hatshepsut's. But it is impossible to say whether there is any link between these two events, nor how the war-goddess figures in this.

111 A. McDowell, *op. cit. [note 31]*, p. 100.

112 A. McDowell, *op. cit. [note 31]*, p. 100; D. B. Redford, *op. cit. [note 31]*, pp. 160, 235-236.

113 R. Ventura, "Snefru in Sinai and Amenophis I at Deir el-Medina", in *Pharaonic Egypt. The Bible and Christianity*, ed. S. Israelit-Groll, Jerusalem (1985). pp. 278-288, 372-374.

114 Cf. *BAR* I §731 "Never had the like been done since the time of the King of Upper and Lower Egypt, Sneferu, triumphant" (said about a 12th Dynasty project in the Sinai). The memory of the activities of Sneferu in the east had also been kept alive by the fact that stations and roads in the north-east Delta, on the way to Sinai, bore his name; cf. *BAR* I §312, §493.

the area, must have been a conscious political move to strengthen the prestige of the monarchy in the border area, after the troubles during the end of the Old Kingdom and the First Intermediate Period[115].

As a concluding remark it may be said that it is difficult to reconstruct what the ancient Egyptians exactly thought of those deified kings – did they consider them 'saints' (i.e. humans with divine powers and/or intermediaries between man and god) or real gods? As J. Baines suggests, despite the wealth of information that can be obtained from the documents dealing with deification, deification itself remains problematic and its significance for personal religion hard to assess[116].

Amr Aly Aly Gaber
Alexandria University, Egypt

Appendix: "God X of King Y" during the New Kingdom

In our research we encountered the goddess "Sekhmet of Sahure" (*Sḫmt S3ḥ-w(i)-Rʿ*) who had a cult from the New Kingdom into the Ptolemaic Period. The question of whether this cult tells us something about the deification of the Old Kingdom king in question could perhaps be answered if we would know what the meaning is of the other combinations of deity and king that occurred in the New Kingdom. Unfortunately, this is a topic that has not yet been explored; only the deities of king Ramesses II have thus far received attention. In the hope of stimulating future debate on the topic, an overview of such combinations ("GN of PN") is presented below.

Amun of Thutmosis I

In the tomb of *Amunnedjem* (TT 84) at *Sheikh abd el-Quraneh* (Thutmosis III's reign), see K. Sethe, *Urk.* IV, p. 136 (54); *PM* I, p. 169 (14), I.

Amun of Thutmosis III

See R. Engelbach, *A Supplement to The Topographical Catalogue of the Private tombs of Thebes (Nos. 253-334)*, pp. 22-23; G. Legrain, "Notes d'inspection: § XXXVII-XXXVIII", *ASAE* 7 (1906), p. 187. For his priest, see *ibid.* p. 187; see also G. Lefebvre, *Histoire des grandes prêtres d'Amon de Karnak jusqu' à la XXIe dynastie*, Paris (1929), p. 110.

A possible god of queen Tiy

See C. Ziegler, "Notes sur la reine Tiy", *Hommages à Jean Leclant*, BdÉ 104/I (1994) p. 536; C. F. Nims, "Another Geographical List from Medinet Habu", *JEA* 38 (1952), pp. 42-43, fig. 2, E 111.

Amun of Ramesses II

See A. Barasanti and H. Gauthier, "Stèles trouvées à Ouadi Es-Sabouà (Nubie)", *ASAE* 11 (1911), p. 80; H. Gauthier, *Le Temple de Ouadi es-Sebouâ*, II, *TIN* (1912), pl. 66A; K. Kitchen, "The Great Biographical Stela of Setau Viceroy of Nubia", *OLP* 6/7 (1975/1976), p. 295; W. Helck, "Die groβe Stele des Vizekönigs *st3w* aus Wadi es-Sabua", *SAK* 3 (1975), p. 90; *LD* III, 180 b; H. Gauthier, *Le Temple de Ouadi es-Sebouâ*, I, *TIN* (1912), p. 204, pl. LXA; *PM* VII, p. 62 (118); A. Barasanti and H. Gauthier, *op. cit.*, pp. 77-81; *LD* III, 191i; *PM* VII, p. 106 (67)-(68); P. Fuscaldo, "Aksha (Serra West): The Stela of Nakht", *ASAE* 73 (1998), pp. 61-69, pl. I; A. Hamada, "A Stele from Manshiyet Es-Sadr", *ASAE* 38 (1938), p. 220, pl. XXX; see also P. Montet, *Les nouvelles fouilles de Tanis (1929-1932)*, Paris (1933) pl. XLIX; A. H. Gardiner, "The Delta Residence of the Ramessides", *JEA* 5 (1918), p. 179; M. Sandman-Holmberg, *The God Ptah*, Lund (1946), p. 236, pl. 60, no. 307; A. Rosenvasser, "The Stele Aksha 505 and the Cult of Ramesses II as a God in the Army", *RIHAO* 1 (1972), p. 108, fig. 3, pl. 4; *KRI* III,

[115] Cf. S. Lupo de Ferriol, "Snefru en la tradición egipcia", *REE* 4 (1993), pp. 67-93. Note that in later times deified kings of the 12th Dynasty became themselves the object of cults in a border area, because of their exploits in the area in question, namely Amenemhet III in the Fayyum and Senusret III in Nubia; see H. Bonnet, *Reallexikon der ägyptischen Religionsgeschichte*, Berlin (1953), pp. 596, 697, 858. The kings of the 12th Dynasty took a keen interest in reviving the glorious past of which they thought themselves the heirs; they restored the cults of the ancestors of the Old Kingdom (notably in the Memphis-Saqqara area), recorded tales about them, and rededicated statues of Old Kingdom kings. See D. B. Redford, *op. cit. [note 31]*, pp. 151-163. Our document (III.4) likely must be seen in this light.

[116] J. Baines, "Practical Religion and Piety", *JEA* 73 (1987), p. 88.

p. 257, 15; see also *LD* III, 146; A. H. Gardiner, *JEA* 5, p. 181; M. Sandman-Holmberg, *op. cit.*, p. 236; see also W. Helck, *Die Ritualszenen auf der Umfassungmauer Rameses' II. in Karnak*, ÄgAbh. 18 (1968) p. 27, fig. 37, p. 17, fig. 16, p. 21, fig. 24, p. 29, fig. 41, p. 30, fig. 43.

Ptah of Ramesses II

See *LD* III, 147 b; dealing with this item, E. Hornung considered that Ramesses II is offering to Ptah, see *id.*, *Conceptions of God in Ancient Egypt: The One and The Many,* trans. by J. Baines, London (1983), p. 215, fig. 19; W. Helck, ÄgAbh. 18, p. 27, fig. 36; see also M. Sandman-Holmberg, *op. cit.*, p. 236, pl. 60, no. 308; see also A. Gardiner, *JEA* 5, p. 257; E. Naville, *Bubastis (1887-1889)*, EEF 8 (1891), pls. XLVI B, XXXVI C. It appeared at Saqqara, see *KRI* III, p. 181, 15. For Wadi El-Sebua, see H. Gauthier, *Le temple de Ouadi es-Sebouâ*, I, p. 151, p. 182, pl. LIVa; *LD* III, 182 C; *PM* VII, p. 60 (84); *KRI* II, p. 734; see also P. Montet, "Les dieux de Ramsés-Aimé-d'Amon à Tanis", in *Studies Presented to F. Ll. Griffith*, Oxford (1932), p. 408; *id., Les nouvelles fouilles de Tanis (1929-1932),* pl. LII; *KRI* II, p. 459.; P. Montet, *Les constructions et le tombeau d'Osorkon II à Tanis*, Paris (1947), p. 51, fig. 12. For Manshiyt es-Sadr, see A. Hamada, *op. cit.*, p. 220, pl. XXX; M. Sandman-Holmberg, *op. cit.*, p. 236, pl. 60, no. 307; A. H. Gardiner, *JEA* 5, p. 179; *KRI* II, p. 361, 10; see also *LD*, Text, V, p. 55 (top); *LD* III, 178b; *PM* VII, p. 36 (40); *KRI* II, p. 725, 9. For Tell Basta, see L. Habachi, *Tell Basta, CASAE* 22 (1957), p. 113. For Mit Rahineh, see L. Habachi, "The Discovery of the Northern Tower of the Pylon and its Inscriptions", in R. Anthes, *Mit Rahineh 1956*, Philadelphia (1965), p. 62, fig. 4 S6. Ptah of Ramesses II was invoked in the *ḥtp-di-nsw* formula on a palette, see A. Erman, "Historische Nachlese", *ZÄS* 30 (1892), p. 44. Also on papyrus Anastasi IV (verso BI), see M. Sandman-Holmberg, *op. cit.*, p. 236, pl. 61, no. 309; see also H. Gauthier, *LdR*, MIFAO 19/I, p. 121, XXXII.

Seth of Ramesses II

See P. Grandet, *Le papyrus Harris I*, BdÉ 109/I (1994), pp. 3-4. Seth, see *id.*, *Le papyrus Harris II,* BdÉ 109/II, p. 48, no. 183, p. 196, no. 806. It designates a statue for which Ramesses II founded a cult at Pi-Ramesse, see *ibid.*, p. 196. It occurred six times in Tanis, see W. M. F. Petrie, *Tanis*, part I, EEF 2 (1885), pl. IV 25A; P. Montet, *Studies Presented to F. Ll. Griffith*, p. 407; see also E. Naville, *op. cit.*, pl. XXXVI, I; *KRI* II, p. 460, 4. On the stele of 400 years, see JE 60539; P. Montet, "La stele de l'an 400 retrouvée", *Kêmi* 4 (1931), pp. 191-215, pls. XII-XV; *KRI* II, p. 287, 8. For Tell Basta, see L. Habachi, *Tell Basta*, p. 113. For Tell Nebesheh, see F. Ll. Griffith, "The God Seth of Ramesses II and an Egypto-Syrian Deity", *PSBA* 16 (1894), p. 88.

Re of Ramesses II

See P. Montet, in *Studies Presented to F. Ll. Griffith*, p. 408; *id.*, *Les nouvelles fouilles de Tanis (1929-1932)*, pl. II; see *LD* III, 148a (left); W. Helck, ÄgAbh. 18, p. 25, fig. 32. For Silsilis, see A. H. Gardiner, *JEA* 5, p. 133; see also E. Naville, *op. cit.*, pl. XLVI B; A. H. Gardiner, *JEA* 5, p. 257.

Atum of Ramesses II

See *LD* III, 147b; W. Helck, ÄgAbh. 18, p. 24, fig. 31; see P. Montet, in *Studies Presented to F. Ll. Griffith*, p. 408; *id.*, *Les nouvelles fouilles de Tanis (1929-1932)*, pl. XLIX.

Horus of Ramesses II

See JE 41407; A. Barasanti and H. Gauthier, *op. cit.*, p. 85, pl. V; *KRI* III, p. 96, 4.

Anath of Ramesses II

See P. Montet, in *Studies Presented to F. Ll. Griffith*, p. 408.

Anukis of Ramesses II

See *LD* III, 178b; *PM* VII, p. 34 (10); *KRI* II, p. 721, 3.

Mut of Ramesses II

See *LD* III, 178a; *PM* VII, p. 35 (14); *KRI* II, p. 723, 3.

Hathor of Ramesses II

See G. A. Gaballa, "Some Nineteenth Dynasty Monuments in Cairo Museum", *BIFAO* 71 (1972), p. 129.

Wadjet of Ramesses II

See P. Montet, *Les nouvelles fouilles de Tanis (1929-1932)*, pl. L. It is worthy of note that there is a fortified military post on the road from Egypt to Syria situated to the north of Migdol / Tell el-Her which was called Wadjet of Seti I. This fort was erected by Seti I and was mentioned in the Karnak inscriptions which listed his military campaigns. Clédat identified this place with Maam or Bir el-Abd, see H. Gauthier, *Dictionnaire des noms géographiques contenus dans les texts hiéroglyphiques*, I, Cairo (1912), p. 181. Also see J. Clédat, "Notes sur l'isthme de Suez", *BIFAO* 21 (1923), pp. 69, 148, 155; A. H. Gardiner believed that during the reign of Ramesses II his name replaced that of Seti I, as can be seen in the papyrus Anastasi I, and that the modern location could be "Katia", also for the Wadjet of Seti I in the Karnak temple, see A. H. Gardiner, "The Ancient Military Road between Egypt and Palestine", *JEA* 6 (1920), p. 110, pl. XL.

Thot of Ramesses II

See G. Roeder, "Zwei hieroglyphische Inschriften aus Hermopolis (Ober-Ägypten)", *ASAE* 52 (1952-1954), p. 325.

Heryeshef of Ramesses II

See W. M. F. Petrie, *Ehnasya 1904*, London (1905), pls. VIII, XVIII, XIX.

Ptah of Merenptah

See J. E. Quibell, "Lintel of Merenptah at Mit Rahineh", *ASAE* 8 (1907), p. 121; *KRI* IV, p. 53, 14; *PM* III, p. 223; H. Sourouzian, *Les monuments des roi Merenptah*, SDAIK 22 (1989), p. 35, fig. 11; W. M. F. Petrie, *Hyksos and Israelite Cities*, London (1906), p. 17, pl. XVI; *KRI* IV, p. 54, 9; *PM* III, p. 223.

Seth of Merenptah

See W. M. F. Petrie, *Tanis* I, pl. II5A.

The gods of Ramesses III

See P. Grandet, *Ramsès III histoire d'un règne*, Paris (1993), p. 284; *id.*, P. Grandet, "Deux établissements de Ramsès III en Nubie et en Palestine", *JEA* 69 (1983), p. 109.

As indicated above, the deities of Ramesses II have received some attention, generating a wide spectrum of interpretation. E. Naville[117] believed that Ramesses II was trying to attribute to himself a special claim to the protection of the gods in coupling his name with theirs. A. Rosenvasser[118] explained "Amun of Ramesses II" as a pretension of Ramesses II to deify himself as Amun without losing his own personality. L. Habachi[119] suggested that Ramesses II was connected in his aspect as a god to these gods and that he was assimilated or identified with them. D. Wildung[120] comments that Anat of Ramesses II is a divine form of Ramesses II himself in the shape of the goddess. P. Fuscaldo[121] believes that these gods are forms of the deified Ramesses II in relation to the gods. A. H. Gardiner[122] believed that Ramesses II added his name to the names of those gods to whom he dedicated temples at *Tanis* as "a display of egotism", but he didn't explain what to do with the rest of the gods in places other than *Tanis*. H. te Velde[123] suggested that the addition of Ramesses II to the names of the gods did not have a geographical meaning (i.e. gods of Ramesses-town), but indicated a special relation between these gods and Ramesses II. The opposite view was held by P. Montet[124], who suggested that the name of Ramesses II when appearing in these combinations should be considered as abbreviation or substitute of the name of the city of *Pi-Ramesse,* so that these names would always refer to gods in that city. M. Sandman-Holmberg[125] thinks it is more probable that we should take the gods of Ramesses II as the images of gods in the temples that were erected by Ramesses II. E. P. Uphill[126] believes that they are figures of the personal cult of Ramesses II. H. Goedicke[127] sees in these formulations the expression of the personal religious attachment of the king to a god. An inscription on a stele that was discovered in the Delta may shed some light on this problem, as it deals with a temple of "Amun of *Usermaatre Setepenre*". A. H. Gardiner believed that Queen Tawosret dedicated the temple to the god in question. This stele, a roughly circular slab of hard crystalline sandstone, was discovered in the little village of Bilgai close to Mansoura Governorate[128]. So the "gods of Ramesses" were not exclusively connected with Ramesses II, but their cult continued after his death.

[117] E. Naville, *Bubastis (1887-1889)*, EEF 8 (1891), p. 42.

[118] A. Rosenvasser, "The Stele Aksha 505 and the Cult of Ramesses II as a God in the Army", *RIHAO* 1 (1972), p. 104.

[119] L. Habachi, *Features of the Deification of Ramesses II*, ADAIK 5 (1969), pp. 35, 44.

[120] D. Wildung, *Egyptian Saints. Deification in Pharaonic Egypt*, New York (1977), p. 27.

[121] P. Fuscaldo, *ASAE* 73 (1998), p. 68.

[122] A. H. Gardiner, "Tanis and Pi-Raʿmesse: a Retraction", *JEA* 19 (1933), p. 127.

[123] H. te Velde, *Seth, The God of Confusion*, Leiden (1977), p. 131.

[124] M. Sandman-Holmberg, *The God Ptah*, Lund (1946), p. 237.

[125] *Ibid.*, p. 237.

[126] E. P. Uphill, *The Temples of Per Ramesses*, Warminster (1984), pp. 235-236.

[127] H. Goedicke, "Some Remarks on the 400-year-Stela", *CdÉ* 41 (1966), pp. 23-39.

[128] A. H. Gardiner, "The Stele of Bilgai", *ZÄS* 50 (1912), pp. 49-57.

The First Prophet of Amenhotep IV / Akhenaten

Federico Rocchi

Introduction

This paper discusses the deification of Amenhotep IV / Akhenaten by focussing attention on a set of documents that demonstrates the existence of a cult for that king while he was still alive. Evidence for a priest of the god Amenhotep IV is collected from the Karnak talatats and similar documents from Amarna. Further analyses draw a diachronic picture of the situation in the 18th dynasty, with particular attention to the reign of Tutankhamun. Summary and conclusions are preceded by a section that deals with the available textual evidence relating to the possible existence of a temple dedicated to Akhenaten at Amarna, and by a section with suggestions concerning the historical and religious developments of the cult of the living Amenhotep IV / Akhenaten, as well as comments on the previous underestimation of the role of the king's priest described here. An appendix deals with some related prosopographical material from the 18th dynasty.

The First Prophet of the King: evidence from the Karnak talatats

Among the inscribed talatats found reused in Karnak and dating back to the earliest years of Amenhotep IV's reign, there are many attestations of the existence of a "First Prophet of Neferkheperura Waenra", *ḥm-nṯr tpy n nfr-ḫprw-rꜥ wꜥ-n-rꜥ*. The man holding this title is always shown in jubilee scenes[1], in particular in a series of kiosk-offering scenes which depict rituals introduced in the *ḥb-sd* festival by Amenhotep IV. Essentially these scenes show the king, in typical jubilee garments, making offerings to the Aten, inside a roofless open-sky structure delimited by walls with portals and doorways which has been termed a *kiosk* (hence the name of the scenes). In these offerings, the standing king is always accompanied by three officiants who are invariably depicted on the talatats in the same place, attitude, attire and position[2]. The two men in front of the king are described by the accompanying inscriptions as the "Greatest of Seers of Ra-Harakhte in the temple of Aten in Southern Heliopolis", *wr mȝw n rꜥ-ḥr-ȝḫty m pr itn m iwnw šmꜥw*, and as the "Chief Lector Priest", *ḫry-ḥbt ḥry-tp*, respectively. In the offering scenes under examination, the Greatest of Seers carries in his hands a small spouted vessel and a *ḫrp*-scepter. The lector priest carries a papyrus roll, probably containing the ritual formula he was asked to read. The third officiant depicted in these scenes is always behind the king, who is almost always barefoot. This person invariably carries his majesty's sandals, a staff, and a small box[3]. He is titled the "First Prophet of Neferkheperura Waenra", *ḥm-nṯr tpy n nfr-ḫprw-rꜥ wꜥ-n-rꜥ*[4]. In many instances he also has the title of "Chamberlain", *imy-ḫnt*.

This First Prophet was identified about 14 times among the reliefs discovered at Karnak by the Akhenaten Temple Project (ATP)[5]. Two more instances found at Medamud were described by

[1] J. Gohary, *Akhenaten's Sed Festival at Karnak*, London 1992.
[2] Gohary, op. cit., 68-86.
[3] The theme of the sandal bearer is reminiscent of the similar one on the Narmer Palette.
[4] For a description of the images and various inscriptional writing variants of this title, see S. Tawfik, "Religious Titles on Blocks from the Aten Temple(s) at Thebes", in R. W. Smith and D. B. Redford, *The Akhenaten Temple Project* (hereafter: ATP), I, Warminster 1976, 95-101. The title *ḥm-nṯr tpy n nfr-ḫprw-rꜥ* should not be confused with the simple and more general term *ḥm-nṯr* mentioned in the Karnak talatats, which probably refers to a priest of a god and not of the king; Tawfik, op. cit., 99.
[5] Gohary, op. cit., 68-86, 117, 132.

Cottevieille-Giraudet[6]. Two additional blocks with this scene, published by Clère, were acquired around 1968 by the Louvre[7]. Two other jubilee reliefs showing the First Prophet known to the present author, but not included in the catalogue compiled by Gohary, are the Gayer-Anderson Jubilee Relief in the Fitzwilliam Museum (EGA 2300.1943) in Cambridge[8] and a relief found at Karnak and, at least up to 1975, stored in the storerooms of the Centre Franco-Égyptien des Temples de Karnak under number 3588. As of 1975, relief 3588 was still unpublished and, to the best of the present author's knowledge, it still is, apart from a black-and-white photo and a general description by Claude Traunecker for the catalogue of the Brussels exhibition of 1975[9]. On this relief only the title, not the figure, of the First Prophet, here bearing also the epithet of Chamberlain, is left. Comparison of n° 3588 with the Gayer-Anderson relief shows striking similarities of composition.

Two important facts must be stressed about this First Prophet: (1) he is never given a personal name in the inscriptions; (2) he is never found or attested in Akhetaten. The first point also applies to the Greatest of Seers and the Lector Priest on the Karnak talatats. As to the second point, the Lector Priest title also seems to disappear in Amarna[10], with a couple of possible exceptions that may date back to the early years of the reign: Meryneith, whose tomb was recently rediscovered at Saqqara[11], and Patwa, whose stela is kept in Berlin[12].

The title "Greatest of Seers" was held by the High Priests of Ra at Heliopolis and was subsequently adopted by the High Priests of Aten in the fashion described by the inscriptions on the talatats. After the move to Amarna, the title continued to be used by the High Priests of Aten but in the shorter form of "Greatest of Seers of the Aten", *wr m3w n p3 itn*[13]. These changes may reflect a change parallel to the theological evolution of the concept of the god Aten from Ra-Harakhte to later forms. By analogy a similar evolution could perhaps be envisioned for the title of First Prophet, as will be shown in the next section.

The First Servant of the King: evidence from the private tombs at Amarna

From the corpus of inscriptions on the walls of the rock-cut private tombs at Amarna, at least six instances attest the existence of a Servant of the living king. They are:

---

[6] Medamud 5427 and 5434 – see R. Cottevieille-Giraudet, *Les reliefs d'Amenophis IV Akhenaton (Medamoud 1932)*, Cairo 1936 = FIFAO 13; non vidi.

[7] Louvre 26013 and 26014 – see J. J. Clère, "Noveaux Fragments de Scènes du Jubilé d'Amenophis IV", *RdÉ* 20 (1968), 51-54.

[8] F. Ll. Griffith, "The Jubilee of Akhenaton", *JEA* 5 (1918), 61-63; id., "The Gayer-Anderson Jubilee Relief", *JEA* 8 (1922), 199-200; C. Aldred, *Akhenaten and Nefertiti*, New York 1973, #11; E. Vassilika, *Egyptian Art*, Cambridge 1995, 60-61. Whether the Cambridge relief comes from Memphis or from the Theban region is basically of no relevance to the present discussion.

[9] *Le Règne du Soleil, Akhnaton et Nefertiti. Exposition organisée par les Ministères de la Culture aux Musée Royaux d'Art et d'Histoire*, Bruxelles, 1975. See number 42 of the catalogue, 108-109.

[10] R. Hari, *Répertoire onomastique amarnien*, Geneva 1976; J. A. Taylor, *An Index of Male Non-royal Egyptian Titles, Epithets & Phrases of the 18th Dynasty*, London 2001.

[11] Meryneith changed his name into Meryra or Meryaten during the reign of Amenhotep IV / Akhenaten, but nothing is known about the fate of his titles after the change of the name. The rediscovery of his tomb at Saqqara may cast more light on his life. See M. Raven, "The tomb of Meryneith at Saqqara", *EA* 20 (2002), 26-28. Cf. URL: http://www.let.leidenuniv.nl/saqqara/homepage.htm

[12] Stela Berlin 9610; B. Porter and R. Moss, *Topographical Bibliography* (hereafter: PM) I, Oxford 1964, 797. This stela is usually dated to the beginning of the reign of Amenhotep IV. However, in the present writer's opinion the stela is, for many reasons, to be dated to the first half of the 18th dynasty.

[13] E.g. N. de G. Davies, *The Rock Tombs of El-Amarna* (hereafter: RTA), I, London 1903-8, pl. VI (tomb of Meryra).

- *b3k n nb t3wy nfr-ḫprw-rˁ wˁ-n-rˁ*

"Servant of the Lord of the Two Lands Neferkheperura Waenra"
(tomb of Panehesy (n° 6))[14]

- *b3k n nb t3wy nfr-ḫprw-rˁ wˁ-n-rˁ m pr itn*

"Servant of the Lord of the Two Lands Neferkheperura Waenra in the temple of Aten"
(tomb of Panehesy (n° 6))[15]

- *b3k tpy n nfr-ḫprw-rˁ wˁ-n-rˁ*

"First Servant of Neferkheperura Waenra"
(tomb of Tutu (n° 8))[16]

- *b3k tpy n nfr-ḫprw-rˁ wˁ-n-rˁ m pr itn m 3ḫt-itn*

"First Servant of Neferkheperura Waenra in the temple of Aten in Akhetaten"
(tomb of Tutu (n° 8))[17]

- *b3k tpy n nfr-ḫprw-rˁ wˁ-n-rˁ m wi3*

"First Servant of Neferkheperura Waenra in the Barque"
(tomb of Tutu (n° 8))[18]

Whether "Neferkheperura Waenra in the Barque" is the private name of a person who had his own cult or a peculiar name of Akhenaten is difficult to say; however the second alternative seems more likely in the eyes of the present author.

- *b3k tpy n [nb t3wy] nfr-ḫprw-rˁ wˁ-n-rˁ*

"First Servant of the [Lord of the Two Lands] Neferkheperura Waenra"
(tomb of Tutu (n° 8))[19]

At Amarna, Tutu is not only named as First Servant of Neferkheperura Waenra, but he also held the title of Chamberlain (*imy-ḫnt*)[20]. This combination of titles could imply that Tutu was also the Chamberlain and First Prophet of Neferkheperura Waenra on the Karnak talatats in the early phase of the reign. If so, then this would be a further indication that after the move from Thebes to Akhetaten a few persons very near to the king continued to serve his majesty in the new capital, even if with slightly altered titles. The similarity of the titles, and the combination with the Chamberlain title, warrants the hypothesis that the First Prophet title of the Karnak talatats became the First Servant title in Amarna.

A possible objection to this hypothesis is the fact that Tutu was appointed First Servant directly by the king at Amarna, as can be deduced from the king's speech in Tutu's tomb[21]:

> "Behold, I appoint him for me, to be the First Servant of Neferkheperura Waenra in the temple of the Aten in Akhetaten."

[14] Davies, RTA II, pl. IV; cf. Taylor, op. cit., #991.
[15] Davies, RTA II, pl. XXI; M. Sandman, *Texts from the Time of Akhenaten*, Brussels 1938, 18, [12, 16]; 20, [10]. Cf. Taylor, op. cit., #992.
[16] Davies, RTA VI, pl. XXIV; cf. Taylor, op. cit., #1001.
[17] Davies, RTA VI, pl. XIX-XX; Sandman, op. cit., 80, [16]; 82, [17-18].
[18] Davies, RTA VI, pl. XIV; Sandman, op. cit., 72, [11-12].
[19] Davies, RTA VI, pl. XVII-XVIII; Sandman, op. cit., 79, [8].
[20] Hari, op. cit., #312.
[21] B. G. Davies, *Historical Records of the Late Eighteenth Dynasty* (hereafter: HRLED), VI, Warminster 1995, 23, partly reconstructed.

He was also appointed Chamberlain by the king at the same time[22]. This may be a sign that the two titles had to belong to the same man. However, the fact that these titles were given to Tutu at Amarna does not in itself speak against the idea that he was the man depicted on the Karnak talatats. Not enough is known about the ways in which titles were given or taken away in the transition period around the move of the capital to allow definite conclusions to be drawn.

A second possible objection to the hypothesis is that a Second Prophet existed at Amarna, in the person of Panehesy:

- *ḥm-nṯr snw n nb t3wy nfr-ḫprw-rꜥ wꜥ-n-rꜥ*
  "Second Prophet of the Lord of the Two Lands Neferkheperura"
  (tomb of Panehesy (n° 6))[23]

He, however, did not hold the title of Chamberlain, but the title of Seal-bearer of Lower Egypt[24]. The existence of a Second Prophet of the King at Amarna would normally imply the existence of a First Prophet, but no evidence for that office exists at all. It seems unlikely that the First Prophet position was just abolished, or the Second Prophet would logically have been renamed. The problem disappears with the hypothesis that the place of the First Prophet of the King was filled by the First Servant of the King.

No reason for the proposed change of the title from First Prophet to First Servant is at hand at the moment, but the reasons may be similar to those which led to the change in the Greatest of Seers title mentioned previously.

When did the First Prophet of the King operate? Comparisons

It is interesting to compare the usage of this title under Akhenaten with that of similar titles relating to kings from other periods of the 18th dynasty. *Table I*, largely based on the Index compiled by Taylor, reports some other occurrences of the title of First Prophet of kings or queens[25].

Firstly, the cases involving kings before the Amarna era must be analyzed. The datable instances clearly show that the title was held by a person who lived after the death of the king of whom he was First Prophet[26]. The case of Piay is less clear and needs a closer look. In this case, the title is found in the left part of the rear wall of the hall of the now lost Theban tomb C6 of Ipy, Piay's father, a tomb which PM dates to the reign of Tuthmosis IV. However, there are indications that the tomb might be later: inside the tomb, three cartouches of Amenhotep III were found[27]. This suggests that Ipy might

---

[22] B. G. Davies, ibid., 23.

[23] Davies, RTA II, pl. IX; Sandman, op. cit., 26, [16]. Cf. Taylor, op. cit., #1466.

[24] See Hari, op. cit., #96, who adds that Panehesy also was First Servant of the Aten in the Temple of Aten at Akhetaten, Overseer of the double granary of Aten at Akhetaten, and Overseer of the cattle of the Aten. Unfortunately, it is not clear from the relief of tomb n° 6 who is the "Servant of the Lord of the Two Lands Neferkheperura Waenra" (cf. main text and n. 16). It seems unlikely that it was Panehesy, because in all other instances (in the Amarna tombs and elsewhere) he is never called thus. In view of the attestations of the First Servant of the King title for Tutu, it is seems to be the general opinion that Panehesy did not hold that title (e.g., Hari, op. cit., #96). If further research would prove that he was both First Servant of the King and Second Prophet of the King, then that would pose a possible objection to the present hypothesis that these titles were part of the same hierarchy.

[25] J. A. Taylor, *An Index of Male Non-royal Egyptian Titles, Epithets & Phrases of the 18th Dynasty*, London 2001. *Table I* does not pretend to be exhaustive.

[26] Another example relating to an 18th dynasty king, but from the 19th dynasty TT31, is the *ḥm-nṯr tpy n mn-ḫpr-rꜥ*, Khons called To, First Prophet of Menkheperra, in the reign of Ramses II (PM I, 47).

[27] B. M. Bryan, *The Reign of Thutmose IV*, London 1991, 302.

Title	Translation	Primary Instance & Title Owner[28]	Number in Taylor's Index	Date according to PM
ḥm-nṯr tpy n imn n mn-ḫpr-rʿ m ḥnkt-ʿnḫ	First Prophet of Amun and of Menkheperra in Henket-ankh[29]	TT72, Ra[30]	1483	reign of Amenhotep II
ḥm-nṯr tpy n ʿȝ-ḫpr-kȝ-rʿ	First Prophet of Aakheperkara	Shrine 15, Aakheperkaraseneb(?)	1494	?[31]
ḥm-nṯr tpy n ʿȝ-ḫpr-kȝ-rʿ m ḫnmt-ʿnḫ	First Prophet of Aakheperkara in Khnemet-ankh[32]	FC/DM 605, Enta (*ntȝ*)	1495	?
ḥm-nṯr tpy n mn-ḫpr-rʿ mȝʿ ḫrw	First Prophet of Menkheperra true of voice	TT72, Ra	1503a	reign of Amenhotep II
ḥm-nṯr tpy n mn-ḫprw-rʿ	First Prophet of Menkheperura	TT C6, Piay	1504	reign of Tuthmosis IV ?
ḥm-nṯr tpy n mryt-imn mȝʿt ḫrw	First Prophet of Merytamun true of voice	FC/DM 226, Meh(y)	1507	?
ḥm-nṯr tpy nbt tȝwy iʿḥ-ms nfrt-iry	First Prophet of the mistress of the Two Lands Ahmose-Nefertari	TT255, Djehuty	1508	reign of Horemheb
ḥm-nṯr tpy n nfrt-iry	First Prophet of Nefertari	FC/DM 210, Amenhotep	1509	?
ḥm-nṯr tpy n nb-ḫprw-rʿ	First Prophet of Nebkheperura[33]	TT40, Khay	not included	reign of Tutankhamun

Table I: First Prophets of the King in the 18th Dynasty

have outlived Tuthmosis IV who did not have a long reign[34]. Lise Manniche suggested that the tomb was decorated after the reign of Tuthmosis IV[35]. In her book on the private Theban tombs, Friederike Kampp writes "Datierung: T.IV./A.III."[36]. Therefore, it is very likely that Piay, Ipy's son, held the title mentioned in *Table I* after the death of Tuthmosis IV. Everything thus points towards a preliminary

---

[28] "TT" means Theban Tomb; "Shrine" means shrines at Gebel-Silsilah, as recorded by Caminos & James; "FC/DM" means a funerary cone in the Davies & Macadam corpus.
[29] Henkhet-ankh is part of the name of the funerary temple of Tuthmosis III at Thebes West.
[30] See recently P. A. Piccione, "Theban Tombs Publication Project: Tombs no. 72 (Rây) and 121 (Ahmose), Season Winter 1990", at URL: http://www.cofc.edu/~piccione/t2p2/1990report.html
[31] Reign of Tuthmosis III (?), according to W. Helck, *Zur Verwaltung des Mittleren und Neuen Reiches*, Leiden-Köln 1958, 435.
[32] Khnemet-ankh is part of the name of the funerary temple of Tuthmosis I at Thebes West.
[33] As reconstructed by Helck, *Urk.* IV, 2068, 14.
[34] The most recent assessment by Bryan (cf. note 27) indicates about 10-12 years.
[35] L. Manniche, *Lost Tombs. A Study of Certain Eighteenth Dynasty Monuments in the Theban Necropolis*, London 1988, 54.
[36] F. Kampp, *Die Thebanische Nekropole, Theben XIII*, Mainz 1996, Teil 2, 620.

conclusion that, before the Amarna period, the title of First Prophet of a king referred to the posthumous cult of a deified dead pharaoh[37].

The post-Amarna case of Tutankhamun also needs a careful analysis. It is certainly true that Khay (see *Table I*) lived during Tutankhamun's reign[38], as did Merymes, who was probably Second Prophet of Nebkheperura and a *wab*-priest of this king[39]. However, one should be certain about the exact time in the lives of these men at which they were appointed First Prophet before drawing precise chronological conclusions from these facts. Another case that needs to be looked at is Userhat's stela in the Metropolitan Museum of New York[40]. Userhat is described not as a First Prophet of a king but as a First Prophet of Amun. However he served in a temple of Tutankhamun that is described as *ḥwt nb-ḫprw-rᶜ* – a formula used in the same stela for a temple of Amenhotep III, *ḥwt nb-m3ᶜt-rᶜ*. If one supposes that this is a funerary temple, and if, with Hayes, one dates the stela to the reign of Tutankamun[41], then one could have a possible indication of a funerary cult for him during his lifetime. But such a conclusion is very speculative. Firstly, it is impossible to date the stela with any certainty by relying only on stylistic grounds as was done by Hayes. Secondly, an identification of the *ḥwt nb-ḫprw-rᶜ* with Tutankhamun's funerary temple remains uncertain. The history of the funerary temple of pharaoh Tutankhamun, its name, fate and whereabouts are unknown[42]. And thirdly, even if the *ḥwt nb-ḫprw-rᶜ* was a funerary temple of the king, and even if Userhat lived during his reign, then strictly speaking that would only indicate that the funerary temple was completed during the life of the king, something which will have happened regularly[43]. But that does not in itself imply a cult being in effect, as Userhat was not a First Prophet of the King. It may perhaps be hypothesized that, until the king died, a First Prophet of Amun oversaw the operation of the completed royal funerary temple.

In contrast to the above examples, the case of Amenhotep IV analyzed in the previous paragraphs shows that Akhenaten certainly overcame the praxis of a posthumous cult, since the First Prophet of the Karnak talatats undoubtedly served while the king was still alive. The same obviously applies for the First Servant found in the private tombs at Amarna.

Did a temple dedicated to Akhenaten exist at Amarna?

It is noteworthy that in several instances (cf. tombs of Tutu and Panehesy), the phrase "in the House of Aten" is added to the First Servant of Akhenaten title. It is not fully clear what this indicates. Did the king not have a funerary temple of his own, but a chapel or cult-statue in the Aten temple, to stress the link between god and king?[44] Or does the *m pr itn* phrase tie in with a common element in the

[37] The First Prophet in this case should have had different roles and tasks from the *Ka* Servant.

[38] B. G. Davies, HRLED VI, 49, partly reconstructed.

[39] Cf. note 38. A certain Pairy, *wab*-priest of Tutankhamun, is known from an ushabti of his, now in the British Museum. See R. H. Wilkinson, *The Complete Temples of Ancient Egypt*, London 2000, 192.

[40] N. 05.4.2, gift of the Egypt Exploration Fund in 1905; W. C. Hayes, *The Scepter of Egypt*, New York 1959, vol. II, 306.

[41] Cf. note 40.

[42] For the *ḥwt nb-ḫprw-rᶜ* temples in Thebes, see M. Eaton-Krauss, "Tutankhamun at Karnak", *MDAIK* 44 (1988), 1-11, and the references cited there. The opinion of Haring (B. J. J. Haring, *Divine Households*, Leiden 1997, 20-30), that a *ḥwt* temple of a king was always a royal memorial temple, is debatable. If his opinion is accurate then how do we explain that Tutankhamun had two different *ḥwt* temples in Thebes? Strangely, Haring (op. cit., 421) does not even mention the existence of the two *ḥwt* temples, the *ḥwt nb-ḫprw-rᶜ mrj imn grg w3st* and the *ḥwt nb-ḫprw-rᶜ m w3st*. According to Eaton-Krauss (op. cit., 11), whose paper is not quoted by Haring, both were autonomous structures of respectable proportions, unequivocally associated with Tutankhamun.

[43] For examples, see Haring, op. cit., 24, 26-29.

[44] Could "First Servant of the King in the House of Aten" be the mirror image of the non-Amarna cases in which we find a "First Prophet of Amun in the funerary temple of the king" (see, e.g., the case of Ra in *Table I*)? In other words, the national god would no longer have a cult in the memorial temple of the king, but the king would have a memorial cult in the temple of the national god. The strong identification of Akhenaten with the living Aten could perhaps have done away with the need to have two separate temples, at least at Amarna.

names of funerary temples? For Theban funerary temples were often named *ḥwt* + king's name + *m pr* + god's name, and it is usually presumed that the *m pr* phrase in these cases is meant to express the administrative inclusion of the royal temple in the estate of the main local temple of a god, although an economic dependence or a religious implication cannot be ruled out[45].

As no certain answer to the question of whether Akhenaten had his own temple at Amarna or not is currently available, a discussion about the nature (funerary, cultic, or both) of this temple would be premature. However, for completeness' sake, it is at least necessary to analyze the archaeological and textual evidence suggesting the presence of a temple of the king at Amarna. From the many scattered, recorded inscriptions, four expressions may have been used to designate a temple dedicated to Akhenaten[46]:

- *pr imn-ḥtp nṯr ḥkꜣ wꜣst* which reports the early name of Akhenaten; this has been found on two hieratic dockets[47] and on a jar sealing[48]. The dockets probably referred to the place of origin of wines. But not all the wines available at court were from the Delta region, as is often found, and it is possible that wine was imported and later labelled as "belonging to", rather than "originating from" an estate.
- *pr ꜣḫ-n-itn* found at Amarna[49], on hieratic dockets[50] and on jar sealings[51]. Another highly noteworthy example, from Lower Egypt, is the inscription on the famous block published by Nicholson and found near Mit-Rainah[52].
- *pr nfr-ḫprw-rʿ* found only once[53].
- *ḥwt nfr-ḫprw-rʿ* found on a hieratic docket[54] and on a jar sealing[55].

As Fairman noted[56], it is highly difficult to pinpoint the exact meaning of the various *pr* and *ḥwt* terms appearing in the inscriptions. For the first word the meaning can range from "house" to "residence", from "estate" to "storehouse" or to "temple"; or the meaning could even be a combination of all these. Haring argues that we should consider the principal meaning of *ḥwt* to be "funerary temple"[57], which would confirm the existence of a temple of Akhenaten. However, if Haring's statement is incorrect, we are left with the vague choice between the meanings of "temple", "mansion", "estate" and "administrative district". Be that as it may, the idea of Akhenaten having at least a funerary temple is not unlikely, since at Amarna we have his tomb but not the structures for his cult (and perhaps these are not to be found at Thebes West near the other funerary temples of the 18th dynasty). Additionally, we don't know which features of the funerary cult of the old religion were kept

[45] For the different opinions about this unsettled matter, cf. Haring, op. cit., 30-34.

[46] The documents that mention the *pr sḥtp itn* are intentionally excluded here, since many doubts still exist about the identity of this *sḥtp itn*. The present writer nonetheless believes this was Akhenaten; cf. *The City of Akhenaten* (hereafter: COA), III, London 1923-1951, 198-199.

[47] W. M. F. Petrie, *Tell el Amarna*, London 1894, pl. XXII.12; COA I, pl. LXIV.22.

[48] Petrie, op. cit. pl. XXI.2.

[49] Davies, III, pl. XXVII.

[50] COA III, dockets 22-27; Petrie, op. cit., pl. XXII.8, pl. XXIV.88; COA I, pl. LXIV.66-69.

[51] COA III, pl. LXXXI.23; Petrie, op. cit., pl. XXI.3-5; COA I, pl. LV.I.

[52] C. Nicholson, "On Some Remains of the Disk Worshippers Discovered at Memphis", pl. II, in *Aegyptiaca*, London 1891. Now the block is in the Museum of Antiquities of the University of Sydney (Nicholson Museum 1143). For the enormously long bibliography on this item, see also B. Löhr, "Ahanjati in Memphis", *SAK* 2 (1975), 139-187 (1975) and W. J. Murnane, *Texts from the Amarna Period in Egypt*, Atlanta 1995, 97, n. 50.

[53] Petrie, pl. XXII.10.

[54] COA II, pl. LVIII.24.

[55] COA III, pl. LXXXI.20.

[56] COA III, 197-198.

[57] Cf. note 42.

and maintained into the monotheistic one. It must also be added that many other *pr* and *ḥwt* have been attested by documents found at Amarna, not belonging to Akhenaten but to Amenhotep III, Nefertiti, Meritaten, Maketaten, Ankhesenpaaten, Smenkhkara, Tiye, Baketaten, Tuthmosis I, Amenhotep II and Tuthmosis IV. So the situation is far from being clear and definite. Further extension of the excavations at Amarna could help in clarifying the state of affairs about these structures. It is beyond doubt that many areas at Amarna, both north and south of the presently known sites, still need extensive excavation[58].

Further thoughts on the cult of the deified king

The preceding analysis of the jubilee scenes in the Karnak talatats suggests that the rites introduced by Amenhotep IV in the jubilee ceremonies required at least the presence of the High Priest of the solar deity and of his own priest. It is possible that, after the change of the capital, the requirement to have these two men present in the later jubilees imposed the creation of two newer priestly titles, the Greatest of Seers of the Aten and the First Servant. Why the First Prophet/Servant also had to be Chamberlain is presently not known. However, given the nature of the Heb Sed rituals, it seems reasonable to suggest that the presence of a priest of Amenhotep IV / Akhenaten was felt necessary when the king, once dead and thus having become a venerable god, underwent rebirth. Consequently, a parallel can be established between the sun-god and his High Priest on one side, and the reborn king and his own priest on the other. This notion may also suggest why the priest of the king was also his chamberlain: he simultaneously served a man (the living king) as chamberlain, and a god (the dead and resurrected king) as priest. In ancient Egyptian thought, parallels derived from religion always lead to identifications. This particular parallel may well be a statement of strong kinship between Amenhotep IV and the sun-god. How closely this idea is linked to the deification program of Amenhotep III proposed recently by Johnson[59] is presently a matter of speculation which will not be further pursued here. What is certain is that the association of the titles of Chamberlain and First Prophet/Servant in the same person in jubilees was first introduced by Amenhotep IV. While a Chamberlain is present in jubilee scenes even from the Old Kingdom[60], the participation of a Prophet of the King seems to be peculiar to the ceremonies invented by Amenhotep IV[61]. Another possible cause for the introduction of a non-posthumous cult by Amenhotep IV / Akhenaten, apart from the desire to establishing a strong kinship with the sun-god, could be his policy of going against the ancient religious conceptions and beliefs of the Osirian tradition and myths in which only the dead king was identified with the supreme deity.

Finally, the opinion of Gohary should be noted, who suggests that the First Prophet on the Karnak talatats was not really a funerary priest[62]:

> "It is possible however, that, in spite of his title, this man was simply an acolyte who carried the king's equipment, a theory strengthened by the fact that he is never shown on the Karnak talatats

[58] For the recently investigated limestone quarries to the North of El Till, see J. Harrell, "Ancient Quarries near Amarna", *EA* 19, 2001, 36-38. For the recently discovered areas with two Amarnian cemeteries for common people, see B. Kemp, "Resuming the Amarna Survey", *EA* 20, 2002, 10-12.

[59] E.g. R. W. Johnson, "Amenhotep III and Amarna: Some New Considerations", *JEA* 82 (1996), 65-82.

[60] See, e.g., L. Borchardt and H. Kees, *Das Re-Heiligtum des Königs Ne-woser-re III*, Berlin-Leipzig 1905-28, 23-24; W. Helck, *Untersuchungen zu den Beamtentiteln des ägyptischen alten Reiches*, Glückstadt 1954, 29; A. H. Gardiner, *Ancient Egyptian Onomastica* I, Oxford 1947, 23* (83).

[61] Clère, op. cit., 54.

[62] Gohary, op. cit., 224-225, note 23.

carrying anything of a religious nature like the Greatest of Seers, but only the king's sandals and a small chest."[63]

In the opinion of the present author, the existence of a cult of the deified living Amenhotep IV / Akhenaten is attested in so many documents that it seems more realistic to take the evidence of the title at face value.

Summary and conclusions

In this paper attention is drawn to various talatats from Karnak that testify to the existence of an always unnamed First Prophet of Amenhotep IV. This prophet operated exclusively, as far as we can tell from what has been so far discovered, in certain ceremonies performed during the first jubilee of Amenhotep IV. In these ceremonies the king is always accompanied by two other, likewise unnamed, men: the "Greatest of Seers of Ra-Harakhte in the temple of Aten in Southern Heliopolis" and the "Chief Lector Priest". After the move to Amarna these three men disappear from the scene. While the Lector Priest is apparently absent from texts from Amarna, the Greatest of Seers of Ra-Harakhte is replaced by the Greatest of Seers of the Aten. In the light of this substitution, due to theological reasons, the idea is advanced that the First Prophet appearing in the Karnak talatats has also been replaced at Amarna, by the First Servant of the King. This idea is strengthened by the fact that the First Prophet also held the title of Chamberlain; in fact at Amarna this title is held by the person who also had the role of First Servant of Akhenaten: Tutu. It cannot be excluded that Tutu was the man represented on the Karnak talatats, even if we are told from the biography in his tomb at Akhetaten that he was appointed First Servant and Chamberlain by the king in the new capital city. Finally, from a preliminary analysis, it seems that Akhenaten was the first pharaoh to introduce his own funerary and/or divine cult while still living. It is unclear whether this innovation was continued by his immediate successors, but it seems unlikely.

Federico Rocchi
rocfed@tin.it
Reggio Emilia, Italy

Appendix: Other Prophets of the King from the 18th Dynasty

In this appendix, attention is paid to Second Prophets of the King who lived before the Amarna Period[64], to see if their data confirms the conclusion of the main text, namely that prophets of a king before the Amarna period only served after the death of this king. The instances identified during the research for this paper are listed in *Table II*. The following remarks survey the prosopographical evidence related to most of these men.

[63] Gohary, op. cit., 225, end of note 23. Note that, for some reason, Gohary overlooks the staff.

[64] Note that in the main text two Second Prophets were mentioned who lived during and after the Amarna Period, namely Panehesy and Merymes. As there is no evidence for other titles it seems likely that the "Prophet of Aakheperura" title of the first entry of *Table II* refers to Neferhebef being Second Prophet of Amenhotep II (cf. second entry in *Table II*), rather than to, e.g., First Prophet of the King later in his life.

Title	Translation	Primary Instance & Title Owner	Number in Taylor's Index	Date
ḥm-nṯr n ꜥꜣ-ḫprw-rꜥ	Prophet of Aakheperura	FC/DM 54[65], Neferhebef	1454	reign of Amenhotep II or later
ḥm-nṯr snnw n ꜥꜣ-ḫprw-rꜥ	Second Prophet of Aakheperura	Statue BM 31, Neferhebef[66]	1455	reign of Amenhotep II or later
ḥm-nṯr snnw n mn-ḫpr-rꜥ	Second Prophet of Menkheperra	FC/DM 228, Kaemamen[67], father of Seqed	not included	reign of Amenhotep II
ḥm-nṯr snnw n mn-ḫpr-rꜥ	Second Prophet of Menkheperra	FC/DM 590, Seqed, son of Kaemamen[68]	1464	reign of Amenhotep II / Tuthmosis IV
ḥm-nṯr snnw n mn-ḫpr-rꜥ	Second Prophet of Menkheperra	Huy[69]	not included	reigns of Tuthmosis III – Amenhotep III ?
ḥm-nṯr snnw n mn-ḫpr-rꜥ	Second Prophet of Menkheperra	Aakheperraseneb, son of Kaemamen[70]	not included	reign of Amenhotep II

Table II

---

[65] FC/DM refers to entries in N. de Garis Davies & M. F. L. Macadam, *A Corpus of Inscribed Egyptian Funerary Cones by the late N. de Garis Davies*, Part I (Oxford, 1957). Funerary cones of type DM 54 in the British Museum are, for instance, cones BM EA 9671-9691, 13863-13866, 13879, 13886-13887, 62692 and 65203. Cf. C. N. Reeves, D. P. Ryan, "Inscribed Egyptian Funerary Cones in Situ: An Early Observation by Henry Salt", *VA* 3 (1987), 47-49. Another cone of this type may be found in Vienna's Kunsthistorisches Museum as AOS 1710, see D. van der Plas (ed.), *Egyptian Treasures in Europe*, vol. 5, CD-ROM by CCER. The only title given on the cones is Prophet of Aakheperura.

[66] The only title listed is Second Prophet of Aakheperura. There is unanimity among Egyptologists in equating this Neferhebef with the Neferhebef of the previous entry, based on name, title, and time period; cf. P. der Manuelian, *Studies in the Reign of Amenophis II*, HÄB 26, Hildesheim 1987, 143.

[67] Funerary cones of Kaemamen (Amenemka) can be found in many museums and private collections around the world. Cf. M. Werbrouck, "Cônes funéraires de Kaemimen", *CdÉ* 33, n. 66 (1958), 223-226, and H. M. Stewart, *Mummy Cases & Inscribed Egyptian Funerary Cones in the Petrie Collection*, Warminster 1986, 45-46. Cones of type DM 228 are, for instance, Louvre CF 37 and Petrie Museum UC 37659. The main study on Kaemamen is A. de Buck, "Een zwerver thuisgebracht", *JEOL* no. 15, Leiden, 1957-1958, 5-11 (I warmly thank Carolien van Zoest, Jacobus van Dijk, and Michael Tilgner for having provided me with copies of the *JEOL* paper). In this work de Buck, basing himself on the inscriptions on the statues he published, reached the conclusion that Kaemamen lived under Amenhotep II; this proposed date was accepted by Peter der Manuelian (op. cit., 145-146).

[68] See previous note. Cones of type DM 590 are, for instance, Louvre CF 83 and Petrie Museum UC 37975. Cones of type DM 246 (like, for instance, Louvre CF 183, Petrie Museum UC 37671, and Bruxelles 152 = E 3989) list the same titles for Kaemamen as DM 590, but do not mention the son. For Bruxelles E 3989, see D. van der Plas (ed.), *Egyptian Treasures in Europe*, vol. 2, CD-ROM by CCER. Photographs of the cones in the Petrie Museum can be seen at the Museum's website, URL: http://www.petrie.ucl.ac.uk/search/index.html

[69] P. der Manuelian, op. cit., 141-147. It is not possible to comment upon this Huy in more detail, as the present writer was not able to consult the Russian source for the one monument on which he is mentioned. Der Manuelian refers to an article by Bogoslowski, and provides no dating criteria, only giving these wide margins ("Tuthmosis III – Amenhotep III?"); even the title "Second Prophet of Tuthmosis III" seems not beyond doubt.

[70] For this person, see the discussion below on the monuments of Kaemamen.

Neferhebef, son of Ithu[71] and Henutweret, is known only through the texts[72] on his parents' statue BM EA 31[73] and from the funerary cones of the type listed as number 54 of the Davies&Macadam corpus. On the cones the name of Neferhebef's wife can be read as Taway or Tataway. The reasons for which BM EA 31 has tentatively been dated to the reign of Amenhotep II are unknown to the present writer. An inquiry directed to the staff of the British Museum[74] revealed that a later dating is much more likely: although Edwards[75] favoured an older dating, recently Wiese[76] preferred to place BM EA 31 in the reign of Tuthmosis IV, and stylistic reasons and the palaeography of the inscriptions on the statue are arguments that urge M. Marée[77] also to choose the reign of Tuthmosis IV. The most recent information would therefore imply that it is highly likely that Neferhebef held his priestly titles after the death of Amenhotep II.

Other monuments exist that bear the name of Neferhebef, but it is difficult to determine whether they all belong to the same person. It may be useful to look at them for prosopographical reasons:

(a) Statue Louvre A57 [N.58][78], of Neferhebef with wife Taiu and son Benermerut, dated in the range from the reign of Amenhotep II to that of Amenhotep III. Unfortunately, the text on this statue does not list any titles for Neferhebef.

(b) The black granite Bologna statue KS 1825[79], from Memphis and from the reign of Amenhotep III, belonging to Amenhotep (also called Huy), the famous chief steward in Memphis, in which he is stated to be the son of the dignitary (*s3b*) Neferhebef and of [Tu]tuya[80]. Strangely enough, Pernigotti[81] gave the name of Amenhotep's father on the Bologna statue (not mentioned among the material published in the fundamental papers on the subject[82]) without any reserve as Nefernebef (and not as Neferhebef), and he gave the name of the mother, unfortunately broken on the statue, as ending with *-tiya (and not as *-tuya), so there seems to be some uncertainty about these readings. The reasons why PM emended the names into Neferhebef and Tutuya are unknown to the present writer, but paleographically the difference between the *nb* and the *ḥb* signs is very small.

(c) The red granite Leiden pyramidion A.M. 6 (K1)[83], of Amenhotep, chief steward in Memphis, which gives the dignitary (*s3b*) Heby as his father and the lady of the house Tutuya (slightly

[71] Cf. P. der Manuelian, op. cit., 125-126 and 143.

[72] *Urk.* IV, 1503-1504.

[73] Painted sandstone, 76 x 33.5 x 54.5 cm. In the British Museum, donated by Henry Salt. The statue is now on a long-term loan to the Antikenmuseum in Basel; a colour photograph can be seen at the following URL: http://www.antikenmuseumbasel.ch/aegypten/aegypten.html

[74] Thanks are due to Marcel Marée, Curator of Ancient Egypt and Sudan in the British Museum; personal communications, August 2002.

[75] I. E. S. Edwards, *Hieroglyphic Texts from Egyptian Stelae etc.*, VIII, London 1939, 6 and pl. 7.

[76] A. Wiese, *Antikenmuseum Basel und Sammlung Ludwig. Die Aegyptische Abteilung*, Mainz am Rhein 2001, 96-97 [58] (non vidi).

[77] Cf. note 74.

[78] PM VIII.2, 488.

[79] The number 157 given in PM VIII.2, 555, as that of Kminek Szedlo's ancient catalogue, is incorrect.

[80] PM VIII.2, 555. The Bologna statue only lists the title "chief steward in Memphis" for Amenhotep. This Amenhotep (Huy) should of course not be confused with his two namesakes of the same period.

[81] S. Pernigotti, *La Statuaria Egiziana nel Museo Civico Archeologico di Bologna*, Bologna 1980, 55-56 and pl. XVI-XVII & LXXII-LXXIV.

[82] G. Bagnani, "Il Primo Intendente del Palazzo, Imenhotpe, detto Huy", *Aegyptus* 14, 33-48; W. C. Hayes, "A Writing-palette of the Chief Steward Amenhotpe and some Notes on its Owner", *JEA* 24, 1938, 9-24.

[83] Bagnani, op. cit., 45, and Hayes, op. cit., 15. See also *Urk.* IV, 1811-1812. The text on the pyramidion gives Amenhotep the following titles: "the hereditary prince and noble, seal-bearer of the King of Upper and Lower Egypt, sole friend, the mouth which makes things peaceful in the entire land, favourite of Horus in his house, the one who conducts the festival of Ptah, South of his Wall, for all the gods of the White Wall, the overseer of works in Khnumet-Ptah, prophet of Great of

different spelling from the ending of the name on the Bologna statue) as his mother. As it is extremely unlikely that there would exist two different Amenhoteps, both chief steward in Memphis during the reign of Amenhotep III, and with mothers with quite similar names, it is generally assumed that the Bologna statue and the Leiden pyramidion refer to one person. This would indicate that Heby was a nickname or shortened form of Neferhebef[84].

(d) A graffito at Aswan[85], of a Heby, chief steward in Memphis. As it mentions the first Nubian campaign of Amenhotep III[86], Heby must have held this office at least into Amenhotep III's 5th year[87].

(e) The depictions and texts in TT 55. Ramose, the famous owner of the tomb, and Amenhotep, in the tomb identified as the chief steward of Memphis, are thought to be close relatives, as Amenhotep is depicted among the family in TT55 and is given the epithet of "brother". In this tomb, the father of Ramose is called "Overseer of the bulls of Amun, overseer of the granary of Amun in the districts which are in Lower Egypt, the scribe Neby"[88]. Under the presumption that Ramose and Amenhotep were real brothers and sons of the same father, born to different wives (of whom the names are not known), it is generally presumed that "Neby" in TT55 is a writing error for Heby, which would make him the same person as the Heby of the Leiden pyramidion A.M. 6 (K1).

(f) A funerary cone of a certain Heby, Florence Museum 6690[89]. It reports the "Revered one with Osiris, scribe and accountant of the bulls of Amun throughout Upper and Lower Egypt, Heby, justified. The son of the scribe and accountant of the bulls of Amun, Senimes, justified, born of the lady of the house Ruia"[90].

The question arises: how are the persons on the above monuments related to each other? As we have seen above, the Neferhebef of KS 1825 is certainly the Heby of A.M. 6, and the latter is with reasonable certainty the "Neby" of TT55. The Louvre statue A 57 and the Bologna statue KS 1825 could well mention the same Neferhebef, given the similarity in the wife's name (Benermerut would then become a brother of Amenhotep). Seeing the similarity in titles (referring to Amun and cattle), the Heby of the Florence cone FM 6690 might be the same person as the Heby of TT55[91]. The chief steward Heby of the Aswan graffito might be the Heby of the Leiden pyramidion, i.e. the father of Amenhotep Huy, if we assume that the son succeeded his father in the same social position[92]. The most intriguing question however is whether the Neferhebef, Second Prophet of Aakheperura (*Table II*), is the same person as any of these men. Obviously, the Neberhebef of BM EA31 cannot be the Heby of the Florence cone, seeing the difference in the name of the parents. If we equate the Heby of the

---

Magic, overseer of the prophets in the temple of Sakhmet, overseer of the two granaries in the entire land, the royal scribe of the recruits, high steward in Memphis".

[84] A shortened form Heby would support the reading Neferhebef on the Bologna statue, versus Pernigotti's Nefernebef. Perhaps this is what motivated the preferred reading of the name in PM.

[85] *Urk.* IV, 1793; cf. PM V, 245 and following; LD, Text IV, 119. The only non-generic title given in the graffito is "chief steward in Memphis"; the other titles are "hereditary prince and noble, the eyes of the King of Upper and Lower Egypt, true royal scribe".

[86] Z. Topozada, "Les Deux Campagnes d'Amenhotep III en Nubie", *BIFAO* 88 (1988), 153-164.

[87] J. Malek, "The Saqqara Statue of Ptahmose, Mayor of the Memphite Suburbs", *RdÉ* 38 (1987), 117-137.

[88] *Urk.* IV, 1784.

[89] Davies & Macadam n° 15.

[90] *Urk.* IV, 1792.

[91] Helck (*Urk.* IV, 1792) states, without providing any motivation, that the Heby of this cone was the same person as Heby, father of Ramose.

[92] Z. Topozada, op. cit., 156, uses the Bologna statue KS 1825 to state that Amenhotep was a son of Heby, the mayor of Memphis. It is not clear how Topozada reached this conclusion; the bibliography quoted is only the exhibition catalogue *L'Egitto Antico nelle Collezioni dell'Italia Settentrionale*, Bologna 1961, edited by S. Curto, pl. 22 (non vidi).

Florence cone with the Heby of TT55, then prophet Neferhebef would not be the father of Amenhotep (KS 1825, A.M. 6, TT55). If we do not make that equation, then it remains possible that the prophet could be the father of Amenhotep, seeing a certain similarity in the names of their wives (Tataway/Taway vs. Tutuya), although the titles would remain very different[93]. Under either option, it remains possible that the Neferhebef of Louvre A57 is to be equated, not with the father of Amenhotep Huy, but with our prophet, again because of a certain similarity in the names of the wives (Taway/Tataway vs. Taiu). But it is difficult, if not impossible, to equate all men called Neferhebef/Heby[94].

A Second Prophet of Tuthmosis III, whose name occurs in two versions, differing in the order of the hieroglyphs, namely Kaemamen (which is similar in pattern to the more common name Kaemwaset) and Amenemka, is mentioned on the following monuments:

(a) A fragmentary seated statue of Kaemamen in the private collection of G. S. Fernhout at Wolfheze, near Arnhem, identified by de Buck[95]. The inscriptions on the statue say that Kaemamen was born of the king's wife (*ḥmt-nsw*) Henuttawy and was "child of the *kap*"[96]. He is given the titles of "Second Prophet of Amun in Henket-ankh", "Second Prophet of Menkheperra in Henket-ankh", "Fourth Prophet of Amun", and "seal-bearer of the king". De Buck is surely correct that the text is meant to indicate that Kaemamen was Second Prophet of Amun in Henket-ankh, the funerary temple of Tuthmosis III in Western Thebes, and Fourth Prophet of Amun in the Amun temple of Karnak[97].

(b) A statue of Kaemamen and his wife, Louvre 10443, published by de Buck in the same study. Kaemamen here only holds the title of "Fourth Prophet of Amun" (although his *ka* is said to

[93] W. J. Murnane, "The Organization of the Government under Amenhotep III", in D. O'Connor and E. Cline (eds.), *Amenhotep III, Perspectives on His Reign*, Ann Arbor 1998, 188ff, says that Ithu, the father of Heby, held the titles of *wab*-priest, overseer of the labor force of Amun, chamberlain and treasurer, and that Heby was allowed to succeed his father in his office of "scribe who counts the cattle of Amun", and later became overseer of the cattle of Amun, overseer of the double granary of Amun throughout the nomes that are in Lower Egypt, and mayor of Memphis; further he says Heby had at least two sons, Ramose and Amenhotep. Murnane gave no reasons or explanations, but it is clear from the listed titles that he equates the Heby, father of Amenhotep (Bologna statue KS 1825, Leiden pyramidion A.M. 6), not only with the Heby of TT55 (overseer titles) and with the Heby of the Aswan graffito (mayor title), but also with the Heby, son of Ithu, of *Table II*. The first four titles listed here for Ithu occur on BM EA 31, but it is completely unclear why Murnane attributed the title "scribe who counts the cattle of Amun" to him. Or to Heby for that matter, as it seems to be taken from FM 6690, which, however, mentions different parents.

[94] What becomes apparent from the material discussed here is that a careful study (a ponderous volume indeed!) of all the many monuments belonging to Ithu's large and important family would be very useful. It would need to include colour photographs, inscriptions, transliterations, translations, commentaries, genealogies, museological data, datings, prosopographical material, etc., ranging from Ithu to his possible great-grandson Ipy, son of Amenhotep Huy, who lived in the reign of Amenhotep IV. In the opinion of the present writer, such a long and difficult study would be very rewarding.

[95] Cf. note 67, A. de Buck, op. cit., 5-11.

[96] As Kaemamen lived during the reign of Amenhotep II, his mother would most likely have been a *ḥmt-nsw* of either Tuthmosis III or Amenhotep II, although it is impossible to say at present what her exact status was. The present writer knows of no other monuments that mention a woman with this name who held this title; she is not listed in L. Troy, *Patterns of Queenship in Ancient Egyptian Myth and History, Uppsala*, 1986, 161-169. That Kaemamen was a "child of the *kap*" would likely have been a consequence of his mother's status. It is important to note that the title of "king's son" is not attested for Kaemamen. He is also not listed among the Dynasty 18 king's sons in A. Dodson, "Crown prince Djhutmose and the royal sons of the Eighteenth Dynasty", *JEA* 76 (1990), 91-96. Whether this is significant, implying, for example, that he was a son of Henuttawy by a man other than the king, is difficult to say. From the reign of Amenhotep II, there is evidence of a woman called Henuttawy who is not called king's wife but held different titles, among them "royal favourite" (*Urk* IV, 1500). It is possible that the two ladies may be equated, under the presumption that the "royal favourite" was first given in marriage to a non-royal person (Kaemamen's father), and after the death of this man became king's wife.

[97] This is also reflected in the *ḥtp-di-nsw* formula on the statue, which mentions "Amun-Re, the king of the gods, and Amun who is in Karnak". A. de Buck, op. cit., 9.

receive offerings from the table of "Amun-Re in Henket-ankh"). His wife Merytra, "the singer of Amun", is mentioned, and their son Aakheperraseneb, who holds the title of "Second Prophet of Menkheperra".

(c) A number of funerary cones belonging to Kaemamen. Cones of the type listed as n° 228 of the Davies/Macadam corpus call Kaemamen (who on certain cones is referred to with the variant name form Amenemka[98]) "Second Prophet of Menkheperra" and mention his wife, the singer Merytra. Cones of the DM 246 type call Kaemamen (Amenemka) "seal-bearer of the king and Fourth Prophet of Amun", so the same titles as on the Wolfheze statue, while cones of the DM 590 type in addition mention his son Seqed who is "Second Prophet of Menkheperra"[99].

The question arises whether Kaemamen had two sons who sequentially held the Second Prophet of Menkheperra title after their father (de Buck opts for this scenario), or whether Seqed was a nickname for Aakheperraseneb. Whatever may be the case, there is little doubt that father and son(s) were Second Prophet of Menkheperra after the death of Tuthmosis III, during the reign of Amenhotep II whose name Aakheperura is written on the *menat* necklace of Merytra (Louvre 10443) and surely inspired the name of (one of) the son(s). The most interesting fact learned from Kaemamen's monuments is that a Prophet of Amun served in the Theban funerary temple of a king (compare also *Table I*, the case of Ra, who may have been Kaemamen's superior).

A possible case that is not included in *Table II* but deserves mention and analysis is that of Meryptah from the reign of Amenhotep III. The monuments of this man are of particular interest since they mention the *ḥwt-nb-mꜣꜥt-rꜥ* which has been encountered above, in the main article, when dealing with the stela of Userhat. Important inscriptions mentioning him are the following:

(a) A funerary cone (Davies & Macadam n° 412). The text on this cone has been translated by Helck[100] as "The honoured one with Osiris, the hereditary prince and noble, the *sem*-priest in the temple of Ptah, great one who controls the craftsmen in Thebes, the prophet in the temple of Nebmaatra, Meryptah, justified". It is important to note that this text doesn't say that Meryptah was Prophet of Nebmaatra, but that he was *ḥm-nṯr m ḥwt*, the preposition *m* being completely certain. As to the owner of this *ḥwt* temple, it must be said that the hieroglyphic text given in Davies & Macadam points towards a variant of hieroglyph C6, which would suggest a temple of Anubis (*ḥwt inpw*), but it is rather certain (seeing the inscriptions below) that this must be a misreading of the famous rebus form of the name Nebmaatra, and that the text should read *ḥwt nb-mꜣꜥt-rꜥ*.

(b) A few lines from TT 55, the tomb of Meryptah's relative Ramose. These lines are translated by Helck[101] as "...*sem*-priest in the temple of [Nebmaatra]...", the portion between square brackets being reconstructed by Helck. However, judging from the text on the funerary cone mentioned above, it is much more probable that the text is to be restored as "*sem*-priest in the temple of Ptah".

[98] Cf. M. Werbrouck, "Cônes funéraires de Kaemimen", *CdÉ* 33, n. 66 (1958), 223-226. See also notes 67 and 68 above.

[99] In TT192, the father of the tomb owner Kheruef is called Seqed. Although no titles for this man survive, it is probable that he was the same man as the Second Prophet of that name in *Table II*, seeing the match in time period and the rarity of the name. If Kaemamen was Kheruef's grandfather, then it seems that the Second Prophet title was inherited from father to son, but not to grandson, as Kheruef does not seem to have held such a title.

[100] *Urk.* IV, 1954; HRLED, V, 77.

[101] *Urk.* IV, 1787; HRLED, V, 5.

(c) A few hieratic inscriptions from Malkata. Two hieratic inscriptions from Malkata were found on wine jar labels from the Middle Palace. Their texts read as follows[102]: "Year 34: wine for the Repeating of the Sed Festival of His Majesty, l. p. h., made by the prophet and steward Meryptah of the temple of the pharaoh, l. p. h." (three examples survive); "Year 37: wine for the third Sed Festival of His Majesty, l. p. h., made by the prophet and steward Meryptah, of the temple of the pharaoh, l. p. h." (six or seven examples survive). In both cases the expression *n tꜣ ḥwt pr-ꜥꜣ* has been used.

(d) The Memphite Leiden stela V 14[103]. In the inscriptions on this stela[104], Meryptah is called, in three different places: *ḥm-nṯr imy-r pr n tꜣ ḥwt nb-mꜣꜥt-rꜥ* "the prophet and steward of the House of Nebmaatra".

(e) A couple of lines from the famous lower portion of a seated limestone statue of Nebnefer found in the temple of prince Wadjimose, a son of Tutmosis I, now in Brussels[105]. The text[106] from this statue is dated to year 20 of Nebmaatra and gives Meryptah only as the First Prophet of Amun.

The first four monuments clearly refer to the same person, Meryptah, *sem*-priest in the temple of Ptah and prophet in the temple of Nebmaatra, who lived during the reign of Amenhotep III. Whether the fifth monument mentions the same Meryptah is not fully certain, seeing the difference in titles, but it seems likely, and if it is so then the Prophet of Amun title could have a particular significance, as will be indicated below. The question arises in which temple Meryptah served as prophet, as there is evidence for at least two *ḥwt* temples of Amenhotep III: the Theban funerary temple in Kom el Hettan and a *ḥwt* in Memphis[107]. As two other hieratic inscriptions from wine jar labels[108] mention a *tꜣ ḥwt* as being in Memphis, it seems likely that the Malkata labels listed above also refer to the Memphite temple. And according to Hayes[109], the temple mentioned on the Leiden stela V 14, is another name of the Memphite temple rather than of the Western Theban funerary temple, as the stela was probably found in Memphis. Whatever may be the case, in all these texts pertaining to Meryptah there is evidence of a *ḥwt* temple of Amenhotep III being active while the king was still alive, and of Meryptah acting as prophet in that temple. Haring's thesis, that the phrases *ḥwt* + king's name and *tꜣ ḥwt pr-ꜥꜣ* always refers to a memorial temple of a king, has already been mentioned above[110]. Should his opinion be invalid, then Meryptah could just have been a prophet in a (Memphite) temple of whatever nature, and thus be irrelevant for the Prophet of the King issue. But even if Haring is correct, and Meryptah did serve in a memorial temple of the king in Memphis, then it is still important to note that not a single inscription testifies to Meryptah actually being Prophet of the King, i.e. a title of the pattern of

[102] W. C. Hayes, "Inscriptions from the Palace of Amenhotep III", *JNES* 10 (1951), fig. 5, #34 (year 34), fig. 7, #59 (year 37); *Urk.* IV, 1954; HRLED, V, 76-77. Circa 25 or 26 other inscriptions from Malkata mention Meryptah without reporting his full titles. See M. A. Leahy, *Excavations at Malkata and the Birket Habu 1971-74: The Inscriptions*, Warminster 1978, 7.

[103] The stela is actually divided in two parts, an upper one in Leiden, and a lower one in the Petrie Museum of London (UC 14463). For the latter fragment, see H. M. Stewart, *Egyptian Stelae, Reliefs and Paintings from the Petrie Collection*, vol. I, Warminster 1976, 26-27 and pl. 16. It can be seen online at URL: http://www.petrie.ucl.ac.uk/search/index.html. The subject of the many monuments of Ptahmose will not be dealt with here; to these monuments, statue Florence 1791 (M. Saleh Ali, *Arte Sublime nell'Antico Egitto*, Milano 1999, pl. 62; the catalogue of the Florence Exhibition, March 6 - July 4, 1999), possibly overlooked by Bosse-Griffiths, could be added.

[104] *Urk.* IV, 1910-1911; HRLED, V, 56-57; K. Bosse-Griffiths, "The Memphite Stela of Merptah and Ptahmose", *JEA* 41 (1955), 56-63; K. Bosse-Griffiths, *Amarna Studies*, Fribourg and Goettingen 2001 (OBO 182), 15-26.

[105] E 1103; see D. van der Plas (ed.), *Egyptian Treasures in Europe*, vol. 2, CD-ROM by CCER.

[106] *Urk.* IV, 1885; HRLED, V, 44-45.

[107] For the temple in Memphis, see *Urk* IV, 1795.

[108] W. C. Hayes, op. cit., 98-101, fig. 4, #3 (year 26) and fig. 6, #58 (year 37).

[109] W. C. Hayes, op. cit., 99.

[110] B. J. J. Haring, op. cit., 26-29. Cf. note 42 above.

those in *Table II*, *ḥm-nṯr n* + king's name, is not found. It is therefore the opinion of the present writer that the case of Meryptah does not argue in favour of Amenhotep III having his own priests while still living. As in the case of Userhat (see main text), it could instead be evidence of a temple being built and administered by a First Prophet of Amun while the king, to whom it was intended to be a memorial, still lived. As we have seen above, from the case of Kaemamen, the Prophets of Amun continued to serve in the funerary temple after the death of the king, when a Prophet of the King also became active. In other words, it is the hypothesis of the present writer that while the king was still alive, Prophets of Amun already served in the royal memorial temple (cf. the cases of Userhat and Meryptah), but that Prophets of the King only started to operate after the death of the king – the case of Akhenaten being an exception.

Enigmatic Kiya

Arris H. Kramer

Much has been romanticized about Akhenaten and Nefertiti, but Kiya, another spouse of Akhenaten, remains less well-known. She was publicly revealed as one of Akhenaten's harem-women in 1959. Knowledge about her has steadily increased since then, and it is now believed that she was more than just a secondary wife at Akhenaten's court. This literature survey summarises the main threads of evidence about the enigmatic Kiya.[1]

1. Kiya's name

In 1959 Aldred[2] was the first to notice the association of the *name* Kiya with the Amarna era. On part of a stela[3] from Nekhen he observed an inscription of presumably sixteen columns. These columns consist of a hymn to Re-Harakhty-Aten at his rising, inscribed on behalf of a man whose name is almost destroyed, except for the last group [hieroglyphs], but which may have been Kia. It is still uncertain whether or not this Kia is connected in any way to Kiya, spouse of Akhenaten. It seems hard to prove, as not even a little part of the female Kiya's titulary is present on the stela. Kiya is a name that apparently could be used by both men and women. However, according to Ranke[4] Kiya as a masculine proper name is attested in the Old Kingdom only.

In 1920, the Metropolitan Museum of Art (MMA) in New York bought an almost complete small calcite cosmetic vase[5] from Howard Carter. This vase was then, with the exception of the inscriptions, mended and restored in this museum. Hayes[6] discussed a translation of these inscriptions, among which he identified Kiya's name as 'Keya', in 1959. Thus, Hayes was the first to reveal Kiya publicly as an apparently unknown member of Akhenaten's harem. Kiya has since been the subject of much scholarly debate.[7]

The British Museum (BM) in London acquired two fragments of other calcite cosmetic vases[8] in December 1959. One was inscribed with text related to Kiya while the other one was without text related to her. The latter fragment is inscribed with the earlier form of the Aten cartouches[9] and the prenomen of Akhenaten only. The size and form of these two vases seem originally to have been virtually identical, as are the layout, content, and cutting of their surviving portions. Except for a few relatively unimportant details, the BM fragmentary cosmetic vase related to Kiya, EA 65901 (see *Fig. 1*), shows the same text as the cosmetic vase from the MMA. On the left, EA 65901 shows the

---

[1] This paper is an expanded version of A. H. Kramer, 'Kiya', *NATP* 4 (1999) 1-4.

[2] C. Aldred, 'The beginning of the El-'Amârna Period', *JEA* 45 (1959) 19-22. Though J. Ray in 'The parentage of Tutankhamun', *Antiquity* 49 (1975) 46, thought that R. Engelbach in 'Material for a revision of the history of the heresy period of the XVIIIth dynasty', *ASAE* 40 (1940) 133-165, had mentioned 'Kia' as an unknown wife of Akhenaten, this is an error.

[3] Royal Museum, Edinburgh, 1956.347, 59.7 x 31.8 x 7.6 cm.

[4] H. Ranke, *Die Ägyptischen Personennamen I: Verzeichnis der Namen*, Glückstadt (1935) 343.

[5] Metropolitan Museum of Art, New York, MMA 20.2.11, unknown provenance; H. W. Fairman, 'Once again the so-called coffin of Akhenaten', *JEA* 47 (1961) 29-30.

[6] W. C. Hayes, *The Scepter of Egypt* II, Cambridge/Mass. (1959) 294.

[7] Bibliography at the end of this paper.

[8] (i) British Museum, London, EA 65901, unknown provenance; (ii) *ibid.*, EA 65900, 6.7 x 3.9 x 1.1 cm., unknown provenance; C. N. Reeves, 'New light on Kiya from texts in the British Museum', *JEA* 74 (1988) 91, pl. XVII, n° 2; H. W. Fairman, 'Once again the so-called coffin of Akhenaten', *JEA* 47 (1961) 29-30.

[9] Cf. J. Assmann, 'Aton', *LÄ* I, Wiesbaden (1975) 526-540; H. A. Schlögl, *Echnaton – Tutanchamun*, Wiesbaden (1993) 32-33.

cartouches of the Aten in the early form of the name, and the nomen and prenomen of Akhenaten. The three vertical columns on the right side show: '*Greatly beloved wife of the King of Upper and Lower Egypt, living on Truth, Lord of the Two Lands,*[10] (*Neferkheprure-waenre*)|, *the beautiful child of the living Aten, who shall be living for ever and ever, Kiya*'. This characteristic formula was uniformly employed on Kiya's monuments and was never associated with Nefertiti.[11] The titulary appears in both 'full' and 'shortened' forms accompanying Kiya's name.

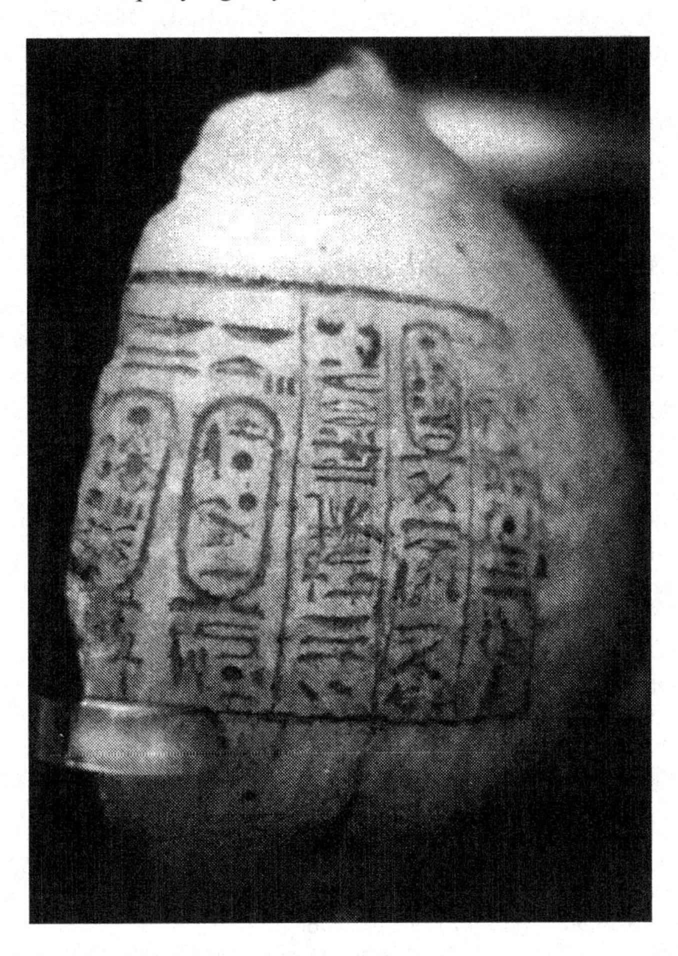

Figure 1: EA 65901 [12]

According to Reeves[13] Kiya's existence is confirmed by the aforementioned cosmetic vases and by a series of fragmentary text-models in gypsum from the Maruaten complex,[14] which outline her characteristic titulary. Aldred[15] argued that these text-models were prepared for the benefit of artisans carving inscriptions on the monuments. Furthermore, her existence is attested by several palimpsest fragments from the Maruaten,[16] where her name and titulary have been partially erased and replaced by

[10] Metropolitan Museum of Art, New York, MMA 20.2.11 omits *nb t3wy*.
[11] M. Birrell, 'Was Ay the Father of Kiya?', *BACE* 8 (1997) 12.
[12] Photograph of EA 65901 taken by the author (March 1988), with permission of the British Museum.
[13] C. N. Reeves, 'New light on Kiya from texts in the British Museum', *JEA* 74 (1988) 91.
[14] T. E. Peet and C. L. Woolley, *The City of Akhenaten* I, Oxford (1984) pl. 32-3; G. Perepelkin, *The Secret of the Gold Coffin*, Moscow (1978) 73-75.
[15] C. Aldred, *Akhenaten, King of Egypt*, London (1988) 204.
[16] J. R. Harris, 'Kiya', *CdÉ* 49 (1974) 27-29 with accompanying footnotes.

texts relating to Akhenaten's daughter Meritaten, and by several blocks[17] from Hermopolis, of likely Akhetaten provenance, which show similar changes.[18] She is also attested by four fragmentary kohl-tubes[19] from Akhetaten, by a small-inscribed wooden panel,[20] and by the canopic jars from tomb KV55 in the Valley of the Kings (discussed in chapter 5, *infra*).[21] Less formal references to Kiya have been recognized on two wine-jar dockets[22] from Akhetaten and on a clay funerary cone[23] of presumed Theban origin.[24] Kiya's formal appearance is also known from a number of representations associated with texts that refer to her (see chapter 3, *infra*). In addition, Reeves[25] discussed five other BM fragments: a portion of an offering slab,[26] three pieces of a column[27] and another fragment of a kohl-tube.[28] On these objects, Kiya's titles were written in her characteristic titulary or in a shortened form. Reeves argued that the portion of the offering slab is of a type that is held by a standing figure in a way only attested for Akhenaten and Nefertiti. If this offering slab is indeed connected to Kiya, she must have had a high status at Akhenaten's court.

Hanke[29] proposed that "Kiya" was a nickname for Nefertiti, but Helck[30] objected that not even a trace of Kiya's name has yet been found on the many talatats[31] at Karnak. Redford,[32] Aldred,[33] and Birrell[34] all agreed that Kiya's name has all the earmarks of a pet name, and suggested that it is a shortened form of a longer and more formal name. Redford wondered whether her name was a contraction of the Mitannian name Gilukhepa, who he supposed to have been a little girl when she entered the harem of Amenhotep III some thirty-five years before. Aldred also thought it was a contraction or abbreviation of a foreign name, probably a Mitannian name such as Tadukhepa; contrary to Redford, he ruled out Gilukhepa, because she would have been too old to be nubile at the start of Akhenaten's reign. Van Dijk,[35] however, argued that, if foreign princesses received Egyptian names, these were more likely to be official court names such as Henutempet or Maat-Hor-neferu-Re (the Egyptian name given to Ramesses II's Hittite princess), rather than a hypocoristicon of the Kiya

[17] J. R. Harris, *ibid.*, 26-29 with accompanying footnotes. These blocks were later reused as building material for a temple of Ramesses II at Hermopolis.

[18] C. N. Reeves, 'New light on Kiya from texts in the British Museum', *JEA* 74 (1988) 91-92.

[19] (i) Egyptian Museum, Berlin 22173; R. Krauss, *Das Ende der Amarnazeit*, HÄB 7, Hildesheim (1981) 109, 285, fig. 4 (Krauss refers to Berlin 22171 in the text, this is an error); (ii) Petrie Museum of the University College, London, UC 585; (iii) *ibid.*, UC 601; (iv) *ibid.*, UC 603.

[20] Petrie Museum of the University College, London, UC 24382, probably found by Petrie (1891-92) at Akhetaten, wood, 3.7 x 9.7 cm; J. R. Harris, 'Kiya', *CdÉ* 49 (1974) 26, 28 n. 5, 28 fig. A; J. Samson, *Amarna: City of Akhenaten and Nefertiti; Nefertiti as Pharaoh*, Warminster (1978) 119-120, pl. 60.

[21] C. N. Reeves, 'New light on Kiya from texts in the British Museum', *JEA* 74 (1988) 91-92.

[22] (i) Year 6 or [1]6: H. Frankfort and J. D. S. Pendlebury, *The City of Akhenaten* II, London (1972) pl. 58, n° 16; (ii) Year 11: British Museum, London EA 59951; W. M. Flinders Petrie, *Tell El Amarna*, London (1894) pl. 25, n° 95.

[23] N. de Garis Davies, *A Corpus of Inscribed Egyptian Funerary Cones I - Plates*, ed. by M. F. Laming Macadam, Oxford (1957), n° 527.

[24] C. N. Reeves, 'New light on Kiya from texts in the British Museum', *JEA* 74 (1988) 92.

[25] C. N. Reeves, *ibid.*, 91-101.

[26] British Museum, London, EA 26814, 12.5 x 12.7 x 4.9 cm., probably from Akhetaten, bought by the museum in 1891.

[27] (i) British Museum, London, EA 58179, 46.5 x 31.5 x 6.5 cm.; (ii) *ibid.*, EA 59165, 15.5 x 10.5 x 7.0 cm.; (iii) *ibid.*, EA 58180; all limestone, from the northern palace at Akhetaten.

[28] British Museum, London, EA 69719, 2.2 x 2.2 x 0.4 cm., from Akhetaten.

[29] R. Hanke, 'Änderungen von Bildern und Inschriften während der Amarnazeit', *SAK* 2 (1975) 93.

[30] W. Helck, 'Einige bemerkungen zu Artikeln in SAK 2', *SAK* 4 (1976) 117.

[31] R. W. Smith and D. B. Redford, *The Akhenaten Temple Project* I, Warminster (1976); A. H. Kramer, 'De reconstructie van Aton-heiligdommen in Karnak', *De Ibis* 16:3 (1991) 112-125.

[32] D. B. Redford, *Akhenaten: The Heretic King*, Princeton (1987) 150.

[33] C. Aldred, *Akhenaten, King of Egypt*, London (1988) 285-287.

[34] M. Birrell, 'Was Ay the Father of Kiya?', *BACE* 8 (1997) 14.

[35] J. van Dijk, 'The Noble Lady of Mitanni and Other Royal Favourites of the Eighteenth Dynasty', in J. van Dijk (ed.), *Essays on Ancient Egypt in Honour of Herman te Velde*, Egyptological Memoirs 1, Groningen (1997) 35-36.

type. Moreover, Birrell, and Harris[36] before him, argued that although the name Kiya is not commonly found in Egypt, it is by no means unique and does not by itself suggest a foreign origin.

Aldred[37] also discussed the possibility that the name Kiya might not have been of foreign origin but Egyptian. Her name could be derived from *ky*, the Egyptian word for 'monkey'. Helck[38] argued that though Kiya's name – like most names ending with – is a short name, the pet name 'monkey' would be less probable if the name had been derived from a foreign name. However, based on her appearance on plaster sculptures[39] from the workshop of the sculptor Tuthmose at Akhetaten, Reeves[40] also called her 'monkey'. In this context, it is worth mentioning Reeves' comment on the general disillusionment felt by the end of Akhenaten's reign. This feeling seems to be reflected on a satirical limestone relief recovered from the ruins of Akhetaten.[41] The relief shows a prestige vehicle, customarily driven by the king: the chariot symbolized the daily course of the sun through the sky. Here, however, a monkey occupies the place of the pharaoh.

On most objects Kiya's name was written as .[42] The apparently uniform transcription of her name is in a sense misleading. As well as *kyiꜣ*, examples have been found in which her name was written as *kiyꜣ*, *kꜣiꜣ*, *kiꜣ* and *kiw*.[43] Sometimes her name appears next to the names of the Aten – in the earlier or the later form[44] – or of Akhenaten, just as is frequently the case with Nefertiti.[45]

2. Kiya's presumed Mitannian origin

Referring to the 'Tale of Two Brothers',[46] Manniche[47] was the first to propose that Kiya came from Naharin (Mitanni). She might have been a princess, perhaps none other than Tadukhepa, the daughter of Tushratta, king of Mitanni. The scribe Inena wrote the only surviving copy of this folk tale some time in the reign of Seti II.[48] The story was made up of at least two parts, originally separate, each with a plot concerning an unfaithful wife. The first section tells how a conscientious young man called Bata was falsely accused of a proposal of adultery by the spouse of his elder brother Anubis, after he had actually rejected her advances. The second part describes the exile of Bata. The woman who appears in the latter part is the wife of Bata, eventually to become queen of Egypt. The king of Egypt first comes to know of her through a lock of hair, which he is told belongs to a daughter of Re-Harakhty. Envoys are sent to search for her in all foreign lands, but especially in the Valley of the Pine Tree, which was presumably somewhere in northern Syria or Asia Minor. When she is brought to the king, he loves her exceedingly and appoints her as 'great noble lady'; she is referred to as 'Ta-Shepset'. The creator of the story must have had a specific reason for calling this woman Ta-Shepset, as the expression was no

[36] J. R. Harris, 'Kiya', *CdÉ* 49 (1974) 26.
[37] C. Aldred, *Akhenaten, King of Egypt*, London (1988) 285-287.
[38] W. Helck, 'Kijê', *MDAIK* 40 (1984) 159-160.
[39] C. Aldred, *Akhenaten and Nefertiti*, New York (1973) 46, fig. 27; *ibid.*, 182, figs. 111, 112.
[40] C. N. Reeves, *Akhenaten: Egypt's False Prophet*, London (2001) 157-161, 174.
[41] Petrie Museum of the University College, London, UC 029, from Akhetaten, limestone, 8.5 x 8.0 x 2.0 cm.; J. Samson, *Amarna: City of Akhenaten and Nefertiti; Key pieces from the Petrie Collection*, London (1972) 37-38, pl. 16.
[42] J. R. Harris, 'Kiya', *CdÉ* 49 (1974) 26-27.
[43] J. R. Harris, *ibid.*, 26.
[44] Cf. J. Assmann, 'Aton', *LÄ* I, Wiesbaden (1975) 526-540; H. A. Schlögl, *Echnaton – Tutanchamun*, Wiesbaden (1993) 32-33.
[45] W. Helck, 'Kijê', *MDAIK* 40 (1984) 163.
[46] J. A. Wilson, 'Story of Two Brothers', in J. B. Pritchard (ed.), *Ancient Near Eastern Texts Relating to the Old Testament*, Princeton (1950), 23-25; a more in depth study is S. Tower Hollis, *The Ancient Egyptian 'Tale of Two Brothers'*, Oklahoma (1990).
[47] L. Manniche, 'The Wife of Bata', *GM* 18 (1975) 33.
[48] Inena, 'The Tale of the Two Brothers', Pap. D'Orbiney, British Museum, London, EA 10183; cf. the following URL: http://www.thebritishmuseum.ac.uk/egyptian/ea/gall/ea10183.html.

longer in ordinary use as a title. His choice of the term may well have been prompted by the knowledge that this part of the story reflected actual events, in which the lady involved had held the title, or had been known as *(tꜣ) šps t*. The designation is indeed so exceedingly rare in the New Kingdom that it is tempting to identify the wife of Bata with the only other woman known to have borne it in what Inena would have seen as the recent past: Kiya.[49]

Several wine-jar dockets have been recovered from excavations at Akhetaten. According to Aldred,[50] these objects are an uncertain source of information. He argued that, in addition to difficulties in reading some of these damaged and ill-written labels, doubt could be cast upon the completeness of the archive. Interruptions to the wine supply from a particular source in certain years could also have occurred for reasons that are now unknown. Moreover, the dates cannot be accepted as absolute, since wine might have been bottled for a time after the owner of an estate had died. Only two wine-jar dockets which may be related to Kiya have been revealed so far. One is dated to Year 11 of Akhenaten's reign[51] and the other apparently to Year 6. The latter date, however, is damaged. According to Aldred the number should probably be amended to 16, since the designation *ḥry bꜥḥ* for the vintner mentioned in the docket is in a form that is not generally employed until Year 13; before that time the customary *ḥry kꜣmw* was used.[52] However, at this time it is generally agreed that the label is from Year 6.[53] The label only mentions *pr (tꜣ) špst*, without any name, so it could refer to a different 'noble lady' from Akhenaten's harem. However, on the docket relating to Kiya's estate at Akhetaten – Year 11 – Kiya's name (surviving in part) appears together with *(tꜣ) špst*.[54]

In view of its rarity, it seems probable that the title *špst <n> nhrn*, i.e. 'noble lady of Naharin', known from a clay funerary cone of Year 6,[55] again refers to Kiya. This appellation was never used for Nefertiti or her daughters. Helck[56] argued that this funerary cone, belonging to a major-domo called Bengay, overseer of the domain (*ꜥꜣ n pr*) of *(tꜣ) špst nhrn*, indicated that Kiya already lived at Akhenaten's court in the beginning of his reign, when the king still lived in Thebes. However, on another funerary cone of this Bengay,[57] from the very same Theban tomb, the god Amen is mentioned. Partly for this reason, van Dijk[58] recently argued that Bengay's 'noble lady' could not be Kiya. In his opinion, Bengay was a contemporary of Tuthmose IV, and the lady from Naharin was the king's wife, Henutempet, named on a third cone of Bengay.[59] It is possible, nonetheless, that during the 18th Dynasty the term *(tꜣ) špst* was used in a restricted sense as an official title designating certain types of foreign princesses, so the association of Kiya with Naharin cannot necessarily be ruled out.

---

[49] G. Perepelkin, *The Secret of the Gold Coffin*, Moscow (1978) 117-119.

[50] C. Aldred, *Akhenaten, King of Egypt*, London (1988) 227.

[51] British Museum, London, EA 59951.

[52] J. van Dijk, 'The Noble Lady of Mitanni and Other Royal Favourites of the Eighteenth Dynasty', in J. van Dijk (ed.), *Essays on Ancient Egypt in Honour of Herman te Velde*, Egyptological Memoirs 1, Groningen (1997) 36.

[53] C. Vandersleyen, *L'Égypte et la vallée du Nil* II, Paris (1995) 444.

[54] W. M. Flinders Petrie, *Tell El Amarna*, London (1894) pl. 25, n° 95; W. Helck, 'Kijê', *MDAIK* 40 (1984) 160; C. N. Reeves, 'New light on Kiya from texts in the British Museum', *JEA* 74 (1988) 92.

[55] N. de Garis Davies, *A Corpus of Inscribed Egyptian Funerary Cones I - Plates*, ed. by M. F. Laming Macadam, Oxford (1957) n° 527; W. Helck, 'Kija', *LÄ* III, Wiesbaden (1980) 422-423; M. Birrell, 'Was Ay the Father of Kiya?', *BACE* 8 (1997) 16-17.

[56] W. Helck, 'Kijê', *MDAIK* 40 (1984) 160.

[57] N. de Garis Davies, *A Corpus of Inscribed Egyptian Funerary Cones I - Plates*, ed. by M. F. Laming Macadam, Oxford (1957) n° 528; M. Birrell, 'Was Ay the Father of Kiya?', *BACE* 8 (1997) 13-14.

[58] J. van Dijk, 'The Noble Lady of Mitanni and Other Royal Favourites of the Eighteenth Dynasty', in J. van Dijk (ed.), *Essays on Ancient Egypt in Honour of Herman te Velde*, Egyptological Memoirs 1, Groningen (1997) 33-35.

[59] N. de Garis Davies, *A Corpus of Inscribed Egyptian Funerary Cones I - Plates*, ed. by M. F. Laming Macadam, Oxford (1957) n° 260; J. van Dijk, *ibid.*, 33.

In this context, it is worth mentioning Krauss' suggestion[60] that Kiya may have returned to Mitanni at the end of Akhenaten's reign, and Vandersleyen's related suggestion[61] that she may have disappeared as a result of diplomatic problems with Mitanni.

However, it is still unknown when and how Kiya died, her mummy has not been identified, and her burial place has not yet been found. Helck[62] thought that Kiya died under violent circumstances shortly after Akhenaten's death. Nowadays, it is generally assumed that Kiya died in the final years of Akhenaten's reign,[63] perhaps because of childbirth in Year 12.[64] According to Reeves[65] an original interment within the royal tomb at Akhetaten may be assumed.

3. Kiya's portrayal

In the 1920s, texts and reliefs of Meritaten were found in the Maruaten. A closer examination, however, showed that Meritaten's name had replaced the original name, and that the hairstyle and the shape of the head had been altered in the reliefs.[66] Until 1968, most scholars thought that the name that had been altered in favour of Meritaten was Nefertiti's. In that year, however, Perepelkin[67] proposed that the original name was Kiya's. By a comparison with Kiya's titulary as it occurs on the cosmetic vases mentioned above and upon a series of plaster casts from Akhetaten,[68] Perepelkin was able to establish that the erasures on blocks from the Maruaten and from Hermopolis, rather than having any connection with Nefertiti, are in fact erasures of Kiya's name and titulary.[69] In 1974, Harris[70] strongly supported Perepelkin's findings. In 1975, Hanke[71] pointed out that the original name in the Maruaten texts could not have been Nefertiti's since her name is too long to fit in the text. Moreover, the text with Kiya's name on the fragment of the cosmetic vase (see chapter 1, *supra*) is compatible with the inscriptions of the Maruaten. A closer examination of the name proved that 'Kiya' fits perfectly in the Maruaten texts. Hanke further stated that in spite of the alterations her name is still recognizable. Moreover, Kiya's name seems to appear more often in the Maruaten than the names of other females. Not all the alterations were in favour of Meritaten: several other texts and depictions seem to have been altered in favour of Ankhesenpaaten.[72]

Hanke[73] stated that Kiya had a 'Sunshade temple' in the Per-Hay as well as in the Maruaten. In addition, she may have had a 'chapel' in this Per-Hay. When queen Tiye got a 'Sunshade temple' this was considered to be very important. Nobody in Akhetaten apart from Tiye and Kiya is known to have had such a temple. Kiya's possession of a 'Sunshade temple' and other buildings seems to point to her high position at Akhenaten's court. According to the altered inscriptions, Meritaten and

---

[60] R. Krauss, 'Kija – ursprüngliche Besitzerin der Kanopen aus KV 55', *MDAIK* 42 (1986) 79.

[61] C. Vandersleyen, *L'Égypte et la vallée du Nil* II, Paris (1995) 446.

[62] W. Helck, 'Probleme der Königsfolge in der Übergangszeit von 18. zu 19. Dyn.', *MDAIK* 37 (1981) 208-209; W. Helck, 'Kijê', *MDAIK* 40 (1984) 166.

[63] D. B. Redford, *Akhenaten: The Heretic King*, Princeton (1987) 186-192; M. Birrell, 'Was Ay the Father of Kiya?', *BACE* 8 (1997) 15.

[64] G. T. Martin, *The Royal Tomb at El-'Amarna* II, London (1989) 37-41, fig. 7, pl. 58-61; C. N. Reeves, *Akhenaten: Egypt's False Prophet*, London (2001) 160.

[65] C. N. Reeves, *Akhenaten: Egypt's False Prophet*, London (2001) 160.

[66] R. Hanke, 'Änderungen von Bildern und Inschriften während der Amarnazeit', *SAK* 2 (1975) 79-93.

[67] G. Perepelkin, *The Secret of the Gold Coffin*, Moscow (1978) [English translation of *Taina Zolotogo Groba*, Moskva (1968)] 58-73

[68] G. Perepelkin, *ibid.*, 73-75; T. E. Peet and C. L. Woolley, *The City of Akhenaten* I, Oxford (1984) pl. 32-3.

[69] G. Perepelkin, *ibid.*, 73-85.

[70] J. R. Harris, 'Kiya', *CdÉ* 49 (1974) 27-30.

[71] R. Hanke, 'Änderungen von Bildern und Inschriften während der Amarnazeit', *SAK* 2 (1975) 87-89.

[72] W. Helck, 'Kijê', *MDAIK* 40 (1984) 163; *ibid.*, 166.

[73] R. Hanke, *Amarna-Reliefs aus Hermopolis*, HÄB 2, Hildesheim (1978) 193-195.

Ankhesenpaaten usurped these 'Sunshade temples'.[74] Finally, three fragments of a column[75] from the northern palace of Akhetaten seem to indicate that Kiya also possessed real estate there.

In many scenes on the stone blocks from the Akhetaten buildings found at Hermopolis, Kiya's names and figures seem to have been either erased or over-carved by the names of Meritaten and Ankhesenpaaten.[76] Samson[77] asserts that Maketaten, Tutankhamen (as 'Tutankhaten') and, repeatedly, Neferneferuaten-Nefertiti are also mentioned on these blocks. Samson noted several scenes of a trio on various Hermopolis blocks. For instance, a scene in which the three unnamed people appear consists of a king, wearing a bulging war crown with an unusual heavily beaded edge of uraei that Akhenaten wore in the middle years of his reign, a lady wearing a so-called 'Nubian wig'[78] pushed back off her brow, without the uraeus, and a small child following her. They represent a distinct family group. According to Samson the trio could well be Akhenaten, Kiya, and their child. In these carvings two younger daughters of the next generation are named as Meritaten-ta-sherit and Ankhesenpaaten-ta-sherit. These children are linked with Akhenaten's name as though he fathered them with his own daughters. His daughters' names cover Kiya's name, however, so it seems likely that he actually fathered the children by Kiya. Another Hermopolis block shows two adults and a child, together with a partly removed subtitle that seems to end with Kiya's name.[79] Hanke[80] thought at least one of Kiya's daughters was called ...-ta-sherit.

Akhenaten, Ankhesenpaaten (as queen of Tutankhamen), Nefertiti, and Tiye are the only members of the royal family depicted with the 'Nubian wig'. However, ordinary citizens also used this hairstyle. Kiya wore it very prominently.[81] Although this style seems to have been Kiya's favourite, she also wore other hairstyles, for instance a round wig, a typical hairstyle during the Amarna period, worn by men as well as women. Nefertiti also wore this wig. In addition, the coffin of tomb KV55 shows somebody with the 'Nubian wig'.[82]

Thus far, apart from tomb KV55, Kiya's objects have only been found in Akhetaten, but mostly not in tombs or private houses.[83] An exception is a blue faience kohl tube from house Q46.5 in Akhetaten.[84] Reeves[85] also attributed three very similar plaster sculptures to Kiya based on the earrings that seem to be a characteristic feature of her representations in relief: one of unknown provenance[86] and two from Tuthmose's workshop[87] at Akhetaten.

---

[74] R. Hanke, *ibid.*, 188-190.

[75] (i) British Museum, London, EA 58179, 46.5 x 31.5 x 6.5 cm.; (ii) *ibid.*, EA 59165, 15.5 x 10.5 x 7.0 cm.; (iii) *ibid.*, EA 58180; all limestone, from the northern palace at Akhetaten; C. N. Reeves, 'New light on Kiya from texts in the British Museum', *JEA* 74 (1988) 93-100.

[76] G. Perepelkin, *The Secret of the Gold Coffin*, Moscow (1978) 100-107.

[77] J. Samson, *Nefertiti and Cleopatra. Queen-Monarchs of Ancient Egypt*, London (1985) 70.

[78] M. Eaton-Krauss, 'Miscellanea Amarnensia', *CdÉ* 56 (1981) 252-258; cf. J. Samson, 'Amarna crowns and wigs', *JEA* 59 (1973) 47-59.

[79] G. Roeder, *Amarna-Reliefs aus Hermopolis* II, Hildesheim (1969) fig. 29-632/VIIIA; R. Hanke, *Amarna-Reliefs aus Hermopolis*, HÄB 2, Hildesheim (1978) 154, pl. 56-632/VIIIA.

[80] R. Hanke, *Amarna-Reliefs aus Hermopolis*, HÄB 2, Hildesheim (1978) 154, pl. 56-610/VII.

[81] M. Eaton-Krauss, 'Miscellanea Amarnensia', *CdÉ* 56 (1981) 254-255.

[82] C. Aldred, *Akhenaten, King of Egypt*, London (1988) 205.

[83] R. Hanke, *Amarna-Reliefs aus Hermopolis*, HÄB 2, Hildesheim (1978) 195-196.

[84] Egyptian Museum, Berlin, 22173, 2.2 x 4.7 cm.; R. Krauss, *Das Ende der Amarnazeit*, HÄB 7, Hildesheim (1981) 109, 285 fig. 4. Krauss refers to Berlin 22171 in the text, this is an error.

[85] C. N. Reeves, 'New light on Kiya from texts in the British Museum', *JEA* 74 (1988) 92; C. N. Reeves, *Akhenaten: Egypt's False Prophet*, London (2001) 158; J. R. Harris, 'Kiya', *CdÉ* 49 (1974) 28 n. 5; cf. C. Aldred, *Akhenaten and Nefertiti*, New York (1973) 182, figs. 111, 112; cf. D. Arnold, *The Royal Women of Amarna. Images of Beauty from Ancient Egypt*, New York (1996) 46-48, figs. 37-38.

[86] Pushkin State Museum of Fine Arts, Moscow, 2141.

[87] (i) Egyptian Museum, Berlin, 21239; (ii) *ibid.*, 21341.

Helck[88] proposed that Kiya and her daughter(s) became subordinate to Nefertiti's daughters after Akhenaten's death; they subsequently usurped Kiya's monuments. Redford[89] assumed that shortly after Kiya's death her monuments and most of the reliefs relating to her were either re-used or destroyed. Her image on a Hermopolis block now in the Ny Carlsberg Glyptotek of Copenhagen[90] is a striking example of the transformation of Kiya's image into Meritaten's. On this block, the two columns of clearly superimposed inscription can be read: 'daughter of the king, of his flesh, his beloved... Meritaten'. However, faint remains of the original inscription show the beginning of Kiya's titulary. The wig of the woman in the relief was originally a 'Nubian wig' that had been changed into what is called a 'modified Nubian wig'. This means that the hair above the forehead and on the back of the head was removed, and, with the help of an added layer of plaster, the hairstyle was transformed into a rather broad sidelock to signify the status of Meritaten as a princess. However, the face of the woman on the relief was not touched; only the eyes were damaged during the destruction. Despite alterations and destructions, an impressive image of Kiya's facial features still exists. Compared with Nefertiti, Kiya's nose seems fleshier, her chin is definitely longer and her cheekbone is less prominent. In several reliefs Kiya is seen accompanying Akhenaten at offerings and official ceremonies, in the same size and manner as Nefertiti, both women being much smaller than Akhenaten.[91]

4. Kiya and the royal family

Akhenaten showed interest in other women besides Kiya and Nefertiti. Krauss[92] and Schulman[93] named the Mitannian princess Tadukhepa as his wife. Schulman further considered a Babylonian princess – the daughter of Burnaburias II – as a consort of Akhenaten; he supposed that several Amarna letters attest to this princess.[94] And Krauss mentioned a certain Py (*pjj*) as a consort of Akhenaten. The name of Py occurs in a text on an ushabti, found by Legrain and reproduced by Sandman.[95] According to Graefe[96] the title and name of the owner of this ushabti can be read as 'Beloved Love of Waenre, royal concubine, Py'. However, Graefe also noted that there is no hard evidence for a Py as a secondary wife of Akhenaten. Finally, in 1988 Aldred[97] mentioned a 'royal ornament' Ipy, a 'true favourite' of the king. Van Dijk[98] discussed the name Ipy more extensively and suggested that it is actually the name of Ay's wife Ty.

---

[88] W. Helck, 'Kijê', *MDAIK* 40 (1984) 166.

[89] D. B. Redford, *Akhenaten: The Heretic King*, Princeton (1987) 186-192.

[90] J. D. Cooney, *Amarna Reliefs from Hermopolis in American Collections*, New York (1965) 34; D. Arnold, *The Royal Women of Amarna. Images of Beauty from Ancient Egypt*, New York (1996) 105.

[91] R. Hanke, *Amarna-Reliefs aus Hermopolis*, HÄB 2, Hildesheim (1978) 193-195; D. Arnold, *The Royal Women of Amarna. Images of Beauty from Ancient Egypt*, New York (1996) 15.

[92] R. Krauss, *Das Ende der Amarnazeit*, HÄB 7, Hildesheim (1981) 45 n. 4.

[93] A. R. Schulman, 'Diplomatic marriage in the Egyptian New Kingdom', *JNES* 38 (1979) 185.

[94] A. R. Schulman, *ibid.*, 185 n. 37; J. A. Knudtzon, *Die El-Amarna Tafeln*, Leipzig (1915) (*EAT* 11, *EAT* 12); W. L. Moran, *The Amarna Letters*, Baltimore (1992) 21-23 (*EAT* 11 ['Proper escort for a betrothed princess']), 24 (*EAT* 12 ['A letter from a princess']). From *EAT* 11, sent by Burnaburias to Akhenaten, Schulman reads: '*Send chariots and people in great numbers, and then Haya will bring the daughter of the king to you. Don't send another envoy. The princess whom you desire, I shall not keep her back with me. But send quickly.*'. *EAT* 12 would then be from the Babylonian princess, and is probably connected with this marriage, since she mentioned 'the gods of Burnaburias' and she addressed the king as 'my lord'. Therefore Schulman assumed that she had already had the oil poured upon her head and was thus already Akhenaten's wife, even though she had not yet come to Egypt. Cf. R. Krauss, *Das Ende der Amarnazeit*, HÄB 7, Hildesheim (1981) 75-78; S. Izre'el, *The Amarna Tablets* (2000), at: http://spinoza.tau.ac.il/hci/dep/semitic/amarna.html.

[95] M. Sandman, *Texts from the Time of Akhenaten*, Bibliotheca Aegyptiaca VIII, Brussels (1938) 177-CCVIII.

[96] E. Graefe, 'Zu *Pjj*, der angeblichen Nebenfrau des Achanjati', *GM* 33 (1979) 17-18.

[97] C. Aldred, *Akhenaten, King of Egypt*, London (1988) 286-287.

[98] J. van Dijk, 'The Noble Lady of Mitanni and Other Royal Favourites of the Eighteenth Dynasty', in J. van Dijk (ed.), *Essays on Ancient Egypt in Honour of Herman te Velde*, Egyptological Memoirs 1, Groningen (1997) 40.

Kiya and Nefertiti never seem to appear together in the same picture.[99] Samson[100] argued that Kiya was not a chief queen like Nefertiti. Kiya was never a 'Great Royal Wife'; Nefertiti had the sole right to this title. Therefore, Kiya had no claim to wear the uraeus, or to enclose her name within a royal cartouche. According to Birrell[101] she had lower standing at the royal court than Nefertiti; her name does not stand alone, but is bound up with an expression of her relationship to Akhenaten. Vandersleyen[102] also noted that Kiya was never depicted wearing a crown or uraeus, and that her name was never written within a cartouche. Robins[103] argued that it is necessary to find an explanation for Kiya's prominence, because earlier in the 18th Dynasty minor wives did not appear on royal monuments and temples, where only *ḥmt nsw wrt*, *mwt nsw*, and occasionally *s3t nsw* were shown. There are thus no parallels to Kiya's position from previous reigns, nor indeed are there any from the succeeding dynasty. Additionally, she wondered why Kiya did not hold the title *ḥmt nsw* like other secondary wives.

Because both women seem to have lived simultaneously at the court of Akhetaten, Reeves[104] thought that Kiya's popularity with Akhenaten would clearly have posed a threat to Nefertiti, and Perepelkin[105] also supposed that Kiya was Nefertiti's rival. In Reeves' opinion it is not by chance that an extraordinary rise in Nefertiti's fortunes began with Kiya's demise. As reviewed in the previous chapter, Kiya is always in a prominent position on pictures with Akhenaten, right behind the king. According to Vandersleyen,[106] this points to Kiya's rank being similar to Nefertiti's.

Sometimes Kiya's presumed daughter is visible behind her. Several Hermopolis blocks show Kiya's name as a subscription.[107] According to Hanke,[108] the text on these blocks is original; no new text has been added. Hanke[109] further reported that Kiya's daughter was allowed to call herself 'daughter of the king, from his body, that he loves', just as with Nefertiti's daughters. Gabolde[110] argued that Kiya's daughter might have been Baketaten, who is shown accompanied by the king's mother queen Tiye. She is often supposed to be a late-born daughter of Tiye, but she does not have the title 'royal sister', only 'royal daughter of his body'. Moreover, she is not described as 'born to the great royal wife Tiye'. According to Gabolde, this means that she was the daughter of Akhenaten, not of Amenhotep III; in his opinion, there is no alternative for Akhenaten as her father. Furthermore, Baketaten cannot be a daughter of Nefertiti, as she never appears with the other daughters of Nefertiti. It is therefore possible that she is the daughter of a secondary wife of Akhenaten; for Gabolde Kiya is a likely candidate. Cabrol[111] argued that Kiya is the same person as Tadukhepa, and that she gave birth to Baketaten as a wife of Amenhotep III; later she would have become a wife of Akhenaten, and when she died, Baketaten would have grown up under the protection of Tiye.

One of the major points at stake in the old dispute about the Amarna co-regency is the parentage of Tutankhamen. In principle Kiya, Nefertiti, Tiye, Tadukhepa or another Mitannian princess, a

---

[99] R. Hanke, *Amarna-Reliefs aus Hermopolis*, HÄB 2, Hildesheim (1978) 188-190; D. Arnold, *The Royal Women of Amarna. Images of Beauty from Ancient Egypt*, New York (1996) 15.

[100] J. Samson, *Nefertiti and Cleopatra. Queen-Monarchs of Ancient Egypt*, London (1985) 69; cf. C. Aldred, *Akhenaten, King of Egypt*, London (1988) 204.

[101] M. Birrell, 'Was Ay the Father of Kiya?', *BACE* 8 (1997) 12.

[102] C. Vandersleyen, *L'Égypte et la vallée du Nil* II, Paris (1995) 445, 460.

[103] G. Robins, 'The Mother of Tutankhamen, 2', *DE* 22 (1992) 26.

[104] C. N. Reeves, *Akhenaten: Egypt's False Prophet*, London (2001) 160.

[105] G. Perepelkin, *The Secret of the Gold Coffin*, Moscow (1978) 108-130.

[106] C. Vandersleyen, *L'Égypte et la vallée du Nil* II, Paris (1995) 445-446.

[107] G. Roeder, *Amarna-Reliefs aus Hermopolis* II, Hildesheim (1969) figs. 11/442-VIII, 15/610-VIII, 23/376-VIII, 29/632-VIII, 32/153-VIII, 57/135-VIII & 148/925-VIII.

[108] R. Hanke, *Amarna-Reliefs aus Hermopolis*, HÄB 2, Hildesheim (1978) 190-191.

[109] R. Hanke, *ibid.*, 192.

[110] M. Gabolde, 'Baketaton, fille de Kiya?', *BSÉG* 16 (1992) 27-40.

[111] A. Cabrol, *Amenhotep III Le magnifique*, Monaco/Paris (2000) 149-155.

Babylonian princess, or an unknown woman could be the mother of Tutankhamen. It seems as if we still do not have any clue. Ray[112] felt that if Tutankhamen was the son of Nefertiti, it is difficult to explain why he never appears on monuments together with the six daughters of the principal queen, as only daughters of Nefertiti are shown on the monuments. Ray and others suggested that the mother of Tutankhamen was Kiya.[113] Van Dijk argued that if Tutankhamen was a son of Kiya he must have been born in or just before Akhenaten's Year 12.[114] This might support the idea that Kiya disappeared from the scene because her production of an heir posed a threat to the position of Nefertiti as Akhenaten's chief queen. It would also imply that Tutankhamen was about 5 or 6 years old when Akhenaten died. This would agree rather well with van Dijk's recent estimate of Tutankhamen's own age at death as about 16 or 17.[115] Robins,[116] on the other hand, argued that surviving evidence suggests that in the 18th Dynasty before the Amarna period, while king's daughters may be shown on royal monuments in ritual scenes, it is highly unusual for king's sons to appear in them. Therefore, it cannot be completely ruled out that Nefertiti was the mother of Tutankhamen.

Tutankhamen associates himself closely with Amenhotep III, calling him *jt*. Aldred[117] concluded that Amenhotep III was indeed the physical father of Tutankhamen. However, Robins noted that while *jt* is the word for 'father', *jt* also means an ascendant, including 'grandfather', and is used by kings to refer to and associate themselves with royal ancestors. Tutankhamen thus uses *jt* in a traditional way to forge an association with his grandfather Amenhotep III, being the direct inheritor of his 'orthodoxy'.[118] In these circumstances, it is unlikely that Tutankhamen would mention his mother, since she would be a link to the otherwise-avoided Akhenaten. Gabolde has also argued that Tiye was simply too old to be the mother of Tutankhamen.[119]

Seele[120] was the first to propose that Ay was the father of Tutankhamen. According to Seele the dismantling of the second pylon of the Karnak temple yielded important information concerning the relations of Tutankhamen and Ay – proof of their co-regency.[121] This proof consisted of the occurrence, on twelve battered architrave blocks belonging to a small temple, of inscriptions in large hieroglyphs recording the names and titles of both kings. The blocks originally belonged to different, probably opposite, colonnades of a temple court, and thus to two separate architraves. One side of each of these architraves contained a single line of hieroglyphs, the other side two lines, one above the other. The one-line inscription on one of the architraves was carved in the name of Ay, the other, in the name of Tutankhamen, in each case with all their royal titles. On the opposite face of each of the two architraves, where the inscriptions occur in two horizontal lines, the upper line is devoted to the royal names and titles of Ay, while the lower one contains those of Tutankhamen. Seele concluded that the building was the joint work of two contemporary kings, the co-regents Ay and Tutankhamen. However, this edifice bears a still more startling revelation. King Ay stated in his inscription that he built the temple 'as his monument for his son' Tutankhamen.[122]

---

[112] J. Ray, 'The parentage of Tutankhamun', *Antiquity* 49 (1975) 46.

[113] E. S. Meltzer, 'The parentage of Tut'ankhamún and Smenkhkaré', *JEA* 64 (1978) 134-135; M. Birrell, 'Was Ay the Father of Kiya?', *BACE* 8 (1997) 11-18; C. N. Reeves, *Akhenaten: Egypt's False Prophet*, London (2001) 8.

[114] J. van Dijk, 'The Noble Lady of Mitanni and Other Royal Favourites of the Eighteenth Dynasty', in J. van Dijk (ed.), *Essays on Ancient Egypt in Honour of Herman te Velde*, Egyptological Memoirs 1, Groningen (1997) 39.

[115] J. van Dijk, *ibid.*, 39 n. 40; cf. F. Filce Leek, 'How old was Tut'ankhamûn?', *JEA* 63 (1977) 115, who set the age at death at '16 or, at the most, 17 years', based on anatomical considerations.

[116] G. Robins, 'The Mother of Tutankhamen', *DE* 20 (1991) 71.

[117] C. Aldred, *Akhenaten, King of Egypt*, London (1988) 293.

[118] G. Robins, 'The Mother of Tutankhamen', *DE* 20 (1991) 73; cf. however Tutankhamen's statement that Tuthmosis IV was his *jt (n) jt (n) jt* (C. N. Reeves, 'Tuthmosis IV as 'great-grandfather' of Tutankhamen', *GM* 56 (1982) 65).

[119] M. Gabolde, 'Baketaton, fille de Kiya?', *BSÉG* 16 (1992) 32.

[120] K. C. Seele, 'King Ay and the close of the Amarna age', *JNES* 14 (1955) 176-180.

[121] K. C. Seele, *ibid.*, 177 n. 57.

[122] O. Schaden, *The God's Father Ay*, dissertation, UMI 78-9739, Ann Arbor (1977) 148 n. 39, blocks 9-2A, 27-2A.

Birrell,[123] inspired by Seele, wondered whether Kiya could have been a daughter of Ay and Ty. A precedent for a king marrying the daughter of a royal nurse (Ty) is found in the example of Sitiah. Sitiah was the daughter of the royal nurse Ipu, and spouse of Tuthmose III. Therefore, Birrell thought that a daughter of Ay and Ty entered the royal harem as a spouse of Akhenaten. This would have justified Ay's use of the title 'God's Father'[124] in the same way the 'God's Father' Yuya had been father-in-law of Amenhotep III. Birrell[125] argued that Ay's title would have associated him with his predecessors, and as propaganda may have justified his right to the throne. Moreover, the title appears to reflect a relationship with both Akhenaten and Tutankhamen, who as kings were considered to be divine. If Tutankhamen was the son of Kiya, Ay could legitimately have retained the title 'God's Father' during the reign of Tutankhamen because he would have been his maternal grandfather.[126]

Martin[127] put forward the hypothesis that the scene on wall F of room Alpha in the royal tomb at Akhetaten might show Tutankhamen's birth; the deceased mother might be Kiya. Gabolde[128] proposed that Nefertiti was actually named as Tutankhamen's mother in room Alpha, based on an examination of Jequier's photographs taken in 1909. In this scene, a woman is shown behind Akhenaten and Nefertiti carrying away a child, presumably Tutankhamen. Robins[129] argued that there is no proof that the woman in room Alpha is Kiya and not some other minor wife of Akhenaten. One thing these scenes do show is that the birth of a royal child was an event of real importance. If Kiya died giving birth to Tutankhamen, this might explain why her name and figures were removed from monuments by changing them into those of Meritaten and Ankhesenpaaten. Furthermore, a relief on a Hermopolis block relating to Kiya now in the Ny Carlsberg Glyptotek of Copenhagen shows part of the head of Akhenaten in high raised relief overlapping the head of a woman in lower relief whose profile juts forward in front of the king's.[130] If, as the hairstyle suggests, the woman should be identified as Kiya, her appearance in this way with the king serves once more to underline her very unusual position. The fact that the two are represented at virtually the same scale testifies to Kiya's importance at court. Reeves[131] guessed that this is a consequence of the birth of a male heir, Tutankhamen.

Harris[132] argued that, if Kiya was the mother of Tutankhamen, her absence in his tomb might be explained in terms of dynastic manoeuvring; it is even possible that Tutankhamen had no knowledge of her.

5. Kiya and tomb KV55

The identity of the skeleton[133] found in tomb KV55 has for many years been the subject of an exasperating controversy, both among anthropologists and archaeologists. This tomb was discovered in January 1907 during excavations sponsored by Theodore M. Davis.[134] One of Davis' personal

[123] M. Birrell, 'Was Ay the Father of Kiya?', *BACE* 8 (1997) 11-18.
[124] O. Schaden, *The God's Father Ay*, dissertation, UMI 78-9739, Ann Arbor (1977).
[125] M. Birrell, 'Was Ay the Father of Kiya?', *BACE* 8 (1997) 11.
[126] M. Birrell, *ibid.*, 15.
[127] G. T. Martin, 'Expedition to the royal tomb of Akhenaten', *The Illustrated London News* 6998 (1981) 67; G. T. Martin, *The Royal Tomb at El-'Amarna* II, London (1989) 37-41, fig. 7, pl. 58-61.
[128] M. Gabolde, *D'Akhenaton à Toutankhamon*, Lyon/Paris (1998) 120 n. 989, 121, 123-124, 124 n. 1011.
[129] G. Robins, 'The Mother of Tutankhamen, 2', *DE* 22 (1992) 25-27.
[130] D. Arnold, *The Royal Women of Amarna. Images of Beauty from Ancient Egypt*, New York (1996) 88, fig. 79.
[131] C. N. Reeves, *Akhenaten: Egypt's False Prophet*, London (2001) 159.
[132] J. R. Harris, *Akhenaten and Nefernefruaten in the Tomb of Tut'ankhamûn*, London/New York (1992) 72 n. 115.
[133] Egyptian Museum, Cairo, CG 61075.
[134] B. Porter and R. L. B. Moss, *Topographical Bibliography of Ancient Egyptian Hieroglyphic Texts, Reliefs, and Paintings* I-2, Oxford (1989) 565-567; M. Rose, 'Who's in Tomb 55?', *Archaeology* 55:2 (2002), with an abstract at the following URL: http://www.archaeology.org/found.php?page=/0203/abstracts/tomb55.html; cf. C. N. Reeves and R. H. Wilkinson, *The Complete Valley of the Kings*, London (1997) 117-121.

ambitions was to locate the final resting place of queen Tiye and, convinced he had found it in tomb KV55, he published the tomb as such in 1910. In 1916, Daressy[135] was the first to make a complete study of the inscriptions on the coffin.[136] He observed that certain parts of the inscriptions had been removed and other words or phrases inserted on patches of gold foil in their place. One of the changes is a bearded seated figure which, in the single place without such alterations, is a seated female figure. Daressy therefore argued that the coffin had originally been made for a woman and then adapted for a man; he concluded that the coffin was originally made for Tiye. In 1931, Engelbach[137] re-examined the coffin and concluded that it was originally prepared for Smenkhkare, who would have been non-royal. According to Engelbach, it was Tutankhamen who had changed the inscriptions on the coffin of Smenkhkare to make them royal, and erased the name of Akhenaten.[138] Helck[139] noticed that the contents of the tomb included items associated with, amongst others, Akhenaten, Kiya, and Tiye, and these have been used in the many attempts to establish the identity of the tomb's occupant. The most recent re-examination by Filer[140] in January 2000 has confirmed the earlier findings of Harrison[141] and Derry[142] that the body of tomb KV55 is, with reasonable certainty, that of a man between the ages of twenty and twenty-five years, probably towards the lower end of the age range. The identity of the person in the coffin is still unclear,[143] but it is certainly not Kiya.

Fairman[144] first pointed out in 1961 that the foot-end inscriptions of the coffin of tomb KV55 show a text nearly identical to the one of cosmetic vase EA 65901 from the BM. In addition, similarities appear between the texts on the coffin and texts on the various Maruaten palimpsest fragments and on the Hermopolis blocks. Perepelkin,[145] however, was the first to propound, in 1968, the theory that the coffin from tomb KV55 was originally Kiya's. Perepelkin has published valuable work about the Amarna period; not all of his work has yet been translated from Russian.[146] He argued for Kiya's original ownership of the coffin, and of the alabaster canopic jars and lids from tomb KV55, based on Kiya's apparently standardized titulary. He further attempted to establish Kiya as a hitherto unrecognised female co-regent of Akhenaten ruling before the accession of Smenkhkare. According to Perepelkin, the body of tomb KV55 lay in Kiya's coffin, refurbished for Akhenaten, in a chamber containing magic bricks inscribed posthumously for the same king. Because the person depicted on the coffin wears the so-called 'Nubian wig', Hanke[147] assumed that the coffin from tomb KV55 was originally meant for Kiya. In 1988, Allen[148] reached the same conclusion as Perepelkin and Hanke: the original name on the coffin was Kiya's. In the lines on the foot, in which Kiya speaks to Akhenaten, a seated god replaced her name. Thus, Allen[149] concluded that the coffin had originally been made for

---

[135] G. Daressy, 'Le cercueil de Khu-n-aten', *BIFAO* 12 (1916) 145-159.

[136] Egyptian Museum, Cairo, JdE 39627.

[137] R. Engelbach, 'The so-called coffin of Akhenaten', *ASAE* (1931) 103-104.

[138] R. Engelbach, *ibid.*, 112.

[139] W. Helck, 'Was geschah in KV55?', *GM* 60 (1982) 43-46.

[140] J. Filer, 'The KV55 body: the facts', *EA* 17 (2000) 13-14; J. Filer, 'Anatomy of a Mummy', *Archaeology* 55:2 (2002), with abstract at URL: http://www.archaeology.org/found.php?page=0203/abstracts/mummy.html.

[141] R. G. Harrison, 'An anatomical examination of the pharaonic remains purported to be Akhenaten', *JEA* 52 (1966) 95-119.

[142] D. E. Derry, 'Note on the skeleton hitherto believed to be that of king Akhenaten', *ASAE* (1931) 115-120.

[143] Cf. J. P. Allen, 'Nefertiti and Smenkh-ka-re', *GM* 141 (1994) 7-8.

[144] H. W. Fairman, 'Once again the so-called coffin of Akhenaten', *JEA* 47 (1961) 30.

[145] G. Perepelkin, *The Secret of the Gold Coffin*, Moscow (1978) 7-36.

[146] E. S. Bogoslovskij, '*Ju. Ja. Perepelkin. Die Revolution von Amen-hotp IV, I, (1967)*, Review', *GM* 61 (1983) 53-63; E. S. Bogoslovskij, '*Yu. Ya. Perepelkin. Amenhotp's revolution, II, (1984)*, Review', *GM* 93 (1986) 85-93; I. Munro, 'Zusammenstellung von Datierungskriterien für Inschriften der Amarna-Zeit nach *J. J. Perepelkin "Die Revolution Amenophis' IV, I, (1967)"*', *GM* 94 (1986) 81-87.

[147] R. Hanke, *Amarna-Reliefs aus Hermopolis*, HÄB 2, Hildesheim (1978) 171-174.

[148] J. P. Allen, 'Two Altered Inscriptions of the Late Amarna Period', *JARCE* 25 (1988) 122-127.

[149] J. P. Allen, 'Nefertiti and Smenkh-ka-re', *GM* 141 (1994) 8.

Kiya. In 2001, a monograph on KV55 by Helck was posthumously published (based on a manuscript of 1992), in which he, too, concluded, based on his own analyses of the sarcophagus texts as well as the work of Perepelkin *et al.*, that the KV55 sarcophagus was originally meant to be used by a female person, Kiya.[150]

In the same year, however, a surprising development occurred. Based on a new and detailed 'autopsy' of the golden sarcophagus trough from KV55, its typology, iconography, and six inscriptions, Grimm concluded that the sarcophagus was originally made for Akhenaten rather than Kiya, and that it was used, after having become obsolete for religious reasons, for Smenkhkare.[151] He found no reason to presume that the uraeus and royal beard on the sarcophagus are secondary, and pointed out that several so-called alterations in the gold foil are optical illusions. The special form of the 'Nubian wig' that appears on the sarcophagus does not point to a female per se, but is also attested for Akhenaten. And the female speaking in the text at the foot-end is, by analogy with other coffins, not a female coffin owner, but the wife of the coffin owner (the dead king), speaking to him in her role as Isis, wife of Osiris. In the other five inscriptions, Akhenaten's epithets occur in the form in which they also occur on Kiya's vases.

Krauss[152] confirmed Perepelkin's findings with respect to the four alabaster canopic jars and lids[153] from tomb KV55: they were made for Kiya and must resemble her. Krauss noticed that all four canopic jars have remnants of texts.[154] Faint traces of the characteristic titulary for Kiya can be read in the rectangular panel (20 x 14.5 cm.) of at least one of the canopic jars. This indicates that she must have had a certain status at Akhenaten's court, elevated above the 'ordinary' harem-women. Furthermore, Krauss argued that the removal of the columns with the text from the right side of the panel made the inscription apply exclusively to Akhenaten instead of Kiya. The face on the lid may show us Kiya. In Krauss' opinion these canopic jars and lids, originally meant for Kiya's canopic remains, were later adapted for a royal person, i.e. for somebody who wore a uraeus. According to Allen[155] two columns originally bearing Kiya's titles and name were ground down on the canopic jars, leaving the cartouches of the Aten (in the earlier form of the name) and Akhenaten to the left. These cartouches were subsequently chiseled out as well. Nothing seems to have been carved in place of Kiya's columns. The new analysis of Grimm confirms that the canopic jars and lids from KV55 were indeed originally made for Kiya.[156]

6. Conclusion

Pharaoh Akhenaten showed strong interest in at least two women, Kiya and Nefertiti. For reasons as yet unknown, Kiya had rare titles, which may indicate that she had a unique place at Akhenaten's

[150] W. Helck, *Das Grab Nr. 55 im Königsgräbertal*, SDAIK 29, Mainz am Rhein (2001) 29-35.

[151] A. Grimm *et al.*, *Das Geheimnis des goldenen Sarges. Echnaton und das Ende der Amarnazeit*, Schriften aus der Ägyptischen Sammlung 10, München (2001) 100-120. Slides of a computer-tomography of the KV55 coffin lid (January 2002, Munich) may be seen at: http://www.aegyptisches-museum-muenchen.de/de/h/h02_in.htm#200202.

[152] R. Krauss, 'Kija – ursprüngliche Besitzerin der Kanopen aus KV 55', *MDAIK* 42 (1986) 67-80; cf. C. Vandersleyen, 'Royal Figures from Tut'ankhamûn's Tomb: Their Historical Usefulness', in C. N. Reeves (ed.), *After Tut'ankhamun. Research and Excavation in the Royal Necropolis at Thebes*, London/New York (1992) 73-81.

[153] Three jars and lids in the Egyptian Museum, Cairo (JdE 39637) and one in the Metropolitan Museum of Art, New York (jar/lid: MMA 07.226.1/30.8.54); B. Porter and R. L. B. Moss, *Topographical Bibliography of Ancient Egyptian Hieroglyphic Texts, Reliefs, and Paintings I. The Theban Necropolis, 2: Royal Tombs and Smaller Cemetries*, Oxford (1989) 566; J. Settgast, *Nofretete – Echnaton*, Berlin (1976) fig. 51; M. Gabolde, *D'Akhenaton à Toutankhamon*, Lyon/Paris (1998) 255-257.

[154] R. Krauss, 'Kija – ursprüngliche Besitzerin der Kanopen aus KV 55', *MDAIK* 42 (1986) 67.

[155] J. P. Allen, 'Two Altered Inscriptions of the Late Amarna Period', *JARCE* 25 (1988) 122.

[156] A. Grimm *et al.*, *Das Geheimnis des goldenen Sarges. Echnaton und das Ende der Amarnazeit*, Schriften aus der Ägyptischen Sammlung 10, München (2001) 117-118.

court. In addition, although not a chief wife, she bids fair to outrival Nefertiti in the amount of attention that was paid to her. Kiya may have given the king at least one daughter and possibly a son, Tutankhamen. It is generally assumed that Kiya died before Akhenaten. Most of the memorials or reliefs in which she was mentioned or portrayed were either usurped or destroyed. There is strong evidence that the canopic jars of tomb KV55 were originally meant for Kiya.

Scarce facts prove the existence of Kiya. The question of who she was cannot yet be fully answered. Further research is needed, which may confirm that she played a prominent role in the Amarna era.

Arris H. Kramer
arrish@ision.nl
Kockengen, The Netherlands

Appendix: Kiya's Chronology

With respect to the presumed chronology of Akhenaten's reign, the following datable items have been held to attest to Kiya:

Year	Object	Selected Literature
6(?) (cf. 16)	Wine-jar docket (CoA II, pl. 58: 16)	H. Frankfort and J. D. S. Pendlebury, *The City of Akhenaten* II, London (1972) pl. 58, n° 16; C. Vandersleyen, *L'Égypte et la vallée du Nil* II, Paris (1995) 444.
6	Funerary cone of Bengay (DM 527)	N. de Garis Davies, *A Corpus of Inscribed Egyptian Funerary Cones I - Plates*, ed. by M. F. Laming Macadam, Oxford (1957), n° 527; W. Helck, 'Kija', *LÄ* III, Wiesbaden (1980) 422-423.
c. 9	Cosmetic vase with earlier Aten name (EA 65901)	H. W. Fairman, 'Once again the so-called coffin of Akhenaten', *JEA* 47 (1961) 29-30; J. Assmann, 'Aton', *LÄ* I, Wiesbaden (1975) 538; W. Helck, 'Kijê', *MDAIK* 40 (1984) 159-160.
c. 9	Canopic jar with earlier Aten name (MMA 07.226.1)	J. Assmann, 'Aton', *LÄ* I, Wiesbaden (1975) 538; R. Krauss, 'Kija – ursprüngliche Besitzerin der Kanopen aus KV 55', *MDAIK* 42 (1986) 71.
11	Wine-jar docket (EA 59951)	W. M. Flinders Petrie, *Tell El Amarna*, London (1894) pl. 25, n° 95; W. Helck, 'Kijê', *MDAIK* 40 (1984) 160; C. N. Reeves, 'New light on Kiya from texts in the British Museum', *JEA* 74 (1988) 92.
12	Wall F of room Alpha in the royal tomb at Akhetaten	G. T. Martin, *The Royal Tomb at El-'Amarna* II, London (1989) 37-41, fig. 7, pl. 58-61; C. N. Reeves, *Akhenaten: Egypt's False Prophet*, London (2001) 160.
[1]6(?) (cf. 6)	Wine-jar docket (CoA II, pl. 58: 16)	C. Aldred, *Akhenaten, King of Egypt*, London (1988) 227.

Bibliography

C. Aldred, 'The beginning of the El-'Amârna Period', *Journal of Egyptian Archaeology* 45 (1959) 19-33
C. Aldred, *Akhenaten and Nefertiti*, New York (1973)
C. Aldred, *Akhenaten, King of Egypt*, London (1988)

J. P. Allen, 'Two Altered Inscriptions of the Late Amarna Period', *Journal of the American Research Center in Egypt* 25 (1988) 117-127

J. P. Allen, 'Nefertiti and Smenkh-ka-re', *Göttinger Miszellen* 141 (1994) 7-17

D. Arnold, *The Royal Women of Amarna. Images of Beauty from Ancient Egypt*, New York (1996)

D. Arnold, 'Nefertiti & the royal women at the Metropolitan Museum', *KMT: A Modern Journal of Ancient Egypt* 7:4 (1996) 18-31

J. Assmann, 'Aton', *Lexikon der Ägyptologie* I, Wiesbaden (1980) 526-540

M. Birrell, 'Was Ay the Father of Kiya?', *Bulletin of the Australian Centre for Egyptology* 8 (1997) 11-18

E. S. Bogoslovskij, '*Ju. Ja. Perepelkin. Die Revolution von Amen-hotp IV, I, Moscow (1967)*, Review', *Göttinger Miszellen* 61 (1983) 53-63

E. S. Bogoslovskij, '*Yu.Ya. Perepelkin. Amenhotp's revolution, II, Moscow (1984)*, Review', *Göttinger Miszellen* 93 (1986) 85-93

J. D. Cooney, *Amarna Reliefs from Hermopolis in American Collections*, New York (1965)

A. Cabrol, *Amenhotep III Le Magnifique*, Monaco/Paris (2000)

G. Daressy, 'Le cercueil de Khu-n-aten', *Bulletin de l'Institut Français d'Archéologie Orientale du Caire* 12 (1916) 145-159

D. E. Derry, 'Note on the skeleton hitherto believed to be that of king Akhenaten', *Annales du Service des Antiquités de l'Égypte* (1931) 115-120

A. Dodson, 'KV 55 and the end of the reign of Akhenaten', *Atti del VI Congresso Internazionale di Egittologia* I, Torino (1993) 135-139

J. van Dijk, 'The Noble Lady of Mitanni and Other Royal Favourites of the Eighteenth Dynasty', in J. van Dijk (ed.), *Essays on Ancient Egypt in Honour of Herman te Velde*, Egyptological Memoirs 1, Groningen (1997) 33-46

M. Eaton-Krauss, 'Miscellanea Amarnensia', *Chronique d'Égypte* 56 (1981) 245-264

R. Engelbach, 'The so-called coffin of Akhenaten', *Annales du Service des Antiquités de l'Égypte* (1931) 98-114

R. Engelbach, 'Material for a revision of the history of the heresy period of the XVIIIth dynasty', *Annales du Service des Antiquités de l'Égypte* 40 (1940) 133-165

H. W. Fairman, 'Once again the so-called coffin of Akhenaten', *Journal of Egyptian Archaeology* 47 (1961) 25-40

F. Filce Leek, 'How old was Tut'ankhamûn?', *Journal of Egyptian Archaeology* 63 (1977) 112-115

J. M. Filer, 'The KV55 body: the facts', *Egyptian Archaeology: Bulletin of the Egypt Exploration Society* 17 (2000) 13-14

J. M. Filer, 'Anatomy of a Mummy', *Archaeology Magazine of the Archaeological Institute of America* 55:2 (2002). Abstract at URL: http://www.archaeology.org/found.php?page=0203/abstracts/mummy.html

W. M. Flinders Petrie, *Tell El Amarna*, London (1894)

H. Frankfort and J. D. S. Pendlebury, *The City of Akhenaten* II, London (1972 reprint; original 1933)

R. E. Freed *et al.*, *Pharaohs of the Sun*, Boston (1999); Dutch translation in *Farao's van de Zon*, Leiden/Amsterdam (2000)

M. Gabolde, 'Baketaton, fille de Kiya?', *Bulletin Société d'Égyptologie Genève* 16 (1992) 27-40

M. Gabolde, *D'Akhenaton à Toutankhamon*, Lyon/Paris (1998)

N. de Garis Davies, *A Corpus of Inscribed Egyptian Funerary Cones I – Plates*, ed. by M. F. Laming Macadam, Oxford (1957)

E. Graefe, 'Zu *Pjj*, der angeblichen Nebenfrau des Achanjati', *Göttinger Miszellen* 33 (1979) 17-18

A. Grimm *et al.*, *Das Geheimnis der goldenen Sarges. Echnaton und das Ende der Amarnazeit*, Schriften aus der Ägyptischen Sammlung 10, München (2001)

R. Hanke, 'Änderungen von Bildern und Inschriften während der Amarnazeit', *Studien zur Altägyptischen Kultur* 2 (1975) 79-93

R. Hanke, *Amarna-Reliefs aus Hermopolis*, Hildersheimer Ägyptologische Beiträge 2, Hildesheim (1978)

J. R. Harris, 'Kiya', *Chronique d'Égypte* 49 (1974) 25-30

J. R. Harris, 'Akhenaten and Nefernefruaten in the Tomb of Tut'ankhamûn', in C. N. Reeves (ed.), *After Tutankhamun: Research and Excavation in the Royal Necropolis of Thebes*, London/New York (1992) 55-72

R. G. Harrison, 'An anatomical examination of the pharaonic remains purported to be Akhenaten', *Journal of Egyptian Archaeology* 52 (1966) 95-119

W. C. Hayes, *The Scepter of Egypt: A Background for the study of the Egyptian Antiquities in the Metropolitan Museum of Art* II, Cambridge/Mass. (1959)

W. Helck, 'Einige Bemerkungen zu Artikeln in SAK 2', *Studien zur Altägyptischen Kultur* 4 (1976) 115-124

W. Helck, 'Kija', *Lexikon der Ägyptologie* III, Wiesbaden (1980) 422-424

W. Helck, 'Probleme der Königsfolge in der Übergangszeit von 18. zu 19. Dyn.', *Mitteilungen des Deutschen Archäologischen Instituts in Kairo* 37 (1981) 207-215

W. Helck, 'Was geschah in KV55?', *Göttinger Miszellen* 60 (1982) 43-46

W. Helck, 'Kijê', *Mitteilungen des Deutschen Archäologischen Instituts in Kairo* 40 (1984) 159-167

W. Helck, *Das Grab Nr. 55 im Königsgräbertal. Sein Inhalt und seine historische Bedeutung*, Sonderdruck des Deutschen Archäologischen Instituts in Kairo 29, Mainz am Rhein (2001)

S. Izre'el, *The Amarna Tablets*, URL: http://spinoza.tau.ac.il/hci/dep/semitic/amarna.html (2000)

Y. Knudsen de Behrensen, 'Pour une identification de la momie du tombeau no. 55 de la vallée des rois', *Göttinger Miszellen* 90 (1986) 51-60

J. A. Knudtzon, *Die El-Amarna Tafeln*, 2 vols., Leipzig (1915)

A. H. Kramer, 'De reconstructie van Aton-heiligdommen in Karnak', *De Ibis* 16:3 (1991) 112-125

A. H. Kramer, 'Kiya', *Newsletter of the Akhenaten Temple Project* 4 (1999) 1-4

R. Krauss, *Das Ende der Amarnazeit*, Hildersheimer Ägyptologische Beiträge 7, Hildesheim (1981)

R. Krauss, 'Kija – ursprüngliche Besitzerin der Kanopen aus KV 55', *Mitteilungen des Deutschen Archäologischen Instituts in Kairo* 42 (1986) 67-80

L. Manniche, 'The Wife of Bata', *Göttinger Miszellen* 18 (1975) 33-38

G. T. Martin, 'Expedition to the royal tomb of Akhenaten', *The Illustrated London News* 6998 (1981) 66-67

G. T. Martin, *The Royal Tomb at El-'Amarna* II, London (1989)

E. S. Meltzer, 'The parentage of Tut'ankhamún and Smenkhkaré', *Journal of Egyptian Archaeology* 64 (1978) 134-135

W. L. Moran, *The Amarna Letters*, Baltimore/London (1992)

I. Munro, 'Zusammenstellung von Datierungskriterien für Inschriften der Amarna-Zeit nach *J. J. Perepelkin "Die Revolution Amenophis' IV, I, Moscow (1967)"*', *Göttinger Miszellen* 94 (1986) 81-87

T. E. Peet and C. L. Woolley, *The City of Akhenaten* I, Oxford (1984 reprint; original 1923)

G. Perepelkin, *The Secret of the Gold Coffin*, Moscow (1978); English translation of *Taina Zolotogo Groba*, Moskva (1968)

B. Porter and R. L. B. Moss, *Topographical Bibliography of Ancient Egyptian Hieroglyphic Texts, Reliefs, and Paintings I. The Theban Necropolis, 2: Royal Tombs and Smaller Cemetries*, Oxford (1989)

H. Ranke, *Die Ägyptischen Personennamen I: Verzeichnis der Namen*, Glückstadt (1935)

J. Ray, 'The parentage of Tutankhamun', *Antiquity* 49 (1975) 45-47

D. B. Redford, *Akhenaten: The Heretic King*, Princeton (1987)

C. N. Reeves, *'G. Perepelkin, The Secret of the Gold Coffin (1978*), English translation of *Taina Zolotogo Groba (1968)*, Review', *Bibliotheca Orientalis* 38 (1981) 293-297

C. N. Reeves, 'Tuthmosis IV as 'great-grandfather' of Tutankhamen', *GM* 56 (1982) 65-69

C. N. Reeves, 'New light on Kiya from texts in the British Museum', *Journal of Egyptian Archaeology* 74 (1988) 91-101

C. N. Reeves and R. H. Wilkinson, *The Complete Valley of the Kings*, London (1997)

C. N. Reeves, *Akhenaten: Egypt's False Prophet*, London (2001)

G. Robins, 'The Mother of Tutankhamen', *Discussion in Egyptology*, Oxford 20 (1991) 71-73

G. Robins, 'The Mother of Tutankhamen, 2', *Discussion in Egyptology*, Oxford 22 (1992) 25-27

G. Roeder, *Amarna-Reliefs aus Hermopolis* II, Hildesheim (1969)

M. Rose, 'Who's in Tomb 55?', *Archaeology Magazine of the Archaeological Institute of America* 55:2 (2002). Abstract at URL: http://www.archaeology.org/found.php?page=/0203/abstracts/tomb55.html

J. Samson, *Amarna: City of Akhenaten and Nefertiti; Key Pieces from the Petrie Collection*, London (1972)

J. Samson, 'Amarna crowns and wigs', *Journal of Egyptian Archaeology* 59 (1973) 47-59

J. Samson, *Amarna: City of Akhenaten and Nefertiti; Nefertiti as Pharaoh*, Warminster (1978)

J. Samson, *Nefertiti and Cleopatra. Queen-Monarchs of Ancient Egypt*, London (1985)

M. Sandman, *Texts from the Time of Akhenaten*, Bibliotheca Aegyptiaca VIII, Brussels (1938)

O. J. Schaden, *The God's Father Ay*, dissertation, UMI 78-9739, Ann Arbor (1977)

H. A. Schlögl, *Echnaton – Tutanchamun*, Wiesbaden (1993)

A. R. Schulman, 'Diplomatic marriage in the Egyptian New Kingdom', *Journal of Near Eastern Studies* 38 (1979) 177-193

K. C. Seele, 'King Ay and the close of the Amarna age', *Journal of Near Eastern Studies* 14 (1955) 168-180

J. Settgast, *Nofretete – Echnaton*, Berlin (1976)

R. W. Smith and D. B. Redford, *The Akhenaten Temple Project* I, Warminster (1976)

A. P. Thomas, 'Some Palimpsest Fragments from the Maru-Aten at Amarna', *Chronique d'Égypte* 57 (1982) 5-13

S. Tower Hollis, *The Ancient Egyptian 'Tale of Two Brothers'*, Oklahoma (1990)

C. Vandersleyen, *Royal Figures from Tut'ankhamûn's Tomb: Their Historical Usefulness*, in C. N. Reeves (ed.), *After Tut'ankhamun. Research and Excavation in the Royal Necropolis at Thebes*, London/New York (1992)

C. Vandersleyen, *L'Égypte et la Vallée du Nil II: De la fin de l'Ancien empire à la fin du Nouvel Empire*, Paris (1995)

B. van der Walle, '*G. Perepelkin, The Secret of the Gold Coffin (1978*), English translation of *Taina Zolotogo Groba (1968)*, Review', *Chronique d'Égypte* 55 (1980) 136-140

J. A. Wilson, 'Story of Two Brothers', in J. B. Pritchard (ed.), *Ancient Near Eastern Texts Relating to the Old Testament*, Princeton (1950), 23-25

Three Notes on Arsinoe I

Chris Bennett

The Egyptian Royal Genealogy Project

This volume is an outgrowth of a project that took place in a very different medium: the Internet. My contribution to the volume has a similar origin. The Egyptian Royal Genealogy website[1] is a project designed to publish the evidence and the issues for the genealogy of the ruling Egyptian families. The first phase of this project, undertaken in 2001, was the publication on the World Wide Web of a detailed Ptolemaic genealogy.

The great advantage of the Web as a publication medium is its dynamism. New data, new arguments and corrections can be added and published almost immediately. Also, as original material becomes available on the Web, the genealogy can be directly linked to it, allowing users to examine the sources for themselves. The Ptolemaic genealogy at the website currently contains links to many classical sources and transcriptions of Greek papyri,[2] and even some images of the source papyri,[3] as well as some numismatic images.[4] Future extensions of the site will certainly draw on resources such as the Howard Carter archives of the Griffith Institute[5] and other Egyptological electronic resources as they become available.

The great disadvantage of the Web as a publication medium is its impermanence. It will be a long time before print is replaced as the medium of record. In compiling the Ptolemaic genealogy, I reached conclusions on a number of topics that differ from those in the literature. This paper is one of several that report these conclusions.

Arsinoe I

The subject of this paper is Arsinoe I, the least visible of Ptolemaic queens. From the classical sources,[6] we know only that she was the daughter of Lysimachus king of Thrace; that she married Ptolemy II and was displaced as his wife by his sister, Lysimachus' widow Arsinoe II; that she was exiled to Coptos in Upper Egypt as a result of a court plot; and that she was the mother of Ptolemy II's legitimate children, namely Ptolemy III, Lysimachus and Berenice. There are no representations of her, and only two possible contemporary references to her are known, both ambiguous.[7]

Modern scholars have made a few additional inferences about her. These are the topic of this paper. It is argued here that something we think we know about her – the likely identity of her mother – is in fact not known. It is next argued that an old dispute – whether she was the mother of Berenice II – has been settled correctly, but for the wrong reason. Finally, it is argued that one of the inscriptions held to refer to her allows us to estimate the date of her exile.

---

[1] Currently hosted at http://www.geocities.com/christopherjbennett, with a mirror site (hosted by Tyndale House) at http://www.tyndale.cam.ac.uk/Egypt/.

[2] At the Duke DataBank of Documentary Papyri, hosted at the following URL:
http://www.perseus.tufts.edu/cgi-bin/perscoll?collection=Perseus:collection:DDBDP.
The electronic publication of demotic and other papyri is unfortunately not as advanced as the publication of Greek papyri.

[3] E.g. pKöln at http://www.uni-koeln.de/phil-fak/ifa/NRWakademie/papyrologie/index.html.

[4] E.g. the ANS image database at http://www.amnumsoc.org/collections/images/imlist.html.

[5] At http://www.ashmol.ox.ac.uk/gri/carter/HomePage.html.

[6] Pausanias 1.7.3, Schol. Theocritus 17.128.

[7] CCG 70031, *KAI* 43, discussed below.

The Mother of Arsinoe I

We do not know the date of Arsinoe I's birth. She was certainly married to Ptolemy II before the death of Lysimachus in February 281,[8] and probably after Ptolemy I made his son coregent at the very end of 285 or the beginning of 284.[9] Assuming, as is usually done, that she was at least 13 or 14 years of age at the time of her marriage, we can estimate from the first date that she was born in or before 296. It seems unlikely, though hardly impossible, that she would have been much more than about 20 years old at the time of her marriage, so she was probably born after c. 305.

Her mother is unnamed in any source, but is universally regarded in modern studies as Nicaea, daughter of Antipater and wife of Lysimachus. Given Ptolemy II's status as a leading Hellenistic king it is safe to assume that his wife was the daughter of one of Lysimachus' principal wives. However, a range of c. 305 - 296 for Arsinoe's birth covers all three known wives.

The three candidates may be considered in reverse order of their marriages.

Arsinoe II. Arsinoe II married Lysimachus shortly after the battle of Ipsus in 300.[10] Macurdy[11] argued against Arsinoe II as the mother of Arsinoe I on two grounds. First, she noted that Arsinoe I would almost certainly have been the oldest daughter of Lysimachus and Arsinoe II, and asserted that in this case she should have been named Berenice after her maternal grandmother. Second, she noted that no ancient writer commented that Arsinoe II had replaced her own daughter as the wife of her brother Ptolemy II. The second objection is cogent, but cannot be regarded as conclusive in view of the almost total absence of sources on the career of Arsinoe I. However, the first is hardly relevant, since Arsinoe II's eldest daughter is more likely to have been named after her paternal grandmother than after her maternal one, and the name of the mother of Lysimachus is unknown. For the same reason, Macurdy's argument that Arsinoe I might have been named after the mother of Lysimachus' first wife Nicaea, whose name is also unknown, is not relevant.

A stronger argument against Arsinoe II is the tightness of the chronology that results from Arsinoe I being assumed to be her daughter. Justin[12] tells us that her younger sons Lysimachus and Philip were 16 and 13 years old at the time of their murder in the winter of 281/0. They were therefore born in 297/6 and 294/3 respectively. Since Arsinoe II married Lysimachus in c. 300, their older son Ptolemy must have been born c. 299/8. It is only possible for Arsinoe I to be the daughter of Arsinoe II if she was born between Lysimachus and Philip, i.e. c. 295. This is right at or after the latest birth date we established above.

Amestris.[13] Amestris was the daughter of Oxathres, the brother of Darius III. She was married by Alexander to Craterus in c. April 324. A few years later she married Dionysios, tyrant of Heracleia in Pontus, by whom she had three sons before his death in 306. She was only married to Lysimachus for a short time, about two years (302-300), and the brevity of the marriage only allows one child to be born from it. According to Polyaenus,[14] she was the mother of a son, Alexander, by Lysimachus, but Pausanias[15] calls Alexander the son of an Odrysian concubine. If Polyaenus is correct then there is little opportunity for Amestris also to have borne Lysimachus a daughter, but Pausanias is generally regarded as more reliable.

---

[8] A. J. Sachs & D. J. Wiseman, *Iraq* 16 (1954) 202. All dates are B.C. unless otherwise indicated.

[9] A. E. Samuel, *Ptolemaic Chronology* (Munich, 1962) 29ff.

[10] Plutarch, *Demetrius* 31. She is not explicitly named by Plutarch, but the chronology of her sons, discussed below, admits of no other solution.

[11] G. H. Macurdy, *Hellenistic Queens* (Baltimore, 1932) 109.

[12] Justin 24.3.

[13] U. Wilcken, *RE* 1 (1894) 1750. The abbreviation *RE* refers herein to the *Pauly-Wissowa Realencyclopädie der classischen Altertumswissenschaft.*

[14] Polyaenus 6.12.

[15] Pausanias 1.10.3.

Nicaea.[16] Nicaea is known to have been nubile c. 321, when her father, the Macedonian regent Antipater, first offered her in marriage to Perdiccas. She married Lysimachus shortly afterwards, and at that point disappears from the record. For all we know, she may have died immediately or may even have stayed married to Lysimachus after his marriage to Amestris, though neither extreme seems likely. Since Lysimachus divorced Amestris to marry Arsinoe II, we may be confident that Nicaea was dead or divorced by 300. Lysimachus' daughter, Eurydice, married the younger Antipater, co-king of Macedon 297-294, and therefore must have been born in or before c. 308. His son and heir, Agathocles, married Ptolemy I's daughter Lysandra c. 292/1, so was presumably born some time before about 310. Since these birth dates are before Lysimachus' marriage to Amestris, Agathocles and Eurydice were almost certainly the children of Nicaea. If Arsinoe I was also her daughter, she was probably several years younger than her uterine siblings.

It can be seen that, while Nicaea is a perfectly possible candidate for Arsinoe I's mother, her contemporary Amestris is equally plausible. The brevity of her marriage to Lysimachus is of little consequence as an objection unless it can be shown that Alexander was her son, as claimed by Polyaenus. As a child of Amestris, Arsinoe would be born c. 301/300, making her between about 15 and 18 years old at the time of her marriage to Ptolemy II.

The Mother of Berenice II

Pausanias tells us that Magas, king of Cyrene, married Apama, daughter of the Seleucid Antiochus I.[17] In modern scholarship, Apama is also considered to be the mother of Berenice II, wife of Ptolemy III, usually without comment. But our only surviving literary sources both state that Berenice II's mother was called Arsinoe. Hyginus names Berenice's parents in passing as Ptolemy and Arsinoe.[18] Justin names them as Magas of Cyrene and Arsinoe, and tells the lurid story of an affair between Arsinoe and Demetrius the Fair, Berenice's fiancé, shortly after Magas' death.[19] In contemporary Egyptian sources Berenice is called the "king's sister" of Ptolemy III.[20] The paternity of Magas rather than Ptolemy is confirmed by an exedra at Thermos, which names queen Berenice as the daughter of king Magas.[21] However, there is no contemporary source that names her mother.

The circumstances detailed by Pausanias[22] imply that Apama married Magas between the accession of Antiochus I in 280 and the outbreak of the First Syrian War in 275. Since Apama was the daughter of Antiochus I by Stratonice, whom he married in 293, she cannot have been born before c. 292, which favours a date towards the end of this range.[23] Callimachus[24] describes Berenice II as showing her heroism as a "young girl" (*parva virgo* in Catullus' translation) in connection with an unspecified incident that is today assumed to be the murder of Demetrius the Fair.[25] Since this event took place shortly after Magas' death, and since Berenice was old enough to have been marriageable at the time, Callimachus' words are taken to indicate that she was born about 14 or 15 years before her father's death.

In order to estimate a date for Berenice II's birth, therefore, we must first establish the date of Magas' death. This date has long been a subject of controversy, with the choices being a 'high'

---

[16] H. Berve, *RE* 17 (1936) 220.

[17] Pausanias 1.7.3.

[18] Hyginus, *Poetica Astronomica* 2.24.

[19] Justin 26.3.

[20] E.g. the Canopus Decree (*OGIS* 56).

[21] *IG* IX I^2 I 56. See C. J. Bennett, *ZPE* 138 (2002) 141.

[22] Pausanias 1.7.3.

[23] Eusebius, *Chron.* I (ed. Schoene) 249; U. Wilcken, *RE* 1 (1894) 2662.

[24] Callimachus, *Coma Berenice*, surviving in the Latin translation of Catullus.

[25] Justin 26.3.

chronology and a 'low' chronology.[26] On the high chronology, Magas died in c. 259/8, resulting in an estimated birth date of c. 274/3 for Berenice II. This is quite soon after the marriage of Magas to Apama. In this case, the probability that Justin's Arsinoe can be equated to Pausanias' Apama must be considered high. Moreover, on this chronology Apama was a little over 30 years old at Magas' death. If she is identified with Justin's Arsinoe, Justin's story of her affair with Demetrius the Fair is perfectly credible at such an age.

On the low chronology, which is more generally accepted today, Magas died in c. 250/49, resulting in an estimated birth date of c. 265/4 for Berenice II. Such a date undermines the chronological case for identifying Pausanias' Apama with Justin's Arsinoe. The difficulty is that the low chronology for Magas inserts around a decade between the marriage of Apama and the birth of Berenice II. This chronology creates plenty of time for Apama to be replaced. While it is perfectly possible that Apama did not bear a child till after some 10 years of marriage, it is also true that there is nothing chronologically implausible, on the low chronology of his reign, in the notion that Magas married twice. Indeed, since no source names Berenice II's mother as Apama while both Justin and Hyginus call her Arsinoe, this would appear, in the absence of other considerations, to be the preferable solution. In addition, the story of an affair between Arsinoe and Demetrius the Fair might be considered somewhat less plausible, though hardly impossible, if 'Arsinoe' (i.e. Apama) was around 40 years old at the time.

The only attempt to answer this objection that I know of was made by Beloch.[27] While favouring the low chronology, he argued that an 'Arsinoe' who was distinct from Apama could only have been a Ptolemaic princess, but objected that a marriage to such a princess would either have been incestuous in itself or would have made the marriage between Berenice II and Ptolemy III incestuous, at least according to the concept of incest as it prevailed amongst Greeks outside Egypt and before the marriage of Ptolemy II and his sister Arsinoe II. Therefore, by a kind of *reductio ad absurdum*, Justin must be in error, and we must equate 'Arsinoe' with Apama.

But there is no obvious reason why this should be true. After all, Magas was only the ruler of a minor provincial kingdom, and for the last two decades of his reign, which (on the low chronology) includes the Second Syrian War, he and Ptolemy II were at peace. This strongly suggests an accommodation had been reached between them. Such an accommodation could easily have been sealed in part by Magas' remarriage to a minor Ptolemaic relative. Distaff relatives certainly existed, e.g. the children of Magas' (probable) sister Theoxena by Agathocles king of Syracuse, who can be identified as Archagathus, epistates of Libya, and a younger Theoxena;[28] Archagathus may well have had official dealings with Magas. A possible source of a wife for Magas would be a daughter or granddaughter of Menelaus, brother of Ptolemy I, whose family is unknown. There is no reason why such a marriage would have been considered incestuous, nor why it should have made the marriage between Ptolemy III and Berenice II incestuous.

In the early nineteenth century, Niebuhr suggested that Justin's Arsinoe was none other than Arsinoe I, divorced from Ptolemy II at the time of her marriage to Magas.[29] Presumably she married him after the death or divorce of Apama. If the standard reasoning for identifying Pausanias' Apama with Justin's Arsinoe is a residue of a false chronology, one might ask whether it is possible that Niebuhr was actually correct, and whether Arsinoe I was not, after all, the mother of Berenice II.

The idea has attractions. Apart from explaining the name of Justin's Arsinoe, it allows Berenice II legitimately to be the (half-)sister of Ptolemy III, as she is described in Egyptian inscriptions. It is clear after the marriage of Ptolemy V to the "king's sister" Cleopatra I, in reality daughter of the

[26] F. Chamoux, *RH* 216 (1956) 18.

[27] K. J. Beloch, *Griechische Geschichte* IV.2 (Leipzig, 1927) 190.

[28] Justin 23.2; *SEG* 18.636; *pOxy* 37.2821; R. S. Bagnall, *Philologus* 120 (1976) 195.

[29] B. G. Niebuhr, *Kleine historische und philologische Schriften* I (Bonn, 1828) 229f. n. 40.

Seleucid king Antiochus III, that the title had no genealogical meaning and had became synonymous with "king's wife", but the only earlier example that can be adduced to show this is Berenice II herself. Beloch's objection that such a marriage would be rejected by Arsinoe I as incestuous has no merit, since in fact an attempt was made to marry Berenice II to her cousin Demetrius the Fair. Also, the proposal is chronologically plausible in that Arsinoe I would still be in her thirties at the time of Berenice's birth even on the low chronology, though she would have been in her late forties or early fifties at the time of the affair with Demetrius.

Nevertheless, Niebuhr's proposal seems most unlikely on circumstantial grounds. Ptolemy II had exiled Arsinoe I to Coptos for attempted murder and treason. He would not then marry her off to Magas, a half-brother with whom he had recently been at war, and it is even less likely that she would have escaped from Coptos across the Libyan desert to Cyrene on her own resources. Therefore, while there is nothing chronologically impossible with this solution, it seems so implausible that it demands positive proof before it can be considered.

There is, moreover, another reason to identify Pausanias' Apama with Justin's Arsinoe that has not, as far as I know, been presented in this context. Appian[30] has Antiochus III tell the Roman ambassadors, shortly before the marriage of Ptolemy V to his daughter Cleopatra I, that he is already a relative of Ptolemy's. If the identity of Apama with Arsinoe is accepted, then the statement has a simple genealogical explanation: they were second cousins once removed.[31] If it is not accepted, then there is no possible way for this statement to be true, except perhaps by invoking otherwise-unknown pre-Alexandrian or even mythological links, or by understanding it only to be metaphorical.

The Exile of Arsinoe I

According to Schol. Theocritus 17.128, Arsinoe I was exiled to Coptos for plotting against Ptolemy II. The exact date of this plot is unknown. The direct evidence is so vague it could even be after the death of Arsinoe II, since Ptolemy II, as a legal fiction, had the children of Arsinoe I adopted by Arsinoe II some unknown time after the death of the latter.[32] However, it is usually assumed that Arsinoe I's plot occurred before Ptolemy's marriage to Arsinoe II, since there is no indication that he was married to both women at the same time. Indeed, if, as Carney and Ogden have plausibly argued,[33] incestuous royal Hellenistic marriage was instituted to circumvent the murderous feuds that arose between the sons of different mothers in polygamous marriages, it is virtually certain that Arsinoe I was exiled before Ptolemy II's marriage to Arsinoe II.

CCG 70031[34] records the career of queen Arsinoe's steward Senu-sher at Coptos. Griffith[35] inferred on several grounds that the queen Arsinoe in this inscription was Arsinoe I in exile, and the identification was generally accepted. However, Quaegebeur[36] challenged this identification, convincingly showing that the stele must refer to Arsinoe II. He argued that CCG 70031 is the only known inscription from Arsinoe II's lifetime.[37]

The earliest dated reference to Arsinoe II as the wife of Ptolemy II is given on CCG 22183, the Pithom stele, erected in his year 21. This names her in connection with a visit to Heroönpolis by the royal couple on 3 Thoth of year 12.[38] However, it is uncertain whether the count of regnal years in this

[30] Appian, *Syriaca* 3.
[31] Antiochus III was the grandson of Apama's brother Antiochus II. Ptolemy V would be her great-grandson.
[32] Macurdy, *op. cit.* (n. 11) 120f.
[33] E. D. Carney, *AHB* 8 (1994) 123; D. Ogden, *Polygamy, Prostitutes and Death* (Swansea, 1999) *passim.*
[34] W. M. F. Petrie, *Koptos* (London, 1896) 19ff.
[35] F. Ll. Griffith in W. M. F. Petrie, *Koptos* (London, 1896) 21.
[36] J. Quaegebeur, *BIFAO* 69 (1971) 191, 215f. nº 47.
[37] Idem in H. Maehler & V. M. Strocka, *Das ptolemäische Ägypten* (Mainz, 1978) 245, 249.
[38] É. Naville, *ZÄS* 40 (1902) 66.

reference is based on Ptolemy II's association on the throne with Ptolemy I in 285/4 ("coregency dating") or on the death of the latter in 282/1 ("accession dating"). We know that Ptolemy II's earliest regnal dates used accession dating, but at some point were switched to coregency dating.[39] For his Macedonian regnal dates, it can be shown that the switch happened very soon after the death of his father,[40] but Egyptian inscriptions with regnal dates based on accession dating are attested at least as late as year 13, and probably as late as year 16.[41] It is generally agreed that by year 21, the date of the Pithom stele, Egyptian regnal dates were based on coregency dating, but it is uncertain whether a retroactive reference, such as the mention of year 12 in the Pithom stele, would have been adjusted from accession to coregency dating or left as the original accession-based date. If the date is accession-based, then the visit occurred on 1 November 272. If it is coregency-based, then the visit occurred on 2 November 274.

Recently, Hazzard has proposed to fix the date of this marriage by accepting at face value the official explanation for it, i.e. that Ptolemy II and Arsinoe II were following the example of Zeus and Hera.[42] He reasonably argues that this implies that the couple must have been deified before the marriage. The dynastic cult first introduces the couple as the Theoi Adelphoi, Sibling Gods, in Macedonian regnal year 14.[43] Hence Hazzard concludes that they must have been deified in year 13 or 14 (Macedonian), and that the marriage occurred about the same time. Since the Macedonian regnal years were coregency-based by this time, this dates the marriage to 273 or 272. This conclusion is completely consistent with the record of the Pithom stele if the Egyptian year 12 reference therein is accession-based.[44]

Additional light on the date of Arsinoe I's exile is given by *KAI* 43.[45] This is a Phoenician inscription from Cyprus which contains the only known, though indirect, dated reference to Arsinoe I. The inscription is dated to year 11 of a king Ptolemy son of king Ptolemy, and refers to a sacrifice instituted by a certain Yatonbaal at Lapethos in Cyprus in year 5 on behalf of "the legitimate scion and his wife". The Ptolemy by whom this inscription is dated has been debated. Oberhummer proposed it was Ptolemy VI, as the "legitimate scion" in opposition to Ptolemy VIII. Berger proposed Ptolemy X in opposition to Ptolemy IX. However, Honeyman presented compelling arguments, universally accepted since, that the inscription dates to Ptolemy II. He also read the essential phrase as referring to "the legitimate scion and his wives" and concluded that it showed Ptolemy II was simultaneously married to both Arsinoe I and Arsinoe II. However, Gibson showed that the singular must be correct.

Honeyman did not attempt to decide whether Ptolemy's regnal years in *KAI* 43 were coregency-based or accession-based. In Cyprus at this time, the regnal dates were certainly calculated according to the Macedonian calendar, not the Egyptian one. If the "year 5" date is accession-based, then the reference to the "legitimate scion" dates to 278/7 and is very difficult to explain. But Hazzard subsequently showed that Ptolemy II adopted coregency-based dating for regnal years in the Macedonian calendar almost immediately after his accession in year 4 (Macedonian). As Teixidor pointed out, a coregency-based year 5 dates the events of *KAI* 43 to 281/0. At this time, Ptolemy II had not yet reached his settlement with his half-brother Ptolemy Ceraunus, who had challenged his

---

[39] Samuel, *op. cit.* (n. 9) 25ff.

[40] R. A. Hazzard, *Phoenix* 41 (Canada, 1987) 140.

[41] Year 13: *iBucheum* 3; year 16: P. W. Pestman, *Chronologie Égyptienne d'après les textes démotiques (332 av. J.-C. - 453 ap. J.-C.)* (Leiden, 1967) 22. But see also B. Muhs in A. M. F. W. Verhoogt & S. P. Vleeming (eds.), *The Two Faces of Graeco-Roman Egypt: Greek and Demotic and Greek-demotic studies presented to P. W. Pestman* (Leiden, 1998), 71.

[42] Theocritus 17.131, Plutarch, *Moralia* 736e; R. A. Hazzard, *The Imagination of a Monarchy: Studies in Ptolemaic Propaganda* (Toronto, 2000) 89f.

[43] *pHibeh* 2.199.

[44] This conclusion has implications for the manner in which Ptolemy II's Egyptian regnal years changed from accession-based to coregency-based dating that I plan to explore elsewhere.

[45] P. Berger, *RdA* 3 (1893) 66; E. Oberhummer, *RE* 12 (1924) 764; A. M. Honeyman, *JEA* 26 (1940) 57; J. C. L. Gibson, *Textbook of Syrian Semitic Inscriptions* III (Oxford, 1982) 140; J. Teixidor, *ZPE* 71 (1988) 188.

succession. The reference to "the legitimate scion" clearly reflects this dispute. Moreover, on this date the wife referred to can only be Arsinoe I.

The significance of a reference to Arsinoe I made in year 11 (Macedonian), i.e. 275/4, has not been fully appreciated. Teixidor, who supposed that the king was by then married to Arsinoe II, says that Yatonbaal "preferred" not to mention Arsinoe I's name. But this raises the question of why Yatonbaal also "preferred" not to name the "legitimate scion", his king. Furthermore, if Ptolemy II's marriage to Arsinoe II is correctly dated to year 13 or 14 then Yatonbaal's alleged motive cannot apply. Since Arsinoe I was disgraced as a traitor, the fact that Yatonbaal felt able to refer to her in a public inscription dated to year 11 strongly suggests that news of her disgrace had not yet reached him. Moreover, if Arsinoe I was still the only wife of Ptolemy II in year 11, then Yatonbaal's reference to her as "the wife" of the "legitimate scion" was completely natural and unambiguous.

We may conclude that Arsinoe I's disgrace and exile took place in year 11 or later, most likely in that year or year 12, so that the ground was cleared for Ptolemy II's marriage to Arsinoe II in year 13.

Chris Bennett
cjbennett@sbcglobal.net
San Diego CA, U.S.A.

The Gayer-Anderson Amenhotep III Commemorative Scarabs in the Portland Art Museum: Their Discovery and Description

John T. Sarr

"A later gift to the museum was the Gayer-Anderson Collection of Egyptian Scarabs."[1] This was the sole reference in a guidebook of the Portland Art Museum in Portland, Oregon (U.S.A.) that I came across in 1993 that would lead me to discover the small but significant collection of Egyptian scarabs held by the museum, including two unpublished Amenhotep III commemoratives.

In my investigation into the whereabouts and contents of the collection, as it was nowhere to be found on display in the museum, I learned that during a major reorganization at the museum over ten years ago the collection had been placed in storage in the museum's vault, where most of it remains today. Being familiar with the Gayer-Anderson collections in Cairo, Cambridge, and Stockholm, I believed the collection had some merit and offered to assess the holdings. To my amazement, the collection contained a great sampling of almost all types of seals, scarabs, and scaraboids. These ranged from Dynasty II cylinder seals to Late Period pectoral scarabs. And among the items were two Amenhotep III lion hunt commemorative scarabs, apparently unknown and unpublished. Given the quality of the collection and the need to have it properly documented, I offered to draw and describe the objects, assemble an informal catalog, and research documentation on the collection and the people involved with it.

The Collection

Briefly, the collection contains about 1350 objects. It was assembled by Major R. G. Gayer-Anderson between 1907 and 1917 in Egypt,[2] loaned to the Ashmolean Museum in Oxford from 1917 to 1925,[3] and sold in 1927[4] to a well-known Portland architect and former Portland Art Museum board chairman, A. E. Doyle. In 1929, one year after Doyle's death, his widow donated the collection in his memory to the Portland Art Museum, Portland, Oregon.

When the objects were on display, shortly after they were donated, most had simple labels that described the period of a grouping of scarabs (Hyksos, New Kingdom, etc.), or the names of pharaohs (Amenophis III) and gods (Amon, Bes, et al.) borne by the objects.[5] No detailed information on the collection appeared in any of the Portland Art Museum's publications other than the fact that the collection existed.

A few American Egyptologists who grew up in the Pacific Northwest have memories of the collection as children,[6] but other than this encounter with the pieces early in the lives of these

---

[1] *A Handbook of the Collections of the Portland Art Museum* (Portland, 1971), 11.

[2] The first page of the 25-page typewritten 'catalog' that Gayer-Anderson included with the collection bears a handwritten inscription: "Collected between the years 1907 and 1917 in Egypt and until recently with the rest of my ancient Egyptian collection on loan to the Ashmolean Museum, Oxford. Signed R. G. G.-A."

[3] In personal correspondence, Dr. Helen Whitehouse confirmed that the Gayer-Anderson scarabs were on loan to the Ashmolean Museum for eight years, from June 1917 to June 1925.

[4] On the last page of the collection's original 'catalog' is handwritten: "R. G. Gayer-Anderson, Major. c/o National Bank of Egypt, 6 King William St., London EC4, 21 Dec. 27." This appears to be the date of the sale.

[5] These labels were found with the scarabs in storage.

[6] Both Ann Macy Roth and Kent Weeks, who grew up in the Pacific Northwest, related to me that they remembered the Egyptian scarabs in the Portland museum. Weeks, in a lecture at the Portland Art Museum in 1998, mentioned that the scarabs played a role in inspiring him to a career in Egyptology.

Egyptologists, the Egyptological community as a whole has known little or nothing of the Gayer-Anderson scarabs in Portland, Oregon.

As a first step in providing details on the collection, I delivered a slide-illustrated presentation at the annual meeting of the American Research Center in Egypt in St. Louis in 1996,[7] wherein I summarized the history of the collection and described several of its most significant pieces. In a short article in 1998, which contained a few photos of the objects, I provided additional details.[8] A temporary but detailed presentation of the scarabs appeared once again before the public when the Portland Art Museum hosted the *Splendors of Ancient Egypt* exhibition from the Pelizaeus-Museum Hildesheim in 1998 and invited me to curate the display of approximately 100 of the masterpieces of the collection.[9] Among the highlights of this small exhibit were the two Amenhotep III commemoratives that are described and illustrated in this article.

Amenhotep commemoratives

To date, there are over 200 documented Amenhotep III commemorative scarabs.[10] These scarabs, which were made to mark events in the life of the pharaoh, were initially studied in detail and classified by C. Blankenberg-van Delden in *The Large Commemorative Scarabs of Amenhotep III.*[11] This seminal book divided the corpus of scarabs into five themes: the marriage of Amenhotep III and his wife Tiye (category A), the king's wild bull hunt (category B), the king's lion hunt (category C), the arrival of the Mitanni Princess Gilukhepa at the court of the king (category D), and the creation of an artificial lake for the king's wife Tiye (category E). In addition, Blankenburg-van Delden listed a small series of commemoratives that were lost[12] (listed as LS plus category letter) or declared forgeries[13] (listed as F plus category letter). As the Portland commemoratives do not figure among the group of lost scarabs, based on inscription details, size, and material, it is safe to assume that the scarabs are authentic lion hunt commemoratives that need to be added to the list of those already known.

Up-to-date accounting of lion hunt commemoratives

The discovery of commemoratives continued after Blankenburg-van Delden's initial publication. More often than not, they were and continue to be brought to light and described when a private art collection is put up for auction, or after a chance discovery of unpublished items in a public collection.

So the question arises: How many lion hunt commemoratives are published? If we take the round figure of 200 for all commemoratives, we find that almost sixty percent of them are lion hunt commemoratives.[14] We have yet to figure out why this category of commemorative scarab is so relatively numerous.

---

[7] The paper was entitled "Rediscovery of the Gayer-Anderson Scarab Collection in the Portland Art Museum" and was presented April 12, 1996 at the 47th annual meeting of the American Research Center in Egypt in St. Louis, Missouri.

[8] "Scarabs in my Backyard: Discovering the Gayer-Anderson Scarab Collection in the Portland Art Museum," *Newsletter of the American Research Center in Egypt* 176 (July 1998).

[9] Details on the exhibit, including text and images of the scarabs on display, can be found at the following URL: http://www.teleport.com/~jsarr/scarabs/.

[10] All the extant and lost scarabs in Blankenberg-van Delden's original work add up to 204 scarabs, with some 20 scarabs published in subsequent articles.

[11] *The Large Commemorative Scarabs of Amenhotep III* (Leiden, 1969).

[12] *Large Commemorative Scarabs,* 149-162: LSA1-9, LSC1-18, LSD1, pls. xxxiii-xxxiv.

[13] *Large Commemorative Scarabs,* 163-165: FA1-5, FC1, pl. xxxv.

[14] Out of the 204 scarabs listed in *The Large Commemorative Scarabs of Amenhotep III*, 126 are lion hunt commemoratives.

In her initial work, Blankenberg-van Delden documented 108 extant lion hunt scarabs (C1-C108) and 18 lost ones (LSC1-18) for a total of 126.[15] In follow-up articles to update her work, Blankenberg-van Delden added five known (C109-C113) and one lost (LSC19)[16] scarab in 1976,[17] and six known (C114-119) scarabs in 1977,[18] bringing the total to 138. Articles by other authors shed light on one in 1979,[19] one in 1985,[20] and two in 1996,[21] for a total of 142 lion hunt commemoratives. The two described in this article would bring the total number to 144, of which 125 are known and 19 lost. But it appears that there are still undocumented but known commemoratives either in public or private collections, and as these make their way into publication these numbers will change.[22]

Other Gayer-Anderson Commemoratives

The two Gayer-Anderson Amenhotep III commemorative scarabs that are described here are not the first ones published from a Gayer-Anderson collection. Two already appear in Blankenberg-van Delden's work and are also lion hunt commemoratives. These include a scarab acquired in 1943 by the Fitzwilliam Museum in Cambridge, England (C21), and a scarab acquired in 1930 by the Medelhavsmuseet in Stockholm, Sweden (C67).[23] It is to be noted that thanks to Major Gayer-Anderson both the Fitzwilliam Museum[24] and the Medelhavsmuseet[25] in Stockholm contain sizeable collections of Egyptian objects, including scarab collections.

Details of the scarabs

For details on the scarabs, I have used the outline that Blankenberg-van Delden created. I have not assigned a number to the scarabs that would continue the C-category of lion hunt commemoratives. I believe that only an author updating the original work in an effort to update the *entire* corpus of commemoratives should give continuing series numbers. Problems of number overlap and confusion by separate authors can result and have occurred.[26]

---

[15] *Large Commemorative Scarabs,* 62-128: C1-C108, 153-160: LSC 1-18.

[16] A mistake by either Blankenberg-van Delden or the article editors lists this scarab as LSC18, although LSC18 appeared previously in *Large Commemorative Scarabs* and referenced a different scarab.

[17] "More Large Commemorative Scarabs of Amenhotep III," *Journal of Egyptian Archaeology* [hereafter: *JEA*] 62 (1976), 74-80, pls. xii-xiii.

[18] "Once Again Some More Large Commemorative Scarabs of Amenophis III," *JEA 63* (1977), 83-87, pls. xiii-xiv.

[19] M. Jones, "The Royal Lion Hunt Scarab of Amenophis III in the Grosvenor Museum, Chester (Chester, no. 429.R/1930)," *JEA* 65 (1979), 165-166, pl. xxviii.

[20] May Trad and Adel Mahmud, "Another Commemorative Lion-hunt Scarab of Amenophis III," *Annales du Service des Antiquités de l'Égypte* 70 (1984-85), 359-361, pl. I, Cairo, JE 97853.

[21] Peter A. Clayton, "Some More 'Fierce Lions,' and a 'Marriage' Scarab: the Large Commemorative Scarabs of Amenophis III," *JEA* 82 (1996), 208-210.

[22] See L. M. Berman, in A. P. Kozloff and B. M. Bryan, *Egypt's Dazzling Sun. Amenophis III and His World* (Cleveland, 1992). The article lists, for example, an unpublished lion hunt scarab loan from a private collection in San Antonio, Texas.

[23] *Large Commemorative Scarabs*, 75-76: C21 pl. xiv; 103-104: C67, pl. xxii. There was some excitement on both sides of the Atlantic when the half Gayer-Anderson lion hunt scarab in Portland (29.16.114.B) was perhaps thought to complement the half Gayer-Anderson lion hunt scarab in Cambridge. This turned out to be not the case as they both contain the lower half of a scarab.

[24] My thanks to Penelope Wilson who at the time of my enquiry provided details of the Gayer-Anderson scarab collection at the Fitzwilliam.

[25] Bengt E. J. Peterson, *Zeichnungen aus einer Totenstadt: Bildostraka aus Theben-West mitsamt einem Katalog des Gayer-Anderson-Sammlung in Stockholm* (Stockholm, 1973).

[26] For example, P. Clayton, in his article in 1996, overlooked Blankenberg-van Delden's 1977 article wherein she listed lion hunt scarabs C114-C119, as well as the subsequent articles of M. Jones in 1979 and M. Trad/A. Mahmud in 1985. By not taking these documents into account, his count of the lion hunt scarabs was off. As a result of the oversight, he assigned a C-category numbers to the two scarabs in his article (C114 and C115) that had already been assigned to other scarabs.

PAM 29.16.114.A – back PAM 29.16.114.A – side PAM 29.16.114.A – front

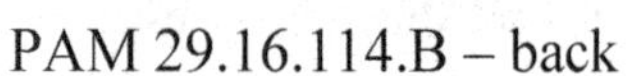

PAM 29.16.114.B – back PAM 29.16.114.B – side PAM 29.16.114.B – front

Plates I-VI

I agree with Geoffrey Martin's early critique of Blankenberg-van Delden concerning the lack of views of the back and sides of the scarabs.[27] The illustration of scarabs with front, back, and side views is crucial if we are to include a study of the whole scarab, which, as Olga Tufnell[28] and W. A. Ward[29] have shown, is needed to allow for the varieties in scarab style and the detection of the emergence of new details. This whole-scarab approach with the Amenhotep III commemoratives, unfortunately

[27] *JEA* 58 (1972), 317.

[28] *Scarab Seals and Their Contribution to History in the Early Second Millennium B.C.* Studies on Scarab Seals II (Warminster, 1984).

[29] *Pre-12th Dynasty Scarab Seals.* Studies on Scarab Seals I (Warminster, 1978); with W. A. Dever, *Scarab Typology and Archaeological Context. An Essay on Middle Bronze Age Chronology*. Studies on Scarab Seals III (San Antonio, 1994).

lacking for the most part in Blankenberg-van Delden's work and subsequent publications, could play a role in substantiating the chronology of the scarabs and perhaps in establishing the dating of patterns of contemporaneous scarabs. As a result, the front, back, and side of the scarabs are illustrated here.

Portland Art Museum, no. 29.16.114.A
Description: steatite; white; traces of green glaze; standard eight lines of text; no cartouche of king on side of scarab; pierced longitudinally.
Dimensions: 7.3 cm. x 5 cm. x 2.5 cm.
Type: single line between the wing cases, single line between the wing cases and prothorax, triangular notches (humeral callosities) below these lines at outer corners.
Preservation: good; first part of lines 2, 3, 4, 7, and 8 and last part of lines 2, 3 chipped; carving is good.
Provenance: location of find unknown; formerly in the Gayer-Anderson scarab collection, purchased by A. E. Doyle in 1927 in London and donated in 1929 to the Portland Art Museum.
Comments on the inscription:[30]
(line 5) the abbreviated writing of *ḥmt nswt* instead of *ḥmt nswt wrt*;
(line 6/7) *š3ʿ-m* written *š3-m*—a similar instance on C41, C60, C78, C88, C93, C100.

Portland Art Museum, no. 29.16.114.B
Description: steatite; reddish brown; traces of green glaze in the grooves; four of eight lines of text remain; no cartouche of king on side of scarab; pierced longitudinally.
Dimensions in present state: 5.5 cm. x 5 cm. x 2.5 cm.
Type: single line between the wing cases, rest of back missing.
Preservation: broken in half with only the last four lines of text remaining; finely carved hieroglyphs that are well preserved.
Provenance: location of find unknown; formerly in the Gayer-Anderson scarab collection, purchased by A. E. Doyle in 1927 in London and donated in 1929 to the Portland Art Museum.
Comments on the inscription:
(line 5) the abbreviated writing of *ḥmt nswt* instead of *ḥmt nswt wrt*;
(line 5) the single instance in the entire corpus of *ʿnḫ.tj* being written with the phonetic complement *j* added to X11;
(line 5) *rḫt* written as *rḫ*—a similar instance on C63 only.

John T. Sarr
jsarr@teleport.com
Portland, Oregon, U.S.A.

[30] For comparison, the usual eight-line transliteration and translation of the lion hunt scarab text reads:

(1) *ḥr ʿnḫ k3 nḫt ḫʿj m m3ʿt* (2) *nbty smn hpw sgrḥ* (3) *t3wy ḥr nbw ʿ3 ḫpš ḥwj sttjw* (4) *nswt-bjtj (nb m3ʿt rʿ)| s3 rʿ (jmnḥtp ḥk3 w3st)|* (5) *dj ʿnḫ ḥmt nswt wrt (ty)| ʿnḫ.tj rḫt m3j* (6) *jn.n ḥm.f m stt.f ḏs.f š3ʿ-m* (7) *rnpt-sp 1 nfryt-r rnpt-sp 10* (8) *m3j ḥ3s 102*

(1) Living Horus: Strong Bull Appearing in Truth; (2) Two Ladies: Establishing Laws, Pacifying (3) the Two Lands; Golden Horus: Great of Valor, Smiting the Asiatics; (4) King of Upper and Lower Egypt (Neb-Maat-Re)|; Son of Re (Amenhotep, Ruler of Thebes)|. (5) Given life. - The Great Royal Wife (Tiy)|. May she live. Number of lions (6) taken by His Majesty by his own shooting, from (7) regnal year 1 to regnal year 10: (8) 102 fierce lions.

The Djed-Ptah-iw-ef-ʿankh Shabti Figurine from the National Museum of Belgrade

Branislav Anđelković and Troy Sagrillo

The ancient Egyptian collection of the National Museum in Belgrade contains, in addition to the Belgrade mummy (Anđelković 1997) and coffin of Nefer-renepet (Panić-Štorh 1997), a number of bronze statuettes, various amulets, scarabs, and shabti figurines (Anđelković, *in press*), one of which is of particular interest.

Description

The shabti figurine (*Fig. 1*) is very schematic, flat-backed, with a light greenish-blue glaze. It is a mold-made faience piece; the arms are not crossed (*cf.* Schneider 1977, fig. 12:10). The eyes and eyebrows are painted in black. The front side bears a vertical text (*Fig. 2*), in black, which reads (↓→):

wsir[1] *ḥm-nṯr 2-nw*[2] *n*[3] *imn*[4] *ḏd-ptḥ-i[w=f]-ʿnḫ*[5]

The Osiris, the 2nd God's Servant of Amen, Djed-Ptah-i[w-ef]-ʿankh.

The dimensions are: *height* 8.2 cm.; max. *width* 2.32 cm.; max. *depth* 1.55 cm. The National Museum Collection number in the inventory book is 8/VI, whereas the previous numbers 8/V and K 396 are written on the statuette itself. The provenance of the object is unfortunately unknown, as well as the date and manner by which it entered the museum.

Figure 1: Shabti figurine, National Museum of Belgrade Collection number 8/VI

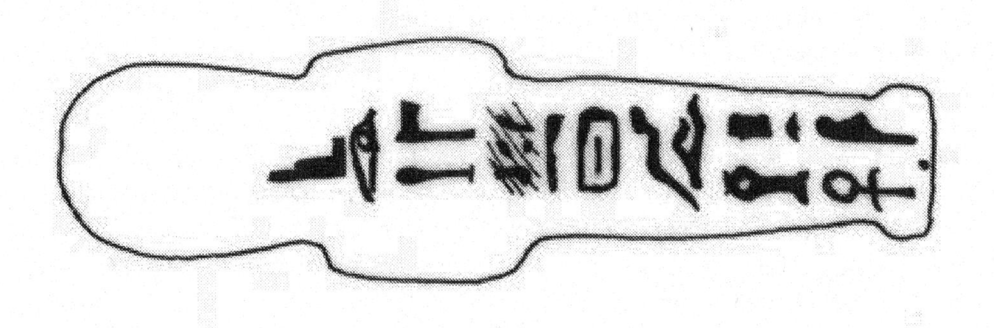

Figure 2: Text on the Djed-Ptah-iw-ef-ʿankh shabti, Belgrade

Notes on orthography

The text on this shabti is written in a very loose, almost hieratic form of hieroglyphs, typical of Third Intermediate Period shabtis. Specific notes on the orthography follow.

1. The writing *wsir* utilised here is most common on shabtis dating to the period after the first part of Dynasty XXII; the writings and are much more often encountered on Dynasty XXI and early Dynasty XXII statuettes.[1] Indeed, *wsir* is unknown in any Dynasty XXII text before the time of Osorkon II and Harsiese A (Leahy 1979, 145, fig. 1). The sole exception cited by Leahy is a canopic jar fragment discovered in the tomb of Sheshenq III at Tanis, which he believed to belong to Sheshenq I. It names a king *ḥd-ḫpr-rʿ stp-n-rʿ šsnq mri-imn sꜣ-bꜣs.t.t nṯr-ḥqꜣ-iwnw* (Dodson 1994, 178/50:2, plate 43b), who is now recognised as Sheshenq 'quartus'—reigning between Sheshenq III and Pimay—and not Sheshenq I (Dodson 1993; Dodson 1994, 93–94). It must be pointed out, however, that the writing does occur in early Dynasty XXII documents (albeit not on shabtis) but not commonly so, beginning with the calcite block of Sheshenq I re-used in the "Apis House" of Memphis (Maystre 1992, 357/1, 4, 13).

2. Although the strokes are located in the break, the text likely reads *ḥm-nṯr 2-nw* '2nd God's Servant.' The *nw*-jar is admittedly somewhat abnormally lengthened, but clearly wider than the proceeding two strokes, making the reading *3-<nw>* '3rd' less likely. Identical orthography is found on shabtis belonging to the 2nd God's Servant of Amen, Djed-Ptah-iw-ef-ʿankh D, now in the Oriental Institute Museum (*vide infra*).

[1] The writing during the Third Intermediate period is a revived, archaising form, encountered initially during the Old and Middle Kingdoms; it was not common during the New Kingdom (Erman 1909, 93).

3. The writing of 𓈖 *n* as —— is taken over wholly from hieratic. This writing does occur both before and after the Third Intermediate Period, but it is especially common during that period, even in monumental hieroglyphic texts (*e.g.*, Jabal al-Silsilah Quarry Stela 100 [Caminos 1952, *passim*]).

4. The cryptographic writing ⊂⊃ *imn* is the hieratic form of 𓏠𓈖. This was used frequently in the later part of the Third Intermediate Period, and became almost a standard in the Late and Græco-Roman Periods (WB 1:84/17; Daumas 1988–1995, 3:470/2018, 2019). It is a calligraphic play on the phonetic values of ⊂⊃ *i* and 𓈖 *n* (*cf.* Fairman 1943, 235:234, 235). For additional examples, which occur on Dynasty XXIII shabti statuettes, see Hölscher and Anthes 1951, 25; Botti 1955 (discussed *infra*). Dynasty XXII writings do not seem to occur earlier than Osorkon II and HPA Harsiese A (*cf.* Jansen-Winkeln 1985, A2/f1, A10/b3, c2). Most Third Intermediate Period texts utilising this writing seem to be connected with the Theban region, but this may only be due to the vagaries of preservation.

5. *i*[*w=f*]-*ʿnḫ:* 𓇋 is falsely written for 𓇋 *i*[*w*] (Jansen-Winkeln 1996, §35); this trend continues into the Græco-Roman Period (Daumas 1988–1995, 2:345/81). The abbreviated writing *ḏd*-DIVINITY-*i*-*ʿnḫ* is frequently encountered in theophoric names of this sort. For a typical example, see the Dynasty XXII shabti of Djed-Khonsu-iw-ef-ʿankh (Aubert and Aubert 1974, 295, plate 51/122–123).

Identification and Date

There are only two known individuals from the Third Intermediate Period bearing the name Djed-Ptah-iw-ef-ʿankh and the title of 2nd God's Servant: Djed-Ptah-iw-ef-ʿankh A (Bierbrier 1975, 64, 74, 135 note 105, 140 note 258; Kitchen [1996], §§157, 244, 245, 266) and Djed-Ptah-iw-ef-ʿankh D (Bierbrier 1975, 96–99, 138 note 186, 140 note 262; Kitchen [1996], §§193, 290, 291, 486).

Djed-Ptah-iw-ef-ʿankh A was buried in the Royal Cachette at Dayr al-Bahrī (TT 320), some time after Regnal Year 10 of Sheshenq I (*c.* 935 B.C.E.), and is known from a number of shabtis that were interred with him (Maspero 1889, 572–574; Yoyotte and Aubert 1987, 144–145/28; see also Botti 1955). However, neither the cryptographic form 𓏠𓈖 *imn*, nor the writing 𓊨𓁹 *wsir*, is used in any of the texts mentioning him, nor in any other known Dynasty XXI or early Dynasty XXII texts. For example, Louvre Shabti E 22085, which belongs to Djed-Ptah-iw-ef-ʿankh A, uses 𓁹𓊨 (see also Botti 1955), as is typical for Dynasty XXI and early Dynasty XXII. (It should be noted, however, that the form 𓁹𓊨 *is* used on his coffin; see Maspero 1889, 572.) It is therefore unlikely that the Belgrade shabti belongs to Djed-Ptah-iw-ef-ʿankh A.

Djed-Ptah-iw-ef-ʿankh D is known from Tübingen Statue 1734 (Brunner-Traut and Brunner 1981, 39–41, pl. 113), as well as shabtis from Tombs 12 and 17 at Madīnat Hābū (Hölscher and Anthes 1951, 25 [OIM 15767–73[2]]). He is also mentioned on the stela and coffins of his great-granddaughter (Porter, Moss, and Burney 1964, 643, 648). The orthography used on all of these documents indicates a connection with the Belgrade shabti. Both the cryptographic form of Amen and the 𓊨𓁹 form of 'Osiris' are used on the Madīnat Hābū shabtis, and the 𓊨𓁹 form of 'Osiris' is standard on the

[2] The authors wish to thank the Oriental Institute Museum, and in particular Ms. Carla C. Hosein, for providing unpublished photographs of these shabtis in order to confirm the orthography employed thereon.

Tübingen statue (lines 4, 4a, 5). Indeed, the general *ductus* and orthography of OIM 15767–73 is identical to the Belgrade shabti under discussion here.

Although originally identified as a son of king Takelot II of Dynasty XXII (e.g., Gauthier 1914, 359/XIX:2; Peterson 1967; Bierbrier 1975, 96–99; Kitchen [1996], §§227, 228, note 478), Vittmann (1978, 89) demonstrated that this individual is better taken as a son of Takelot III of Dynasty XXIII on the basis of genealogical data; this has been accepted by most other scholars working with the data (e.g., Kitchen [1996], §486; Aston and Taylor 1990, 134). The late Third Intermediate Period orthography of the Belgrade shabti makes an identification with Djed-Ptah-iw-ef-ᶜankh D all the more likely given Vittmann's correction.

While it is not entirely certain whence this shabti originated, it is possible it came from the area of Madīnat Hābū, given that other shabtis belonging to Djed-Ptah-iw-ef-ᶜankh D are known from excavations there (Hölscher and Anthes 1951, 25). It is certainly a welcome addition to the growing body of Dynasty XXIII documents.

Branislav Anđelković
University of Belgrade

Troy Leiland Sagrillo
meshwesh@bigfoot.com
University of Colorado at Boulder

References

Anđelković, B. 1997. "The Belgrade Mummy." *Recueil des travaux de la Faculté de philosophie* (Belgrade) 19/A:91–104.

———. *in press*. "The Ancient Egyptian Collection in the National Museum of Belgrade." *Journal of the Serbian Archaeological Society* (Belgrade) 18.

Aston, D. A., and J. H. Taylor. 1990. "The Family of Takeloth III and the 'Theban' Twenty-Third Dynasty." In *Libya and Egypt c. 1300–750 B.C.*, edited by M. A. Leahy. London: School of Oriental and African Studies. 131–154.

Aubert, J.-F., and L. Aubert. 1974. *Statuettes égyptiennes: Chaouabtis, ouchebtis*. Paris: Librairie d'Amérique et d'Orient, Adrien Maisonneuve.

Bierbrier, M. L. 1975. *The Late New Kingdom in Egypt (c. 1300–664 B.C.): A Genealogical and Chronological Investigation*. LMAOS. Warminster: Aris & Phillips Limited.

Botti, G. 1955. "Una statuetta funeraria del principe Djedptahefonkh nel Museo del palazzo Della Silva in Domodossola." *Rendiconti dell'Instituto Lombardo* 88:94–98.

Brunner-Traut, E., and H. Brunner. 1981. *Die ägyptische Sammlung der Universität Tübingen*. Mainz am Rhein: Verlag Philipp von Zabern.

Caminos, R. A. 1952. "Gebel el-Silsilah No. 100." *JEA* 38:46–61.

Daumas, F., ed. 1988–1995. *Valeurs phonétiques des signes hiéroglyphiques d'époque gréco-romaine*. 4 vols. Montpellier: Institut d'égyptologie université Paul-Valéry.

Dodson, A. M. 1993. "A New King Shoshenq Confirmed?" *GM* 137:53–58.

———. 1994. *The Canopic Equipment of the Kings of Egypt*. Studies in Egyptology, ser. ed. A. B. Lloyd. London: Kegan Paul International.

Erman, A. 1909. "Zum Namen des Osiris." *ZÄS* 46: 92–95.

Fairman, H. W. 1943. "Notes on the Alphabetic Signs Employed in the Hieroglyphic Inscriptions of the Temple of Edfu." *ASAE* 43:193–318.

Gauthier, H. 1914. *Le livre des rois d'Égypte*. Volume 3: *De la XIXe à la XXIVe dynastie*. MMAF 19. Cairo: Imprimerie de l'Institut français d'archéologie orientale du Caire.

Hölscher, U., and R. Anthes. 1951. *The Excavation of Medinet Habu*. Volume 4: *The Mortuary Temple of Ramses III: Part 2*. Translated by E. B. Hauser. UCOIP 55. Chicago: University of Chicago Press.

Jansen-Winkeln, K. 1985. *Ägyptische Biographien der 22. und 23. Dynastie*. ÄAT 8/1–2, ser. ed. M. Görg. Wiesbaden: Otto Harrassowitz.

———. 1996. *Spätmittelägyptische Grammatik der Texte der 3. Zwischenzeit*. ÄAT 34, ser. ed. M. Görg. Wiesbaden: Harrassowitz Verlag.

Kitchen, K. A. [1996]. *The Third Intermediate Period in Egypt (1100–650 B.C.)*. 2nd ed. with Supplement. Warminster: Aris & Phillips Limited.

Leahy, M. A. 1979. "The Name Osiris Written ." *SAK* 7:141–153.

Maspero, G. 1889. "Les momies royales de Déir el-Baharî". In *Mémoires publiés par les membres de la Mission archéologique française au Caire*. Vol. 1 (fascicle 4). Paris: Ernest Leroux, éditeur. 511–788.

Maystre, C. 1992. *Les grands prêtres de Ptah de Memphis*. OBO 113, ser. ed. O. Keel. Freiburg and Göttingen: Universitätsverlag Freiburg and Vandenhoeck & Ruprecht.

Panić-Štorh, M. 1997. "Ancient Egyptian Anthropoid Coffin from the National Museum in Belgrade." In *Mélanges d'histoire et d'épigraphie offerts à Fanoula Papazoglou*, edited by M. Mirković. Belgrade: Faculty of Philosophy. 71–87.

Peterson, B. J. 1967. "Djedptahefanch, Sohn des Takeloth II." *ZÄS* 94:128–129.

Porter, B., R. L. Moss, and E. W. Burney. 1964. *Topographical Bibliography of Ancient Egyptian Hieroglyphic Texts, Reliefs, and Paintings*. Volume 1: *The Theban Necropolis*. Part 2: *Royal Tombs and Smaller Cemeteries*. 2nd ed. Oxford: The Griffith Institute.

Schneider, H. D. 1977. *Shabtis: An Introduction to the Study of Ancient Egyptian Funerary Statuettes with a Catalogue of the Collection of Shabtis in the National Museum of Antiquities at Leiden*. 3 vols. Collections of the National Museum of Antiquities at Leiden 2. Leiden: Rijksmuseum van Oudheden te Leiden.

Vittmann, G. 1978. *Priester und Beamte im Theben der Spätzeit: Genealogische und prosopographische Untersuchungen zum thebanischen Priester- und Beamtenum der 25. und 26. Dynastie*. Beiträge zur Ägyptologie 1, ser. eds. H. Mukarovsky, and G. Thausing. Wien: Afro-Pub.

Yoyotte, J., and L. Aubert. 1987. "Statuettes funéraires et personnalités des XXIe et XXIIe dynasties". In *Tanis: L'Or des pharaons*. [Paris]: Ministère des Affaires Étrangères and Association française d'Action artistique. 116–151.

The Coffin and the Cartonnage of Kaipamaw

Igor Uranic

Three Egyptian coffins are kept in the ancient Egyptian collection of the Zagreb Archaeological Museum. The oldest one, a coffin with a mummy in cartonnage, under number 687, is the most interesting. This object represents the Egyptian art and funerary customs of the Late Period (1070-332 B.C.). It was a gift of the Egyptian National Museum in Cairo in the seventies, as an expression of gratitude for the role of some Croatian (and also some other former Yugoslavian) companies in the great international action of saving Nubian monuments. In Cairo the coffin had inventory number 1516. In publishing this Egyptian coffin I received great help from my colleagues professor István Nagy and Éva Liptai, Egyptologists from the Szépművészeti Múzeum in Budapest.

Ancient Egyptian coffins can be divided into two main types according to their shape: architectonic (square form), and anthropoid (human mummy form).[1] The coffin of the woman Kaipamaw belongs to the second type. Among anthropoid type coffins there are numerous variations in style and workmanship, in the shaping and painting of coffins and cartonnages, and of course in the usage of religious iconography and texts connected with beliefs in life after death. For example, Middle Kingdom coffins were inscribed with long chapters of the so-called *Coffin Texts*, while in the New Kingdom and Late Period (to which most of the well preserved objects of this type held in museums belong) texts are usually shorter. They are reduced to a few formulas enumerating food offerings and mentioning the names of the dead persons. However, coffins of the Libyan and Late Periods, as well as cartonnages dressing the mummies, are very copiously furnished with iconography representing various gods and symbols of magical protection for the soul of the dead dwelling in the underworld.

Provenance

The tomb of Kheruef (TT 192)[2] is, if measured by its contents, one of the most important sites in the region of Assasif near Hatshepsut's temple in Deir-el-Bahri. The owner lived during the reigns of Amenhotep III and Amenhotep IV (Akhenaten) and held the title of "servant of queen Teje." Remains of beautiful relief in this tomb testify to the high quality of New Kingdom art. Some of these reliefs show the jubilee of Amenhotep III in which the tomb owner himself took part. An inscription on Kheruef's almost life-size statue reads: "Splendid is the one who fills the heart of the Two Lands. The steward of the palace and the king's festivals."[3] The tomb was already known in the 19th century and one of the first explorers, who entered it in 1885, was the famous German Egyptologist Adolf Erman. The tomb was robbed a few times, and in the beginning of the 20th century its entrance was blocked up again. Clearing and digging of the tomb proceeded in 1957. On that occasion not only was the damaged statue of Kheruef found, but also later burials, dating from the 21st to the 26th dynasties. However, the mummy or coffin of the tomb owner himself was not found. Among a few well preserved burials there was a coffin with the mummy of Kaipamaw, which was most probably placed there during the 22nd dynasty, and is now displayed in the Archaeological Museum in Zagreb. All the mummies found in coffins were burials of priests of Amun and members of their families. The Egyptian Egyptologist Labib Habachi, who tried to determine the relationships between the mummies,

[1] E. Brovarsky, *Lexikon der Ägyptologie*, V, Wiesbaden, 1983, 471.

[2] For this tomb, see: Porter-Moss, *Topographical Bibliography of Ancient Egyptian Hieroglyphic Texts, Reliefs and Paintings,* I/2, Oxford, 1964. *The Theban Necropolis*, II, 627; and also: F. J. Martín Valentín, "La tumba de Kheruef (TT 192): Indicios de una corregencia," *Boletin de la Asociacion Española de Egiptologia* (Madrid) 3, 1991, 224-225.

[3] *Urk.* IV, 1860, 4-5.

suggested that Kaipamaw might be closely related to a female mummy which bears the name Shepenkhonsu and is also buried in TT 192.[4]

After a few centuries of prosperity, the New Kingdom fell into decadence in the times of the Late Ramessides. The documents from this period testify to rebellions and organised tomb robberies. For example, the *Abbott Papyrus* (Pap. British Museum 10221) from the sixteenth year of the reign of Ramses IX Noferkare-Setepenre (1127-1109 B.C.) reports on an investigation and legal proceeding against a man who robbed tombs in Thebes. Such cases induced the Theban priesthood of the 21st dynasty to move the bodies of the sacred pharaohs from their tombs to more securely hidden places.[5] By the end of the 19th century Maspero, Loret and Mariette had succeeded in locating those places near Deir el Bahri as well as in the Valley of the Kings. They were the Deir el Bahri cache and the tomb of Amenhotep II (KV 35). A significant number of known royal mummies was found in these locations. In the same way the priesthood strove to protect their own burials and the burials of their family members. In 1891 Daressy found an underground tomb in the Bab el-Gusus which evidently had assumed the function of a cache. In the Late Period 153 mummies were stored in it. The same situation almost undoubtedly applies to Kheruef's tomb as well. The later burials are probably not related to the original owner but were placed in TT 192 almost 400 years later.

Description of the Coffin and the Cartonnage

The coffin dates to the time of the Libyan kings (dynasty 22-24) who ruled in Lower Egypt after the death of Psusennes II (959-945 B.C.). They were able to take control of the region due to the fact that they held high military positions in the Egyptian army. Comparing the style of anthropoid coffins in the 21st dynasty to that of later periods,[6] and noting gradual changes introduced on similar coffins, it can be concluded that the coffin of Kaipamaw dates from the 22nd Libyan dynasty. In the history of ancient Egyptian art, the 21st dynasty, with its specific taste for archaism, represents a good base for dating. In the case of this object, the use of the yellow background in the painting of the cartonnage is one of the elements inherited from the 21st dynasty. On the other hand, significant changes in iconography took place after the 21st dynasty.[7] However, a painted surface was still divided into a few horizontal registers in which, in the case of Kaipamaw's coffin, winged creatures or gods occur. These seem to protect a fetish of Abydos which extends across the cartonnage. The text on the cartonnage has the shape of the letter T, and together with the fetish of Abydos it makes a cross. The form of the letter T covers the middle of the lower part of the body.

The mummy was placed in a painted cartonnage[8] and in a wooden coffin (cf. *Fig. 1* and *Fig. 2*). The total length of the coffin is 195 cm., the maximum width is 56 cm., and its depth is 62 cm. The dimensions of the cartonnage are: length 162 cm., maximal width 39 cm., and depth 25 cm. Only the

---

[4] L. Habachi, "Clearance of the Tomb of Kheruef at Thebes (1957 - 1958)," *ASAE* 55, 1958, 325-350.

[5] For the hidden tombs and mummies of the kings and priests, see A. Niwiński, "The Bab el-Gusus and Royal Cache in Deir el-Bahri," *JEA* 70, 1984, 73-80.

[6] That development is shown in: *LÄ* V, 445-452 ("Sarg. NR-SpZt"); A. Niwiński, *21st Dynasty Coffins from Thebes,* Theben V, Mainz, 1988, par. 78; S. D'Auria, P. Lacovara, C. H. Roehrig, *Mummies and Magic*, Boston, 1988, 166-171, 220-221; J. H. Taylor, *Egyptian Coffins*, Aylesbury, 1989, 46-52.

[7] For types of the coffins in the 22nd dynasty, see: J. E. Quibell, *The Ramesseum,* Eg. Research Account 1896. Cambridge, 1898, Cartonnage of Nekh-ef-Mut, pl. XVI; R. A. Martin, *Mummies,* Chicago Natural History Museum, 1945, Coffin of a woman named Tinto, pl. 1.; S. D'Auria, P. Lacovara, C. H. Roehrig, *Mummies and Magic*, Boston, 1988, 170, The mummy of Tabes; E. Liptay, "The Cartonage and Coffin of *Js.t-m-3ḫbi.t* in Czartoryski Museum, Cracow," *Studies of Art and Civilization* (Kraków), 1993, 7-26; E. Varga, "Egy késöi mumiakoporsó," *A Szépművészeti Múzeum közlemény* (Budapest), 50-51, 1987, 185-191.

[8] For the Egyptian cartonnage technology, see K-Th. Zauzich, *Lexikon der Ägyptologie*, III, 353.

Fig. 1 - The Coffin of Kaipamaw (Zagreb Archaeological Museum)

Fig. 2 - The Coffin plus Cartonnage of Kaipamaw

top of the coffin is painted, showing the face of the owner. Her head is depicted covered with a traditional scarf in greenish blue, yellow, white and red colours. The wings of the goddess Mut hang down along both sides of the head. The corresponding part of the cartonnage is coloured almost in the same manner but we also find there the legs of the vulture goddess holding a *šnw* symbol on each side of the head. The legs and the wings spring from the fillet on the woman's head. The eyebrows on her face are in the same greenish blue colour that occurs on the scarf. The collar, painted in the same colour, spreads over the chest. Below it there is a white line, inscribed with hieroglyphs, which reaches to the feet and ends with a determinative of a flower. The inscription contains an offering to gods and the name of the owner. The rest of the anthropoid coffin is not painted. Almost the same is the coffin of Shepenkhonsu (which, however, is a double coffin) found in the same tomb as Kaipamaw. The texts on these two coffins are so similar that Habachi concluded the two women were probably related.[9] The main difference in the way in which these two coffins are painted is the representation of a winged scarab on the chest of Shepenkhonsu. The cartonnages of the two mummies are also very similar.

On the bottom of the lower part of the coffin, Isis is shown in a red dress, spreading her arms to protect the dead person (*Fig. 1*). She holds two *ˁnḫ* symbols. The goddesses shown on the bottoms of such coffins are connected with traditional beliefs about protection of the soul, beliefs that date from at least the times of the *Pyramid Texts* where the pharaoh asks for the protection of the goddess of the sky, Nut, claiming she is his mother: "O King, your mother Nut spreads above you to protect you from all evil..."[10] We already find the wings of Nut shown above the name of the dead on an ivory tablet from the grave of the archaic King Djet.[11] There the wings of Nut are spread above Horus, who symbolises the King. Nut appears most often in this role, although in later periods Isis and Nephthys became the traditional protectors.

The iconography of the cartonnage, which in the upper part is similar to that of the coffin, is very rich and nicely coloured (*Fig. 2*). The images are harmoniously divided in four fields by the fetish of Abydos that covers the middle of the body. The bottom of the column of the fetish is shown placed in a *ḏw* hill, held by two lion-headed creatures who fix it with their backs. The symbol reaches to the middle of the body and is crowned with two feathers.[12] Above it there is a spread-winged hawk which probably represents the *bꜣ* soul of Kaipamaw freed from its mummy. The bird, which dominates the whole scene, holds the symbols of life.

Above the bird's wings the four sons of Horus can be found. If we look to the mummy from *enface*, their order from left to right is: Imset, Hapy, Duamutef, and Khebehsenuf.

In the fields below the bird's wings on both sides of the cartonnage, Isis and Nephthys are represented as two winged women and as two birds. They are offering the *wḏꜣt* eye and protect the symbol of Abydos from both sides. Both deities, shown with a Sun disc on their heads, are named in the accompanying text as "Great goddess from Abydos and mistress of the sky." Two Upuauts are painted on the feet of Kaipamaw's cartonnage.

Identity of the Mummy

As was already mentioned, the coffin contains the mummy of a woman named Kaipamaw, who lived in Thebes in the time of the 22nd dynasty (945-715 B.C.). She bears the title of chantress of Amun (*šmˁyt n Ỉmn*) which means she had to be a member of the priestly community. Her other title, *nbt-pr*, does not reveal much except for the fact that she was probably married. The fact that Kaipamaw was

[9] The coffin and a cartonnage of Shepenkhonsu (Luxor 7.106), L. Habachi, "Clearance of the Tomb of Kheruef at Thebes (1957 - 1958)," *ASAE* 55, 1958, pl. XVIII. a.

[10] K. Sethe, *Die altägyptischen Pyramidentexte*, II, Leipzig, 1809, 825.

[11] On the first representation of this theme, see R. Wilkinson, *JSSEA* (Toronto), 9/1973, 88.

[12] For the symbol of Abydos (*ꜣbḏw*) and its origin, see H. Bonnett, *Reallexikon der ägyptischen Religionsgeschichte*, Berlin 1952, 3, 805.

buried near the family of Amun's priest Thaenwaset (there were two groups of three coffins), who bore the title god's father, makes it possible that she was a close relative. Her sarcophagus is made in the same manner as that of a woman called Shepenkhonsu – probably Thaenwaset's wife. It is therefore also possible that Kaipamaw was closely related to Shepenkhonsu – she might have been her sister or her daughter.

The name Kaipamaw was probably of Libyan origin and rare in Egypt as it is not listed in this form in Ranke's *Ägyptische Personennamen*. Originally, Habachi had rendered the name found on the coffin as Kapathaw. This reading was based on the reading of the hieroglyphic sign of the sail, which is usually read *ṯꜥw* or *nfw* and only rarely has the value *mꜣw*. But the German Egyptologist Erhart Graefe demonstrated that the correct reading is Kaipamaw, since the name is written on the cartonnage in three different ways.[13] Besides

which is identical to the writing on the coffin, the cartonnage has another version:

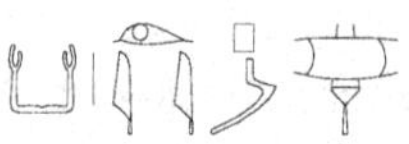

In this second version there is a *kꜣ* instead of *ḳꜣ*, three signs are used to denote a single yod,[14] and in front of the mentioned sail hieroglyph the *mꜣ* sign is written, which discloses the correct reading. Graefe compared this name to what must be a longer version of the same name, Kaipamairdiset,[15] and thought it should be regarded as one of the god's secret names or attributes.

The Text

The text consists of the usual offering formulas and the titles of the dead. There are a few mistakes in the inscription, and the language is quite simple and standard for the period in which the object was made.

The coffin:

Royal offering to Ra-Horakhti, Atum, the Lord of the Two Lands from Heliopolis and to Ptah-Sokaris-Osiris, the Lord of the necropolis, that they might give offering in food to mistress of the house Kaipamaw.

[13] E. Graefe, *Untersuchungen zur Verwaltung und Geschichte der Institution der Gottesgemahlin des Amun vom Beginn des Neuen Reiches bis zur Spätzeit*, ÄA (Wiesbaden) 37, I/1981, 153 (q. 10).

[14] F. Daumas, *Valeurs phonétiqes des signes hiéroglyphiques d`epoque Gréco-romaine,* I, Montpellier 1988, 148 (no. 83).

[15] E. Graefe, *Untersuchungen zur Verwaltung und Geschichte der Institution der Gottesgemahlin des Amun vom Beginn des Neuen Reiches bis zur Spätzeit*, ÄA (Wiesbaden) 37, I/1981, 153 (q. 10).

The cartonnage:

Royal offering to Osiris who is first in Amenti, the Great God from Abydos, that he might give invocation burnt-offering and incense offering to mistress of the house Osiris Kaipamaw.

Blessed before Osiris, mistress of the house, the chantress of Amun Kaipamaw.

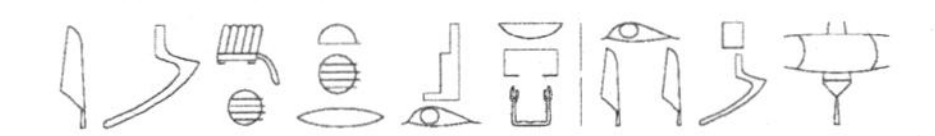

Blessed before Osiris, mistress of the house Kaipamaw.

Igor Uranic
Zagreb Archaeological Museum
N. S. Zrinskog 19, HR – 10 000 Zagreb,
Croatia

The Case of the Misplaced Cow: ROM Cartonnage 1910.10*

Gayle Gibson

The golden-faced cartonnage coffin of Djedmaatiusankh is one of the most recognizable artifacts in the Royal Ontario Museum. The mummy and cartonnage were acquired in Egypt by the Museum's first director, Charles Trick Currelly, between 1907 and 1910. Curator Nicholas Millet wrote about them in a 1973 article, "An Old Mortality."[1] A thorough cleaning in 1974 enhanced the beauty of the decoration,[2] and since then, Djedmaatiusankh has become an icon of the ROM. In this article we will look at one set of images on her cartonnage in particular, a set which is unusual both in being used at all, and in the oddities of its presentation, namely the images of the seven celestial cows and their husband, the bull, from Chapter 148 of the Book of the Dead.

The Owner of the Cartonnage

Djedmaatiusankh's name means "The goddess of Truth has said 'She will live'." Unfortunately, we know very little about her.

In a scene spread across the chest of the mummy (see *Fig. 1a*), the dead lady is introduced to the god Osiris (who is sitting on his throne, behind him the goddesses Isis and Nephthys) by Horus.[3] She is in her finest clothes, a pink gown of pleated linen with a blue fringe. Her long hair, or fine wig, frames her delicate, eternally youthful and healthy face. She is slightly plump (the fashionable shape in the XXII Dynasty) with full thighs visible beneath the transparent robe. She wears a festive collar and a cone of sweet smelling wax and oils on her head.[4] A bud and a blossom of the water lily are attached to the cone, though it is not clear whether the nocturnally-blooming white, or diurnal blue lily was intended; either is appropriate. The whole headdress means to express that she rejoices to meet the gods.

Written in two vertical lines of hieroglyphs directly over her head are her name and titles:

> "Lady of the House, Chantress of Amun-Re King of the Gods,
> Djed-maat-iu.s-ankh, True of Voice."

Djedmaatiusankh's titles are common for the time and place.[5] They tell us that she owned property, and had the right to enter farther into the temple than common people and to take part in some rituals.

---

* This essay is based on a longer (yet unpublished) work which examines the meaning of the iconography of ROM 1910.10, and compares it to several other cartonnage coffins from the same time period, and, I believe, from the same workshop.

[1] Nicholas Millet, "An Old Mortality," *Rotunda*, Volume 5, Number 2, Spring 1972, pp. 18-27.

[2] Susan Wilson and Carol Baum, "A Real Case History: The Coffin of Djed-Ma'at-es-ankh," *Rotunda*, Volume 11, Number 3, Fall 1978, pp. 9-13.

[3] The god Thoth is standing behind Djedmaatiusankh, and behind him we find a Weighing of the Heart scene (cf. *Fig. 1b*), which is a most unusual image on a XXII Dynasty cartonnage. Some details of the scene, like the scale with the heart, can just be seen in the upper part of *Fig. 4*. Seeing the overall design of the coffin, one might have expected a winged figure to appear on the viewer's right, as a pendant opposite to Nut, but, if so, the Weighing scene was felt to be important enough to disregard any such artistic considerations of symmetry.

[4] A wig saturated with these materials can be seen in the Metropolitan Museum of Art in New York.

[5] See Millet, *op. cit.,* for a discussion of these common Theban women's titles.

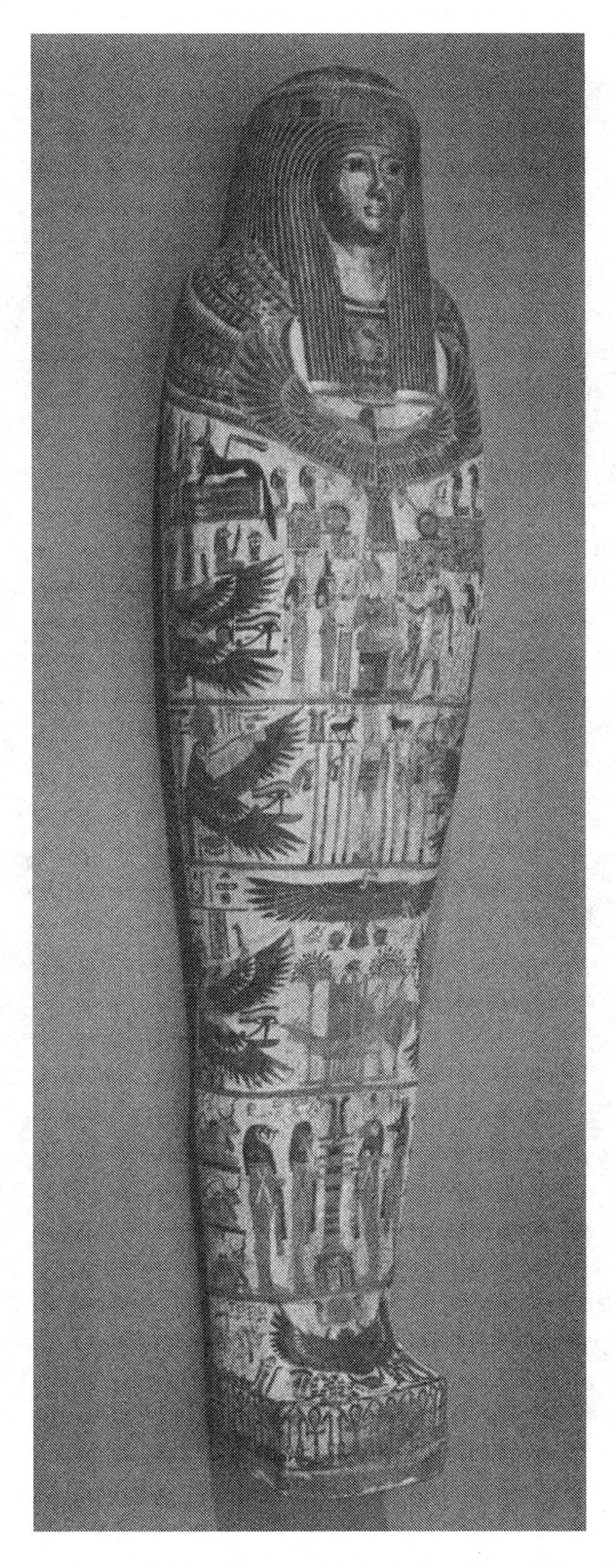

Fig. 1a. An overall view of the front and right side of the coffin. (Photo: ROM)

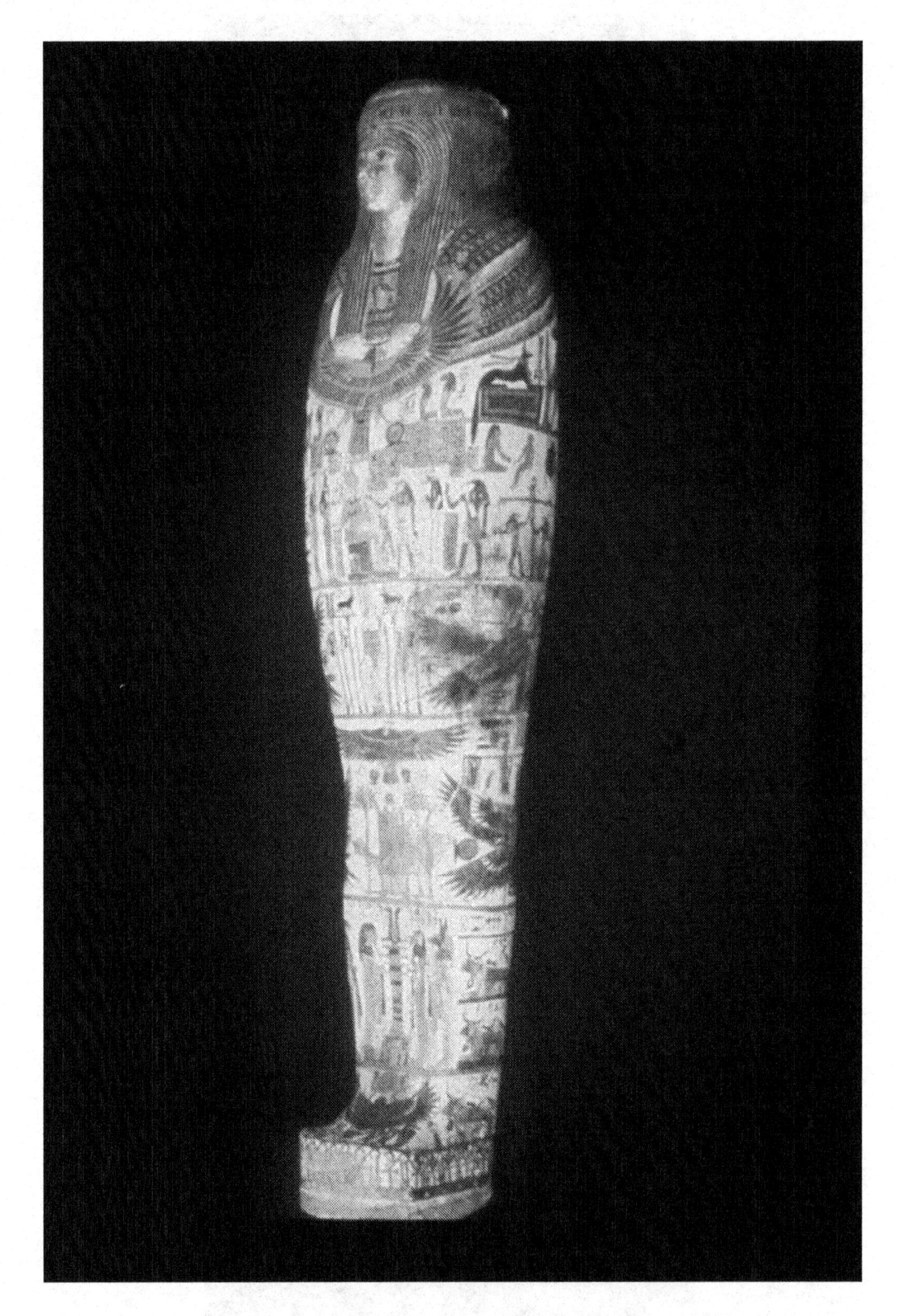

Fig. 1b. An overall view of the front and left side of the coffin. (Photo: ROM)

The pretty, rather bland, face has taken on an added poignancy since a CT-Scan in 1994 revealed that Djedmaatiusankh had suffered from severe dental abscesses.[6] Djedmaatiusankh died in anguish, her jaw discoloured, her cheek distended. But the illustration on the cartonnage proclaims that in the next life, all pain will be forgotten, and she will be again and eternally beautiful.

General Observations about the Cartonnage

By the XXII Dynasty, the coffin had taken the place of an expensive painted tomb for Thebans of all classes,[7] and augmented or replaced hand-written and painted Books of the Dead. XXII Dynasty coffins tend to exhibit a limited number of scenes, generally eliminating mythological images and most specifically Theban references.[8] The calm, spacious designs may reflect the relatively stable political situation under the first Libyan kings.

Like many XXII Dynasty cartonnage coffins, Djedmaatiusankh's is a beautiful object. The artists who decorated it were skillful painters whose repertoire of images included the Book of the Dead, New Kingdom coffin decoration, and the royal imagery of the Journey of the Sun as employed both in the Valley of the Kings and at Tanis. The choice of elements resulted in a cartonnage which appears typical of its time and place, but is also highly idiosyncratic and personal.[9]

Prominent sets of wings, painted a deep green, are the characteristic and dominant feature of XXII Dynasty cartonnages and coffins. A limited number of winged beings protect the front, while other wings reach from the sides, sometimes meeting, or even overlapping, at the centre line. On Djedmaatiusankh's coffin, winged beings are arranged to protect all parts of her body. On the front, from top to bottom: a ram-headed bird spreading his wings across her breast, a vulture stretching across her upper thighs (Nekhbet, in the third register), and a winged scarab beetle protecting her feet. They all carry one or two protective *shen* symbols. Along the sides of the coffin, we find five winged figures who lift their wings in salutation and protection: a goddess (Nut, in the first register, on the right side of the coffin, i.e. on the viewer's left in *Fig. 1a*), two hawks or falcons (Horus the Behdetite, both on the left and the right, in the second register), and two goddesses (Neith, on the viewer's left, and Serket, on the viewer's right, in the third register[10]). Between their wings, Nut and Horus offer an *udjat*-eye to ward off evil, and Serket offers a protective *shen* symbol.

The pictures on a coffin were painted prayers, short-hand images of a whole theological system, which incorporated both the realm of the living, ruled and powered by solar deities, and the chthonic realm of Osiris. For example, the winged scarab at the mummy's feet can be read as a reference to the Solar Religion, but also to the Book of the Dead. In BD 83, which seems particularly appropriate for the winged coffins of the XXII Dynasty, the Deceased proclaims:

---

[6] Lee-Ann Jack, "The Faces of Djed: A CT-Scan of a mummy illuminates a life from Ancient Egypt," *Rotunda,* Winter 1995, Volume 28, Number 3, pp. 30-37, and Kathy A. Svitil, "The Mummy Unwrapped," *Discover*, April 1995, Volume 16, Number 4, pp. 62-65.

[7] This type of coffin is well represented in Museum collections. The still unpublished coffin ROM 910.5.1, though nameless, and rather hastily decorated, illustrates the genre well.

[8] No comprehensive study of XXII Dynasty coffins has yet been published. John Taylor devoted three paragraphs to this type of coffin in *Death and the Afterlife in Ancient Egypt* (Chicago: Univ. of Chicago Press, 2001), pp. 233-236. Aidan Dodson and Selima Ikram dealt with this type in six paragraphs (including the royal examples) in *The Mummy in Ancient Egypt: Equipping the Dead for Eternity* (London: Thames and Hudson, 1998), pp. 233-236.

[9] For purposes of discussion in this article, the coffin is divided into five registers (cf. *Fig. 1*): the first register has the scene of Djedmaatiusankh before Osiris' throne, the fourth register has the Sons of Horus, and Khepri figures in the fifth register. For reasons of space, the central scenes of the second and third register of the coffin will not be discussed here.

[10] Note that all four goddesses who as a rule protect the deceased (and notably the canopic jars), namely Isis, Nephthys, Serket, and Neith, are present on the coffin, the first two not as winged protectresses at the sides of the coffin, but in the throne scene in the first register.

> "I have flown up like the primeval ones, I have become Khepri, I have grown as a plant, I have clad myself with turquoise. I am the essence of every god."[11]

Djedmaatiusankh, like the great kings of old, survives death by identifying with the sun god, Re, in his various manifestations. Khepri is the god of morning, of new beginnings; placed on a coffin, he assures us that death is not the end. Here he is shown with hawk's wings to enable him to soar into the sky. He pushes the sun before him along its daily path, protected by two rearing cobras. Beneath Khepri, two more cobras may represent the royal uraei Wadjet and Nekhbet, or Isis and Nephthys. They function to complete the spell of protection and rebirth. Together with a ram-headed hawk on Djedmaatiusankh's bosom (a *bꜣ* bird), and a scarab on the crown of her head, these images show that the deceased will share in the eternal life of the Sun god, born again each day, dying each night, part of the endless cycle of birth and death.

The wings and other elements of the decoration seem at first glance to be perfectly symmetrical, but examination reveals differences between the two sides of the coffin. In part, this is because Egyptian art is rarely completely symmetrical, with small variations deliberately introduced into even the most balanced compositions. In this case, the artisans were not always in full communication (or, perhaps, not in agreement). However carefully priests and family members or the artists themselves may have planned the decoration, confusions, omissions, and possibly subversion crept in. Every coffin was individually decorated. Constraints of time, wealth, or skill affected each differently. Each coffin documents the process which created it and the world it came from.

BD 148: Celestial Cows and Heavenly Steering Oars

From about knee level downward, along both sides of Djedmaatiusankh's cartonnage, bovines and oars represent Chapter 148 of the Book of the Dead. BD 148 lists a spell that is "to be spoken ... when Re manifests himself" and opens thus:

> *Spell for Making Provision for a spirit in the realm of the dead.*
> "Hail to you, You who shine in your disc, a living soul who goes up from the horizon! I know you, and I know your name; I know the names of the seven cows and their bull who give bread and beer, who are beneficial to the souls and who provide daily portions; may you give bread and beer and make provision for me, so that I may serve under you, and may I come into being under your hinder-parts."

Then the names of the cattle are listed, with a prayer at their address:

> "May you grant bread and beer, offerings and provisions which shall provide for my spirit, for I am a worthy spirit who is in the God's Domain."

This is followed by the names of four steering-oars, with a similar prayer. The spell concludes with general prayers for provisions and protection against evil, and with some prescriptions on how to recite the spell.

The seven cows are usually shown reclining, with a solar disk between their horns. This refers to the myth of the Celestial Cow, who rose from the Primeval Waters, gave birth to the Sun, and raised it

[11] All quotations from the Book of the Dead are from R. O. Faulkner's translation. R.O. Faulkner, *The Ancient Egyptian Book of the Dead* (Austin, Texas: University of Texas Press, 1990).

upon her horns to Heaven.[12] As heavenly mother cow, she suckles and nurses the god Horus, and by extension the pharaoh and the dead. Each day, at sunrise, she gives birth to the Sun anew; BD 17 expresses this idea thus:

> "I have seen this sun-god who was born yesterday from the buttocks
> of the Celestial Cow; if he be well, then will I be well, and vice versa."

As one of the lines of BD 148 ("may I come into being under your hinder-parts") signals, the dead hoped to share this fate, to be born anew from, and be raised up and nurtured by, the Celestial Cow. That there are seven such cows depicted in BD 148 seems to refer to the idea of the Seven Hathors, manifestations of the most important cow-goddess, Hathor. The Seven Hathors fortell the fate of a child at birth,[13] like the fairies in Sleeping Beauty.

To keep the cows happy, and to ensure fertility and energy in the hereafter, a (virile striding) bull completes the herd. The male's name expresses his function: the Bull of Bulls, the Husband of the Cows. There is no ambiguity in the spelling of this name in hieroglyphs which requires the scribe to draw five penises along with four other signs.

The cows' names are more poetic, and there are at least two traditions for naming them.[14] In the tradition that is relevant in our case, the names of all the animals are:[15]

1. Mansion of *Kas*, Mistress of All
2. Silent One who dwells in her place
3. She of Chemmis whom the god ennobled
4. The Much Beloved, red of hair
5. She who protects in life, the particoloured
6. She whose name has power in her craft
7. Storm in the Sky which wafts the god aloft
8. The Bull (of Bulls), husband of the cows.

The seven cows and bull can be seen not only in vignettes to Books of the Dead, but also on many mythological papyri, in the tomb of Nefertari,[16] and on some coffins of the XXI Dynasty.[17] Djedmaatiusankh's artists may have seen the beautiful rendition on the walls of room 24 in the Osiris Complex at Medinet Habu, where all seven cows, the bull, and the steering oars are drawn and named.

The four steering oars of BD 148 represent, via their natural function, the four cardinal directions of Heaven, and help the dead to travel the Afterlife. Their traditional names and epithets are:

---

[12] For the motif of the Heavenly Cow, see Hans Bonnett, *Reallexikon der ägyptischen Religionsgeschichte* (Berlin/New York: Walter de Gruyter, 2000), pp. 402-405, 459. Note that the cows on Djedmaatiusankh's coffin do not carry a solar disk. The motif of the cow raising the sun is also mirrored in the name of the seventh cow ("Storm in the Sky which wafts the god aloft," see below; for the storm motif, cf. PT 336b).

[13] See, for example, the stories of "The Two Brothers" (P. BM 10183) and "The Doomed Prince" (P. Harris 500, verso), in Miriam Lichtheim, *Ancient Egyptian Literature: Volume II, the New Kingdom* (Berkeley: University of California Press, 1975), p. 200 and p. 207.

[14] George Hart, *A Dictionary of Egyptian Gods and Goddesses* (London: Routledge & Kegan Paul, 1986) pp. 79-80.

[15] Faulkner, op. cit. (BD 148).

[16] QV66. In this tomb, the herd is shown in splendid naturalistic detail, with none of the animals having any artificial ornaments, and the cows are depicted striding, like the bull.

[17] A monograph on BD 148 is R. El Sayed, "Les sept vaches célestes, leur taureau et les quatre gouvernails d'après les données de documents divers," *MDAIK* 36 (1980), pp. 357-390. The article lists 72 monuments and documents that carry the motif of the seven cows and their bull; the list contains mostly tombs, temples and funerary papyri, but also three sarcophagi (Cairo CG 41025, CG 29301, CG 29305).

1. Good Power, the good steering oar of the northern sky
2. Wanderer who guides the Two Lands, good steering oar of the western sky
3. Shining One who dwells in the Mansion of Images, good steering oar of the eastern sky
4. Preeminent who dwells in the Mansion of the Red Ones, good steering oar of the southern sky.[18]

It is likely no coincidence that on Djedmaatiusankh's coffin the cows and oars are placed in such a way that they flank the central scene of the coffin's fourth register, the scene with the Four Sons of Horus standing around a Djed Pillar. For the Sons of Horus regularly figure in vignettes of BD 148,[19] and have several functions in common with the cows and oars, besides the number four ruling them all. As guardians of the Canopic jars, which hold the internal organs of the dead person, the Sons of Horus, like the cows, protect the dead person against thirst and hunger (cf. PT 552[20]). At times they also represent the four cardinal directions, just like the steering oars. In that capacity, they herald the king's coronation to the four corners of the earth, lift up the dead person and raise him to heaven (cf. PT 1092, 1338-1340, 1983), and are positioned at the four corners of the tomb or the coffin (cf. the vignette of BD 151).[21]

The BD 148 herd on Djedmaatiusankh's coffin is unusual for a XXII Dynasty coffin. Of similar white background cartonnages known to me, only that of Nebnetjeru in the University of Pennsylvania Museum of Archeology and Anthropology has any cows.[22] Djedmaatiusankh's cows have peculiarities which do not seem attributable to fine points of theology, nor to references to unusual sites. They seem to be, simply, mistakes.

A Misplaced Cow

The bovines at the bottom of Djedmaatiusankh's coffin are divided in two groups, four on the right side of the coffin (*Fig. 1a* and *Fig. 2*) and three on the left side (*Fig. 1b* and *Fig. 3*): three cows on each side, plus a bull at ankle level on the coffin's right side (*Fig. 2*). The herd does not follow the colour scheme indicated by some of their names and shown in Nefertari's tomb and in fine versions of the Book of the Dead, such as that for Maiherperi.[23] Djedmaatiusankh's cows were all painted with a light wash of yellow ochre, then daubed with resin on the face and back to give them what now appears as a golden glow. The three cows on the viewer's right are slightly bigger than those on the left, but all wear red blankets and *menat* necklaces with counterpoises (S18).[24] Bull and cows, except for the last cow on the viewer's left, 'carry' the *nekhakha* flail (S45) on their backs. The six cows, but not the bull, have eyes 'made up' to look like the *udjat*.

---

[18] Faulkner, op. cit. (BD 148).

[19] In the BD 148 vignette of the Papyrus of Nes-min, in the Detroit Institute of Arts (see the following URL: http://dia.org/bulletin/papyrus/), there are four columns divided in four rows, with the eight bovines spread over the first two columns, the steering oars, each accompanied by a large *udjat*-eye, in the third column, and the Sons of Horus in the fourth column. El Sayed, op. cit., p. 381, also lists some examples in which the Sons of Horus accompany the cows. And cf. BD 141 in which the Sons of Horus follow the cows and the oars in a list of divinities.

[20] R. O. Faulkner, *The Ancient Egyptian Pyramid Texts* (Oxford: 1998).

[21] See H. Bonnett, *Reallexikon der ägyptischen Religionsgeschichte* (Berlin/New York: Walter de Gruyter, 2000), pp. 315-316, and B. H. Stricker, *De Hemelvaart des Konings* (Leiden: EOL, 1990), pp. 13-20.

[22] Nebnetjeru's two cows are similar to Djedmaatiusankh's in style, but wear the *atef* crown and sit on decorated baskets as in E4, the *hesat* cow. Like Djedmaatiusankh's they wear red blankets, but instead of a *menat* with counterpoise, they wear *bat* emblems around their necks. Behind each is an unnamed steering oar.

[23] Cairo, CG 24095.

[24] Hieroglyphs are referred to by letter and number according to Sir Alan Gardiner's system, *Egyptian Grammar, Third Edition* (Oxford: Griffin Institute, 1969). The *menat* collar is associated with Hathor. Some BD papyri give the cows a mummiform look by depicting them as being fully wrapped in red cloth, only leaving their heads free, instead of wearing just a red blanket.

Six cows and a bull, seven animals in all on the lower half of Djedmaatiusankh's coffin. But there should be seven cows. Seven is a magic number. Would six cows be able to work the magic? Someone noticed that there were not enough cows. This artist clearly believed that he had to find a space on the coffin for the seventh Hathor. In the second register of the coffin, on the viewer's right, in a space behind the winged image of Horus the Behdetite, there is a seventh cow (see *Fig. 4*).[25] She was drawn on her own blue base line by a different hand. Unlike the others, her head is not tilted back, her horn does not touch her ear, and she has (like the cow above the bull) no *nekhakha* scepter. She does complete the herd, but the errors made had made it impossible for another artist to correct the situation completely.

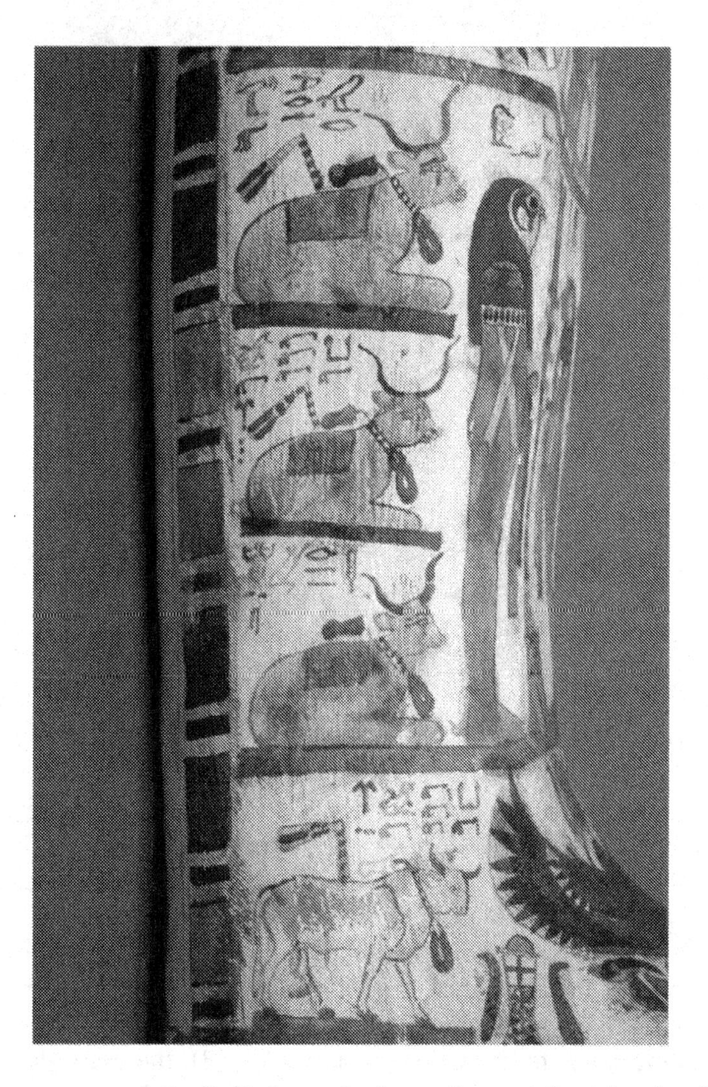

Fig. 2. Three cows and the bull, from the lower legs, right side of coffin. (Photo: ROM)

[25] Below the cow there is an unnamed, seated and knife-bearing guardian with the head of a hippopotamus. In the corresponding space at the opposite side of the coffin, behind the Horus of Behdet at the viewer's left, there are two epithets of the falcon god ("Lord of the Sky" and "Lord of Mesen") and a feather on a standard (symbolizing the West).

Fig. 3. Three cows, from the lower legs, left side of coffin. (Photo: Gayle Gibson)

The names of the members of the herd on Djedmaatiusankh's coffin are also in disarray.

The first cow on the viewer's left (*Fig. 2*), in the fourth register, is "Greatly-beloved, red of hair." The fact that she is the same colour as all the other cows and the bull does not seem to have bothered the scribe or the artist. The name over the second cow cannot be correct. The five penises show that the scribe looked at the little cow, and wrote "the Bull of Bulls, husband of the cows." The third cow is "Her name is powerful in her craft." Finally in the fifth register, on the viewer's left, the Bull himself is depicted, with his proper name.

At the corresponding areas to the viewer's right (*Fig. 3*), there are three cows. The uppermost is named "Shining One who is in the Mansion of Images, beautiful oar of the Western Sky." A fine name, but not for a cow. This is the name of one of the steering oars from BD 148, but the name is usually associated with the oar of the Eastern sky. The middle cow has two names written behind and above her: her own, "Silent One who is foremost in her place," and the name of another oar, "Wanderer who guides, beautiful oar." Traditionally, the "Wanderer" is associated with the Western

Sky. The last cow, in the fifth register, has a purely bovine name: "She of Chemnis,[26] whom the god ennobled."

Several of the magic names were missing. Two were added behind the 'extra' cow in the second register (*Fig. 4*). She is "Mansion of *Ka*s, Lady of All," and has an oar, "the Beautiful steering oar of the Eastern Sky." In the fourth register, immediately under the red register line that divides the three larger cows from the winged image of the goddess Serket above them (*Fig. 3*), is another name: "Rainstorm who raises up the god," written in large glyphs and not connected with any image at all.

In short, several errors were made, both in drawing the figures and in placing the names. Do the errors enable us to see the artists at work, to get an impression of how the work was executed, and to understand how familiar the artists and Djedmaatiusankh's relatives were with the iconography on the coffin?

Fig. 4. The misplaced seventh cow, behind Horus the Behdetite. (Photo: Gayle Gibson)

A Cascade of Errors

The cows representing BD 148 and the Weighing of the Heart scene are both unusual components of a XXII Dynasty cartonnage design. Do they suggest that the family of Djedmaatiusankh knew the importance of the Book of the Dead but could not afford to send a copy with her? Perhaps. But the fact that the scenes were unusual, and had to be ordered specially, would on the one hand have increased the chance of errors in decoration, the common artists not being very familiar with the images and the names, and on the other hand have increased the visibility of the scenes, and thus the need to correct any mistakes so not to displease the commissioning family.

[26] The name of this Delta town means literally "Papyrus Thicket of the Lower Egyptian King," and it was the place in which the god Horus was supposedly born and nursed. So this name brings up the image of the nurturing mother cow.

Although mistakes are only human, and need not have a detectable rationale, it is tempting to see if there could be a logical explanation for the errors on the coffin. One hypothetical scenario could run as follows, presuming that two artists went at work after each other, first an artist for the pictures, and then a (more experienced?) scribe for the hieroglyphs.[27]

The first artist, realizing there were eight bovines (cow's figures 1-7, plus a bull, to be paired with cow's names 1-7, plus bull's name), would have wanted to depict four animals on each side of the coffin, for nice symmetry. For some reason he seems to have started at the bottom of the coffin's right side (*Fig. 2*), first painting the very recognizable bull in the fifth register, and then working upwards, drawing the seventh, sixth and fifth cow's figures on the viewer's left in the fourth register. Then he moved to the viewer's right, again starting at the bottom of the column, with the fourth cow's figure in the fifth register (opposite the bull), and then in the fourth register the third and second cow's figures (*Fig. 3*). But suddenly he found out he had drawn these latter two cows too large, so that there was no longer any room for a figure of the first cow! Why he did not draw all the blue base lines first, to guarantee equal space for all cows, remains puzzling.

Subsequently, the second artist had to put the names with the pictures. He, however, seems to have started at the top of the viewer's right, with the intention to begin properly with the first cow, but made the basic error of not checking the work of his predecessor (i.e., he did not count whether there were eight bovines). He made three further errors. Firstly, when putting names with the cow's figures on the viewer's right (*Fig. 3*), he started out wrongly, perhaps by looking at the wrong list with names, and gave the first cow's figure the name of an oar, and on top of that, he linked that oar with the wrong cardinal direction. He quickly seems to have realized his mistake and gave the second cow's figure the second cow's name from BD 148 (see name list above), and the third cow's figure the third cow's name, so that when arriving at the top of the column on the viewer's left (*Fig. 2*), he gave that fourth cow's figure the fourth cow's name, a logical thing to do. But then he made a third mistake. When looking at his notes, which may have had the bovine names in two rows (as on Nefertari's wall), like this:

4321
8765

instead of moving to the start of the next row, he wrongly looked down, and painted the names as they occurred in his notes, i.e. to the second cow on the viewer's left, below the cow with the fourth cow's name, he gave the eighth bovine name (the name of the bull), instead of the fifth cow's name.[28] Again realizing his error, he gave the next cow's figure the name that was planned, namely the sixth cow's name, only to find out that there was no figure to receive the name of the seventh cow! Pausing, he noticed the cow-less space on the viewer's right (*Fig. 3*), realized that it was the first cow's figure that was missing (rather than the seventh), and proceeded to correct the errors of his predecessor as best he could. Of course he named the bull correctly, and then moved the missing seventh cow's name into the cow-less space on the viewer's right, to keep it as close as possible to the sixth cow's name. Now there was a bovine name just above the cow's figure that had wrongly received the name of an oar (*Fig. 3*). Moreover, he painted an extra cow figure in another spot on the coffin (*Fig. 4*), namely the missing figure of the first cow, and logically gave it the first cow's name (the name he himself had wrongly swapped with an oar's name, when naming the first bovine figure on the viewer's right in the fourth register). Now there were eight bovine figures and eight bovine names. Even though the fifth cow's name ("She who protects in life, the particoloured") was still missing, and there were two bull's names, this was apparently felt to be less a problem than having, for instance, nine bovine names.

[27] Other possible scenarios will be explored elsewhere.

[28] This error, like the error of giving a cow an oar's name, would suggest that the second scribe only had two lists of names, without the figures of bovines and oars.

The plan might have been to put two oars on each side of the coffin, paired with two cows, for symmetry. He had earlier given one oar's name to a cow on the viewer's right, a cow he had thought was the first cow, although why he had picked the name of this particular oar and why he wrote down the wrong cardinal direction (West instead of East), remains unexplained.[29] But, now needing to finish the oars after having completed the cows, and after writing down the name of the oar "Wanderer" on the viewer's right (*Fig. 3*), he realized that this was the real western oar, and settled for an unfinished epithet "beautiful oar" without mentioning the sky, so as not to get two western skies on this side of the coffin. It would be unlikely that someone would recognize the names of the oars, but perhaps someone would recognize the hieroglyph for West? For consistency, he placed the missed epithet "beautiful oar of the Eastern Sky" behind what was, after his earlier corrections, the real first cow, namely the dislocated cow with the first cow's name (*Fig. 4*). Apparently, it was at this moment that he finally realized that the errors could not be fully corrected, with the eastern and western oars now appearing mixed and split up, so he decided not to place the names and designations of the northern and southern oars on the opposite side of the coffin. The fact that there were now two oar's names plus two sky directions seems to have satisfied the requirement of the number four.

Coffins, chests of life, were expensive, high status purchases. What does it mean that there remained several uncorrected errors on the piece?

Firstly, it may have been a matter of time. How long did a team of artists work on decorating a cartonnage? Did the limited time between their reception of the mummified body and the funeral prevent them from correcting certain errors? For example, to correct the error of calling a cow a bull, another layer or two of gesso would have had to be applied over the mistake. Each layer would have to dry, and then the name could be written. Was there no time to do this? Or was the budget so tight that either the extra gesso or the extra time was not available? And thus more ad hoc (and less ideal) solutions had to be found.

Secondly, some mistakes may have been more important than others. The four Sons of Horus, for example, are nicely drawn on Djedmaatiusankh's coffin, but, as often happened, Qebesenuf's name was switched with Duamutef's. It is clear that, while the iconography of the Sons of Horus was something that every artist working on coffins knew, their names were less important. Did the scribe who labelled the Sons of Horus not know or care which was which? The Sons of Horus frequently appear on coffins without labels, known by their iconography, not their name-tags. The scribe could, perhaps, rely upon the funeral guests to be, if not illiterate, too concerned with other matters to notice.[30] So it was not necessary to correct this common name error on the coffin. The cows, however, must have been more important, seeing the efforts to correct previous mistakes. There are many coffins and Books of the Dead on which the cows are not named at all,[31] or only a token handful are shown. But they do not receive such casual attention on Djedmaatiusankh's coffin, on the contrary. Perhaps this was due to the fact that these cows are Hathors: Hathor was the goddess of music, and thus possibly of special religious importance to a chantress like Djedmaatiusankh. This may not only have given them a higher emotional value, their rarity alone would increase the chance that they would receive the attention of the family. Guests at a funeral might not all be literate, but they would certainly be able to count up to seven. And thus someone thought it was vital to add the

[29] Admittedly, El Sayed's overview (op. cit., pp. 376-378) suggests there is considerable variation in the way in which the names of the oars are connected with the directions of the sky.

[30] A good parallel might come from Christian iconography: the four evangelists are represented by a man, a bull, a lion, and an eagle. Which is Mark, which Luke, John, Matthew? Few people can identify all four.

[31] For example, the names are not written beside the images on the Papyrus of Ani, BM EA 10470/5, XIX Dynasty, nor on the Ptolemaic Papyrus of Hor, son of Djedhor and Sebat, BM 10479. In Piankoff's study of the Mythological Papyri (Bollingen Series XL.3, Pathenon Books, 1957), ten of the papyri do use the cow image, though only three, Nes-pa-ka-shuty, Khonsu-mes, and Nesi-ta-Nebet-Taui, show all eight animals, and only the last papyrus actually names all eight bovines and all four oars.

seventh cow figure. But why was it necessary to add the missing cow name in the fourth register (*Fig. 3*), even though there was no image to associate it with? Would someone really notice a name to be missing? Perhaps the scribe or some family member was more familiar than most with the texts of the Book of the Dead, particularly BD 148? So, special attention was given when depicting the cows, not only to their figures but also to their names. Still, even here a perfectly correct rendering apparently was not overly important: the numbers seem to have mattered most, and eight bovine figures and eight bovine names were felt to be sufficient to do the BD 148 magic, even though each cow was not paired with its proper name, one cow's name is missing, and there is a double bull's name. The same with the oars, which received a lot less attention: the number four was completed, via two names and two sky descriptions, even though the result was far from ideal; basically only two oars are represented.

This essay has examined only a few aspects of Djedmaatiusankh's cartonnage. Like every other ancient artifact, it is a message from the past. This coffin speaks to us of the ancient Egyptian longing for immortality, of the accidents of history, and of the skills and fallibility of human beings. And it reminds us of Djedmaatiusankh's all-important name, so that we may pray for her, so that she may hold infinity in the palm of her hand, and enjoy her eternal hours.

Gayle Gibson
Royal Ontario Museum
Toronto Ontario
Canada

Tracing the Origins of the Ancient Egyptian Cattle Cult

Michael Brass

Studies of ancient Egyptian religion have examined texts for evidence of cattle worship, but the picture given by the texts is incomplete. Mortuary patterns, ceremonial buildings, grave goods, ceramics and other remains also contain evidence of cattle worship and underline its importance to early Egypt. The recently discovered cattle tumuli at Nabta Playa in the Western Desert are identified here as a potential source of evidence on the origins of cattle worship in the ancient Egyptian belief system.

Climate and Society in the Palaeolithic and Neolithic of North-East Africa

Throughout the past 250,000 years, North-East Africa has known alternating periods of dry (arid) and moist (pluvial) climatic conditions. During the rainy periods, the eastern Sahara was covered with grassy plains inhabited by wild herd animals, including wild cattle, *Bos primigenius*. Whenever the climate changed to arid, only pools and oases remained and the archaeological and faunal visibility of both humans and herds becomes very scarce.

A northward shift of the African monsoons around 10,000 B.C. saw the end of the Late Palaeolithic[1] period of hyper-aridity[2], and the Western Desert once again became inhabited. We will focus on one site in that Desert, the Nabta Playa basin situated 100 kilometres west of Abu Simbel, and on simultaneous developments in the Nile Valley.

The 10th millennium B.C. yields the evidence of the earliest sediments within the Nabta Playa basin. The first occupational layers are dated to 8,800 B.C. As summarised in *Table 1*, the El Adam industry continued until 7,800 B.C. (Wendorf 1998, pers. comm.). The excavators, Fred Wendorf and Romauld Schild, noted that the artefacts which they assigned to this industry display flaking techniques similar to the near-contemporary Nile Valley Arkinian industry (Hoffman 1979: 102; Wendorf & Schild 1998: 100). On this basis, the first inhabitants of Nabta Playa have been hypothesised to be immigrants from the Nile Valley (Wendorf et al. 1985: 135).

It was during the arid post-El Nabta/Al Jerar phase that the present topography of Sites E-75-8, E-94-2 and E-94-1 was formed (Wendorf et al. 2001: 650), and Nabta Playa became depopulated. The Late Neolithic groups which subsequently re-occupied Nabta Playa, from Saharan areas as yet undetermined, display signs of social systems with a degree of organisational control not present in contemporary Nile Valley communities (Malville et al. 1998: 488).

The desiccation of the Western Desert over the course of the Late Neolithic occurred about the same time as a drop in the flood levels of the Nile River, as documented by Hassan (1988: 146-47). The tempered, black-topped and red-slipped pottery found in the Late Neolithic layers of Nabta Playa sites E-75-8 and E-94-2 are similar to that of the Badarians in the Nile Valley (Nelson 2001: 539-40), suggestive of an interaction between the two areas.

While there is evidence for Eastern Desert influence on the Badarian culture (Majer 1992), Midant-Reynes (2000: 148, 164) convincingly argues for a strong element of Saharan culture:

---

[1] For North-East Africa, the Palaeolithic period divides into: Lower Palaeolithic (c. 500,000-250,000 B.P.), Middle Palaeolithic (ca. 250,000 -70,000 B.P.), Transitional Group (70,000-50,000 B.P.), Upper Palaeolithic (ca. 50,000-24,000 B.P.), Late Palaeolithic (ca. 24,000-10,000 B.P.) and Epipalaeolithic (ca. 10,000-7,000 B.P.).

[2] The Late Palaeolithic was an entirely arid period with no pluvial interruptions.

> "With regards to lithics, [Holmes (1989: 183)] points out that there are some similarities with the post-Palaeolithic culture in the Sahara (an industry based on blades and flakes, in which polished axes and hollow-base arrowheads are not lacking), which means that we cannot exclude the possibility that the semicircle formed by the Bahariya, Farafra, Dakhla and Kharga oases might have been the point of origin of populations who perhaps already pursued a pastoral mode of subsistence; these people might have been pushed eastwards by increasing aridity and would eventually have settled in the region of Asyut and Tahta . . . [It] might even be suggested that the Neolithic cultures of the oases and the Faiyum could be regarded as the eastern fringes of the Sahara Neolithic groups."

Taken together, the two indications above suggest that the population which arrived in Nabta after 5,500 B.C. – apparently pastoralists from the Sahara with a new and higher level of organisation – influenced the developments in the nearby Nile Valley. Wendorf & Schild (1998: 114) also hypothesise that the primary external stimulus for the rise of social complexity in Upper Egypt was contact with the pastoralists of the Western Desert. If this view is correct, social complexity in the Nile Valley was the end product, not only of numerous differentiating factors associated with the rise of craft specialisation, but also of the dynamic interaction between two contrasting lifestyles, pastoral and centralised agricultural economies, existing in close proximity (Wendorf & Schild 1998: 113). The Nile Valley and desert economies were characterised by structural and functional differentiation providing mutual support for each. Yet a tense harmony would also have been present, as well as diffusion of ideas and practices. Rock art from the Predynastic to the east and west of Armant, situated on the west bank of the Nile River, depicts domesticated cattle with artificially deformed horns, indicative of pastoralism (Hoffman 1979: 234-35).

Nabta Playa	Dates	Nile Valley	Dates
El Adam[3]	8,800 – 7,800 B.C.	Makhadma	11,300 – 10,000 B.C.
El Ghorab	7,600 – 7,200 B.C.	Arkinian	c. 7,400 B.C.
El Nabta/Al Jerar	7,100 – 6,300 B.C.	Fayum B	7,100 – 6,000 B.C.
Middle Neolithic	6,100 – 5,600 B.C.	Badarian	5,200 – 4,000 B.C.
Late Neolithic	5,500 – 4,000 B.C.	Fayum A	5,200 – 4,000 B.C.
		Merimde	5,000 – 4,400 B.C.
		Nagada I	4,000 – 3,700 B.C.
		Nagada II	3,700 – 3,300 B.C.
		Nagada III (Terminal Predynastic)	3,300 – 3,100 B.C.

Table 1. Comparative chronology of Nabta Playa and contemporary sites in the Nile Valley.

The contemporary Badarians engaged in a semi-sedentary way of life based on riverine exploitation, crop cultivation and selective herding practices. Assertions by Trigger (1983: 27) that the Badarians possessed a predominantly egalitarian social community have been challenged as a result of mortuary analyses at Armant and Nagada (Bard 1987: 123-25; 1988: 55) and at Badari, Matmar and Mostagedda (Anderson 1992: 51-66). Wendy Anderson examined the distribution of grave goods across age and sex divides. She also quantified which graves were most targeted for plunder, and the different types of goods likely to have been most prevalent within the plundered graves, based on the contents of graves

[3] Wendorf & Schild (1998: 100) termed the El Adam, El Ghorab and El Nabta levels Early Neolithic, although the basis for this claim is unfounded as will be discussed below.

of the same size and shape which were not robbed, and the intact burials of all shapes and sizes. Correlations were found between the contents of certain types of graves and the area in which they were most prevalent, and between the condition of the graves and the sex and age of the inhabitants, although no relationship was found between the sex of the dead and the type of grave goods most prevalent in their grave types. Anderson (1992: 60) found that the plunderers targeted the graves containing predominantly luxury materials, which she terms "sociotechnic" artefacts. The results of Anderson's analysis indicate that the Badarian society was hierarchical, although whether or not inherited status existed remains far from proven; the most that can be stated is that it was a culture which practiced social differentiation.

In regard to this greater social complexity, which has been linked above to the rise of pastoralism, Nabta Playa may be said to predate (cf. Malville above) the developments of Late Predynastic Egypt. The date of cattle domestication and its influence on society and culture are the topics of the next chapters.

Cattle domestication in North-East Africa

Originally, cattle had only been hunted in its wild form (*Bos primigenius*) – in Egypt since at least the Middle Palaeolithic. Recent research of genetic signatures has lent support to the hypothesis of an independent center for cattle domestication in North Africa (Stockstad 2002; Hanotte et al. 2002), versus the old hypothesis of diffusion from the Middle East. But when did domestic cattle reach the Nile Valley?

Some authors have posited that the first signs of cattle domestication can be seen at the sites of Tushka (c. 11,000 B.C.) and Dibeira West (c. 7,400 B.C.). The cattle remains at these two sites are smaller in size than the remains found at Isna (c. 15,000 B.C.), and Wendorf & Schild (1994: 127) believe that the reduced size indicates the first stages of cattle domestication. More than 20 *Bos* bones and teeth were found in the El Adam layers of Nabta Playa. Morphologically, these fall within the size range measurements of Tushka and Dibeira West. Based on these deductions, Wendorf & Schild (1994) identified the Nabta cattle remains as a domesticated descendant of the wild *Bos primigenius*.

However, two observations weaken this conclusion: firstly, those Nabta *Bos* bones which can be measured fall within the size range of the bones of wild *Bos*; and secondly, the *Bos* from Isna are of even greater size than those from Middle Palaeolithic sites (Gautier 1993). Therefore it seems that bone sizes in this case do not allow us to draw conclusions in regard to domestication at such an early date, whether in the Nile Valley or at Nabta Playa.

Wendorf et al. advance several other arguments for an early date for cattle domestication in Nabta Playa. Since the term 'Neolithic' is linked by definition to agriculture and animal husbandry – in North-East Africa and the Sahara notably to pastoralism – the issue of cattle domestication also affects the question of whether the earliest Nabta layers should be considered as Early Neolithic or not. The main additional argument of the authors is that the ruling environmental conditions were too arid to support a wild population of *Bos*, and that the *Bos* bones found would therefore belong to domesticated cattle. Wendorf et al. (1985: 136) said:

> "The early Holocene desert fauna lacks all of these medium-sized animals, indicating an environment too harsh to support anything larger than a small ruminant."

It is also contended (Wendorf & Schild 1994: 126) that the lack of aquatic vertebrates is evidence for the absence of permanent water. From this the authors deduced that any *Bos* in the desert would have needed human assistance to survive. However, any *Bos* under the control of humans would also require adequate grass cover and the water to sustain this nourishment. The desert playas would have held large amounts of water after the rains. Grass cover adequate to sustain domesticated *Bos* would also be capable of sustaining wild *Bos*. It should further be noted that hundreds of gazelle and hare bones have been found in the El Adam layers at Nabta Playa – gazelles are medium-sized animals and hares prefer short grasses. All in all, there is no reason to conclude that the environmental conditions were such that wild *Bos* could not have survived.

Another point advanced by Wendorf et al. is that at Late Palaeolithic sites in the Nile Valley, such as Wadi Kubbaniya and Makhadma (Wetterstrom 1993: 179, 181), wild *Bos* and hartebeest remains are found in association, with Wadi Kubbaniya also yielding the bones of gazelle. As hartebeest remains are absent in El Adam layers at Nabta Playa, the authors suggest that the conditions would probably also be unfit for wild *Bos*, and that the layers should be assigned a post-Palaeolithic date, with *Bos* remains being those of domesticated (or at least semi-wild) cattle. However, as already indicated above, plenty of gazelle bones are found in these layers, as they are in Wadi Kubbaniya, and it is not clear why wild *Bos* and hartebeest have to be considered cohorts per se, just because the species occupy the same ecological niche. Large desert antelopes like *Addax nasomaculatus* are still present today in the Central Sahara and yet their remains are lacking at Nabta Playa, which indicates that the full range of animals is not necessarily represented in a limited faunal record (Smith 1986: 198). Therefore, one should not assign too much weight to the absence of hartebeest remains.

In short, the hypothesis of Wendorf & Schild about early cattle domestication at Nabta Playa, based on presumptions about the ruling environmental conditions, is problematic and best rejected. Their pre-Middle Neolithic phases of Nabta Playa are defined here as falling in the Terminal Pleistocene.

As the arguments about bone sizes and environment cannot be regarded as sufficient to sustain the hypothesis of an early cattle domestication at Nabta Playa, there is no reason to assign the El Adam layers to the Early Neolithic.[4] The fact that only 2% of the faunal assemblage in the El Adam layers was *Bos* also does not suggest specialisation of some kind (e.g. pastoralism), unless one were to explain the scarcity of bones by the suggestion that the cattle were not used for meat but only for milk and blood.

From the evidence presented to date, it can at best be hypothesised that the inhabitants of Nabta Playa had access to a variety of wild animals and that they selectively hunted gazelles, hares and some wild *Bos*. The environmental conditions were generally good during the Great Wet Phase of the Terminal Pleistocene (Muzzolini 1993: 228, figure 11.1) and there would have been little pressure on or incentive for the hunter-gatherers to alter their way of life. Wendorf & Schild's Middle Neolithic phase at Nabta Playa coincides with the beginning of the Great Mid-Holocene Arid Phase (Muzzolini 1993: 228, figure 11.2). This would have placed pressure upon the hunter-gatherer inhabitants to alter their lifestyle accordingly. Fat is extremely important in a hunter-gatherer diet, and the percentage of fat stored in cattle is greater than in other game animals. Reduced availability of water and grass cover would have placed corresponding pressures upon population sizes of *Bos*.

In order to maintain a reasonable level of access to this valuable food source, control measures would have been introduced and this would have had an impact on the numbers of *Bos* bones recovered archaeologically, and there should also be clear signs of size reduction through genetic manipulation as domestication began. Indeed, this is exactly the trend seen. Wendorf & Schild's Middle Neolithic levels at Nabta Playa E-75-8 show an increase in the proportion of *Bos* bones to 7.7% (Smith 1986: 201). Further, the bones are smaller in size. The conclusion that cattle domestication spread relatively late is further reinforced by the central Saharan Fezzan and Tassili rock paintings

[4] Another argument the authors advance in support of an early domestication is the analysis of language by Christopher Ehret (1993). Ehret has bracketed a period around 8,000 B.C. for Proto-Northern Sudanic (part of the Nilo-Saharan language family), which includes words indicative of the exploitation of *Bos*. But such an analysis is risky as most linguistic roots describing wild or domestic *Bos* are neutral and the same words that later were applied to domesticated *Bos* may well have been originally applied to their wild ancestors.

which depict the transition from a hunter-gatherer lifestyle to a pastoralist way of life around 4,500 B.C. during the Neolithic Humid Phase (Muzzolini 1992).

The spatial distribution of artefacts in Wendorf & Schild's pre-Middle Neolithic layers (in this article redefined as being Epipalaeolithic layers) at Nabta Playa does not reveal sophistication beyond that of a tight hunter-gatherer band organisation. It should be noted that the ethos of hunter-gatherers is to share their food and not to accumulate unwarranted surpluses (Smith 1986: 200). The domestication of *Bos* would have initiated a new mode of production, primarily through the introduction of surpluses into their socio-economic lifestyle, with the potential for accumulating wealth and political status.

Cattle burials in Nabta Playa and the Nile Valley

On the western edge of the largest wadi to the north of Nabta is the first of two differing types of stone-covered tumuli marking the burial sites of cattle. Seven out of the nine tumuli examined have been excavated (Applegate et al. 2001: 468). At E-94-1n, at the northern end of the Late Neolithic ceremonial complex, the stones covered the articulated remains of a young cow in a clay-lined chamber. Radiocarbon tests on the wood from the roof returned a date of 5,400 B.C. The poorly preserved cow is around 125 cm. in height, with the spine oriented north-south and its head facing south (Applegate et al. 2001: 468).

The second type of tumulus consists of disarticulated *Bos* bones scattered between unshaped rocks. Sites E-94-1s, E-96-4, E-97-4, E-97-6 and E-97-16 are associated with the remains of 3, 4 (2 sub-adult, 2 young adults), 2 (1 juvenile, 1 sub-adult), 1 and 1 (sub-adult) bovines respectively (Applegate et al. 2001: 473-481). No particular body part was deliberately selected for deposition.

A similar emphasis on the cattle cult can be observed in Egypt. The Nabta cattle burials are paralleled somewhat by the Badarian animal graves. Certain animals, including *Bos*, were revered during Badarian times as witnessed by their burials in select sections of different cemeteries either on their own, or in association with human burials or within human graves (Brunton & Caton-Thompson 1928: 42, 91-94). This does not presuppose that each species was buried for the same purpose, since pattern variation within the burials is well documented by Flores (1999). Some graves had linen and matting which may have covered the animal (Brunton & Caton-Thompson 1928: 42). Cow remains are present in numbers at Hammamiya (Baumgartel 1955: 21-22). Human and *Bos* remains were also sometimes found buried together. The latter practice continued down into late Old Kingdom times, as evidenced by the ox burials at Qau. These later burials show signs that the cattle had been carefully dismembered before burial.

A *Bos* burial from Tomb 19 at Hierakonpolis Locality Hk6 was examined by Sylvia Warman (2000: 8-9). The tomb dates either to the end of Nagada I or to the beginning of Nagada II. Its occupant is a specimen of *Bos primigenius* (the wild ancestor of the domestic cow *Bos taurus*). Reed matting and resin were utilised in the burial of the whole corpse, as indeed they were with human burials in the same locality (Warman 2000: 8-9).

Remains of cattle have also been uncovered at localities Hk11, Hk29 and Hk29A at Hierakonpolis (McArdle 1992). Many of the *Bos* from Hk11 were of mature age, which is indicative of animal husbandry and has led McArdle (1992: 56) to believe "this reflects their use for purposes other than to supply meat (e.g. dairy products, draft animals, religious or social symbols)". Locality Hk29 also displays strong signs of animal husbandry but it is at Hk29A, a ceremonial complex dating to late Nagada II (Friedman 1996), where something unusual occurs in the faunal patterns. There is a discrepancy between the cranial and post-cranial age profiles, wherein the younger *Bos* crania are under-represented (McArdle 1992: 54, 56). The implication of ritual activities involving cattle at Hk29A is given additional weight by Friedman (1996: 30), who states that "representations [on

Predynastic seals and vessels] of fences topped with the impaled heads (mainly cattle) may explain the head to torso discrepancy among the faunal remains at HK29A".

From the Terminal Predynastic period, cattle burials have been found at the Nubian A-Group cemetery at Qustul (Williams 1986: 176) and again at Hierakonpolis. There is one burial at Hierakonpolis of particular interest, that of Tomb 7 in Locality Hk6 (Adams 2000: 33-34). By contrast with Nagada II tombs, this cattle burial grave has an almost square shape. It has a length of 2.5 m., a width of 2.1 m. and a depth of 0.65-0.75 m. It had been lined with stone slabs during the original construction (Hoffman 1982: 56). The community had placed grass matting over the bones of three dead animals, which were each buried in one piece. An intriguing aspect of this burial was the presence of a dark organic substance that sheathed a few of the bones. Hoffman (1982: 56) hypothesised:

> "Since this organic substance was associated with only the ribs and was tightly packed around them, the possibility exists that it was used to fill out the animal's eviscerated abdomen – a practice foreshadowing the mummification of later time."

It has a parallel in the earlier *Bos* burial from Tomb 19, mentioned above. This is the first known *Bos* burial triad. *Bos* triads are also known through representations on predynastic Nagada ceramics.

Fekri Hassan (1992) posits that the cattle beliefs of the dynastic Egyptians had their origins in the Saharan pastoralists in the predynastic period and stemmed from the Saharan pastoralists, which in essence is also the hypothesis of this article. However, Hassan proposes a fertility-women-cattle ideology, which in the view of the present writer is based on several unfounded assumptions. He assumes that only women were associated with fertility and the provisioning of the essential ingredients of life, water and food, but there is no solid evidence to back up this claim. He also suggests that it was the women amongst the Saharan pastoralists who herded the cattle, but the ethnographic records of, for example, the Nuer of the Sudan reveal that teenage males herd the cattle (Hoffman 1979: 242). According to Hassan's hypothesis, the predynastic male king drew power from his association with the female goddesses and the very status of women within Predynastic society – but there is a lack of data for the latter claim. An integral part of Hassan's (1992: 315) hypothesis are the Nagada female figurines shown with their arms curved above their heads in a posture which he has interpreted as representing the bovine horns of a mother goddess. However, Hassan does not account for the presence of male figurines, or the absence of figurines representing a mother and a son as would be expected in a "Mother Goddess" cult, or the lack of exaggerated features (breasts, buttocks) suggesting fertility and divinity, or alternative explanations of the curved arms, such as the invocation of a bird deity (Ucko 1968: 427; Maisels 2001: 46). In short, there is little reason to connect the cattle burials, or cattle symbolism in general, with a "Mother Goddesses" cult, although there certainly are important symbols that have to do with a cow goddess in relation to the king, as will be demonstrated.

Cow heads and the codification of religion

Whether the frequent occurrence of separated heads of cow goddesses (humanised or not), often on pillars or poles, in the Predynastic Nile Valley has any relation with the impaled cattle heads at Hierakonpolis (Hk29A) is hard to say. But it is certain that the cow's head seen front face, associated with a goddess, is an important bovine motif in early Egyptian art.

Dating to the Terminal Predynastic is a sculptured palette that has been found at Gerzeh (Baumgartel 1960: 90). Oval-shaped, it has one blank side with a flat relief covering the whole of the other side. The relief is of a cow's head reduced to a geometrical form with ears and horns curving outwards, all embellished with five stars. It is identical to the Hathor Bowl dating from the 1st Dynasty at Hierakonpolis (Burgess and Arkell 1958). The stars might suggest that the palette refers to the

Heavenly Cow and (Fisher 1962: 11) to the epithet "Mistress of the Stars", which in the later "Tale of Sinuhe" (B, 270-274) is applied to Hathor.

Also on the Narmer Palette, on either side of the top of the palette (as if looking down on the scenes from heaven), there are two frontal, humanised, cow heads. Fischer (1962; 1963) has argued for the identification of these heads with the goddess Bat. Bat was a local goddess of the seventh nome in Upper Egypt, where the *b3.t* fetish appears often on official pendants. One of the earliest written occurrences of the goddess' name is from the 6th Dynasty: "I [Menenre] am *B3t* with her two faces." (Pyramid Text §1096; Faulkner 1969). Fischer (1962: 11) speculates that the first occurrence of Bat's name is on a diorite vase excavated at Hierakonpolis and dated to the 1st Dynasty. The vase displays a human face with cow ears and horns. A *jabiru* stork was found near the vase. Fischer regards the vase and the *jabiru* stork (*b3*) as representing a hieroglyphic construction, so that together they make up *b3.t*. This word is the same as that represented on the shrine of Sesostris I at Karnak (Fischer 1962: 11). A gold amulet from the archaic period at Naga ed-Deir displays in it a pendant of the *b3.t* fetish. As the classic representation of Hathor is with an outward curving pair of horns, Fischer (1962: 12) argues that, if Bat were a later offshoot of Hathor, then she should be expected initially to have adopted Hathor's elegant and outward curving horns, instead of the heavy, ribbed, and inward curving horns that appear in some of the Predynastic cow heads. These heavy archaic horn forms are therefore ascribed to Bat, who supposedly lost this bovine horn structure over time, replacing it with antennae with spiral tips that likewise curl inwards. Based on these arguments, Fischer (1962: 11-12) concludes that the Predynastic representations of cow heads are that of Bat, who consequently also appears at the top of the Narmer Palette, and that Hathor is first mentioned in the 4th Dynasty, a time in which she becomes connected with the *b3.t* fetish.

A close examination of Fischer's points reveals frailties and circular arguments. The Hierakonpolis 1st Dynasty vase was found in uncertain association with the *jabiru* bird. The context in which Bat ("female *ba*") is mentioned in the Pyramid Texts is that of funerary items. Furthermore, Fischer's argument is critically flawed by a point he himself mentions (Fischer 1962: 13-14):

> "And even as late as the Eighteenth dynasty the term *b3.t* is still applied to the pendant worn by court officials. But a Middle Kingdom coffin gives the fetish itself the label 'human-faced,' and the use of this much less distinctive designation, which is later applied to the human-headed *b3*-bird, and so on, advises caution in regarding every appearance of the *b3.t*-fetish as a specific reference to the goddess Bat."

And Jonathan van Lepp (1998 & 2000: pers. comm.) suggests:

> "The inward curved horns on the Narmer Palette have antecedents in Predynastic art where they appear to be the form of surgical horn manipulation originally applied to bulls.[5] This motif is also applied to sheep . . . Thus, the inward curved horns are not species specific and cannot be considered the domain of Bat . . . [If Bat was] important, as the prominent positioning on the Narmer Palette indicates, then there should be some evidence of temples, cults, and priestesses to her . . . There is no evidence of [Old Kingdom] kings associated with temples belonging to Bat. None of the princesses, or noble women, at Memphis have this titulary. If the Goddess on the Narmer Palette is Bat, she should be manifested in some way among the royalty, as clearly there is an important attachment to the deity and the king."

[5] For a short survey of the different types of cow's horns in Predynastic art, see van Lepp 1999: 101-105.

Of interest in this regard is that, in the Armant rock art, as has already been noted, horn manipulation is a mark of early pastoralism in the Nile Valley.

The present author therefore prefers to identify the goddess depicted on the Narmer Palette as Hathor. Noteworthy is Pyramid Text §546, "My kilt which is on me is Hathor" (Faulkner 1969), which reminds one of the four cow heads on Narmer's kilt (van Lepp, 1988 & 2000, pers. comm.), identical to the ones on top of the Palette. A statue of Djoser displays a similar belt which Lana Troy (1986: 54) has identified as displaying Hathor's head. Fisher's idea that Hathor appeared relatively late can also not be maintained: a temple of Hathor at Gebelein dates from the late 2nd dynasty (Wilkinson 1999: 312). While Toby Wilkinson (1999: 283) is probably accurate in saying "it seems likely that, in this area, Egyptian theology was characterised by 'a common substratum of ideas which lent the two goddesses a somewhat similar character' (Fischer 1962: 12)", it is Hathor who is consistently connected with the pharaoh, not Bat. In the Valley Temple of Mycerinus there are triads of Hathor, Mycerinus and a nome diety. Hathor stands in the middle of the figures in two of the triads, with her arm around Mycerinus as if in a protective stance and showing that she is related to him.

The connection of the pharaoh with bovines is apparent from the beginnings of dynastic Egypt. On the Narmer Palette, a bull breaking down the enemy's fortifications symbolises the conquering pharaoh. The name Menes (*mni*) is usually linked with either the pharaohs Narmer or Hor-Aha (Wilkinson 1999: 66-68) and Fairservis (1992: 62) hypothesises it has a possible origin in *mniw*, meaning "herdsman". Hor-Aha's name is proof of a close relationship between the pharaoh and the god Horus, who is sometimes depicted in bull form (Wendorf & Schild 1998: 116). It is during the late Predynastic that Horus (and by extension the king) adopts the cow goddess as his mother, formalised in the form of Hathor (*ḥw.t-ḥr*, "House of Horus"). This event is also evidenced by the dedication of the Narmer Palette in the temple of Horus at Hierakonpolis. The Narmer Macehead has many features in common with the ceremonial complex Hk29A (Friedman 1996: 33) and also displays the pharaoh in association with bovines.

The end products of the codification of such traditions are commemorative objects like the Narmer Palette, which built on the works of earlier commemorative pieces. It was a combination of formal commemorative hieroglyphic writing with the basic iconography of the evolving kingship.

Conclusion

Bos primigenius, originally only hunted, became domesticated in order to protect a valuable source of fat in the hunter-gatherer diet and to enhance chances of survival during changing environmental conditions. At Nabta Playa in the Western Desert, evidence of the domestication of cattle dates from the Middle Neolithic. This brought about socio-economic changes within the desert communities, which is later reflected in the Late Neolithic cattle tumuli and megalithic constructions at Nabta Playa. The *Bos* tumuli are indicative of cattle worship, and the Late Neolithic site as a whole displays evidence of a community with greater social complexity than its contemporaries in the Nile Valley. Prolonged contact with desert pastoralists led to the first socially complex society in the Nile Valley, the Badarian. It introduced a new religious and socio-economic element into the life of the Upper Egyptians, namely ownership and burial of domestic cattle. *Bos* burials are found in Nagada period settlements, in clearly ceremonial contexts. As pastoralism became increasingly fused in the Nile Valley economy with agriculture, religious associations evolved between the cow goddess and the king. These aspects became codified in the artefactual representations dating from the time of Unification.

Michael Brass
mike@antiquityofman.com
Oxford, U.K.

References

Adams, B. 2000. *Excavations in the Locality 6 Cemetery at Hierakonpolis: 1979-1985.* British Archaeological Reports 903. Oxford: Archaeopress

Anderson, W. 1992. Badarian Burials: Evidence of Social Inequality in Middle Egypt During the Early Predynastic Era. *Journal of the American Research Center in Egypt* 29: 51-66

Applegate, A., A. Gautier, and S. Duncan. 2001. The north tumuli of the Nabta Late Neolithic Ceremonial Complex. In Wendorf, F., R. Schild and Associates (eds.) *Settlement of the Egyptian Sahara.* Volume 1: *The Archaeology of Nabta Playa.* New York: Kluwer Academic

Bard, K. 1987. *An Analysis of the Predynastic Cemeteries of Nagada and Armant in Terms of Social Differentiation.* University of Toronto: Unpublished PhD Dissertation

Bard, K. 1988. A Quantitative Analysis of the Predynastic Burials in Armant Cemetery 1400-1500. *Journal of Egyptian Archaeology* 74: 39-55

Baumgartel, E. 1955. *The Cultures of Prehistoric Egypt* (Vol. I). London: Oxford University Press

Baumgartel, E. 1960. *The Cultures of Prehistoric Egypt* (Vol. II). London: Oxford University Press

Brunton, G. & G. Caton-Thompson. 1928. *The Badarian Civilisation and Predynastic Remains near Badari.* London: British School of Archaeology in Egypt

Burgess, E. & A. Arkell. 1958. The Reconstruction of the Hathor Bowl. *Journal of Egyptian Archaeology* 44: 6-11

Edwards, I.E.S. 1993. *The Pyramids of Egypt.* London: Penguin

Ehret, C. 1993. Nilo-Saharans and the Saharo-Sudanese Neolithic. In Shaw, T., Sinclair, P., Andah, B. & Okpoko, A. (eds.) *The Archaeology of Africa: Food, Metals and Towns.* London: Routledge

Fairservis, W. 1992. The Development of Civilization in Egypt and South Asia: A Hoffman-Fairservis Dialectic. In Friedman, R. & Adams, B. (eds.) *The Followers of Horus: Studies Dedicated to Michael Allen Hoffman 1944-1990.* ESA Publication No. 2. Oxford: Oxbow Monograph 20

Faulkner, R. 1969. *The Ancient Egyptian Pyramid Texts.* Oxford: Oxford University Press

Fischer, H.G. 1962. The cult and nome of the goddess Bat. *Journal of the American Research Center in Egypt* 1: 7-18

Fischer, H.G. 1963. Varia Aegyptiaca. *Journal of the American Research Center in Egypt* 2: 17-51

Flores, D. 1999. *The Funerary Sacrifice of Animals during the Predynastic Period.* Ph.D. Dissertation. Toronto: Department of Near and Middle Eastern Civilizations, University of Toronto

Friedman, R. 1996. The Ceremonial Centre at Hierakonpolis Locality HK29A. In Spencer, J. (ed.) *Aspects of Early Egypt.* London: British Museum Press

Gautier, A. 1993. Faunal remains from Site E71K12 and E71K13 near Isna in Upper Egypt. In Mohamed, A. *Two Upper Pleistocene Kill-Butchery Sites in the Nile Valley of Egypt*, pp. 333-344. Ph.D. dissertation. Dallas: Southern Methodist University

Hanotte, O., D. Bradley, J. Ochieng, Y. Verjee, E. Hill, and J. Rege. 2002. African Pastoralism: Genetic Imprints of Origins and Migrations. *Science* 296: 336-9

Hassan, F. 1988. The Predynastic of Egypt. *Journal of World Prehistory* 2(2): 135-175

Hassan, F. 1992. Primeval Goddess to Divine King: The Mythogenesis of Power in the Early Egyptian State. In Friedman, R. & B. Adams. (eds.) *The Followers of Horus: Studies Dedicated to Michael Allen Hoffman 1944-1990.* ESA Publication No. 2. Oxford: Oxbow Monograph 20

Hoffman, M.A. 1979. *Egypt Before the Pharaohs.* New York: Barnes & Nobles

Hoffman, M.A. 1982. *The Predynastic of Hierakonpolis – an interim report.* Giza and Macomb, Illinois: Cairo University Herbarium and Western Illinois University. Egyptian Studies Association Publication No. 1

Holmes, D. 1989. *The Predynastic lithic industries of Upper Egypt: A comparative study of the lithic traditions of Badari, Nagada and Hierakonpolis.* BAR (15) 469. 2 vols. Oxford: Archaeopress

Maisels, C. 2001. *Early Civilizations of the Old World: The Formative Histories of Egypt, the Levant, Mesopotamia, India and China.* London: Routledge

Majer, J. 1992. The Eastern Desert and Egyptian Prehistory. In Friedman, R. & Adams, B. (eds.) *The Followers of Horus: Studies Dedicated to Michael Allen Hoffman 1944-1990.* ESA Publication No. 2. Oxford: Oxbow Monograph 20

Malville, J., F. Wendorf, A. Mazar & R. Schild. 1998. Megaliths and Neolithic astronomy in southern Egypt. *Nature* 292:488-491

McArdle, J. 1992. Preliminary Observations on the Mammalian Fauna from Predynastic Localities at Hierakonpolis. In Friedman, R. & Adams, B. (eds.) *The Followers of Horus: Studies Dedicated to Michael Allen Hoffman 1944-1990.* ESA Publication No. 2. Oxford: Oxbow Monograph 20

Midant-Reynes, B. 2000. *The Prehistory of Egypt: From the First Egyptians to the First Pharaohs.* Oxford: Blackwell Publishers

Muzzolini, A. 1992. Dating the Earliest Central Saharan Rock Art: Archaeological and Linguistic Data. In Friedman, R. & Adams, B. (eds.) *The Followers of Horus: Studies Dedicated to Michael Allen Hoffman 1944-1990*. ESA Publication No. 2. Oxford: Oxbow Monograph 20

Muzzolini, A. 1993. The emergence of a food-producing economy in the Sahara. In Shaw, T., Sinclair, P., Andah, B. & Okpoko, A. (eds.) *The Archaeology of Africa: Food, Metals and Towns*. London: Routledge

Nelson, K. 2001. The pottery of Nabta Playa: A Summary. In Wendorf, F., R. Schild and Associates (eds.) *Settlement of the Egyptian Sahara*. Volume 1: *The Archaeology of Nabta Playa*. New York: Kluwer Academic

Smith, A. 1986. Review article: *Bos* domestication in North Africa. *African Archaeological Review* 4: 197-203

Stockstad, E. 2002. Early Cowboys Herded Cattle in Africa. *Science* 296: 236

Trigger, B. 1983. The Rise of Egyptian Civilization. In Trigger, B., Kemp, B., O'Connor, D. & Lloyd, A. (eds.) *Ancient Egypt: A Social History*. Cambridge: Cambridge University Press

Troy, L. 1986. *Patterns of Queenship in Ancient Egyptian Myth and History*. Uppsala: University of Uppsala. Acta Universitatis Upsaliensis. Boreas. Uppsala Studies in Ancient Mediterranean and Near Eastern Civilizations 14

Ucko, P. 1968. *Anthropomorphic Figurines of Predynastic Egypt and Neolithic Crete with Comparative Material from the Prehistoric Near East and Mainland Greece,* London: Andrew Szmidla

van Lepp, J. 1999. The Misidentification of the Predynastic Egyptian Bull's Head Amulet. *Göttinger Miszellen* 168: 101-111

Warman, S. 2000. How Now, Large Cow? *Nekhen News* 12: 8-9

Wendorf, F., A. Close & R. Schild. 1985. Prehistoric settlements in the Nubian Desert. *American Scientist* 73: 132-141

Wendorf, F. & R. Schild. 1994. Are the Early Holocene Cattle in the Eastern Sahara Domestic or Wild? *Evolutionary Anthropology* 3(4): 118-128

Wendorf, F. & R. Schild. 1998. Nabta Playa and Its Role in Northeastern African Prehistory. *Journal of Anthropological Archaeology* 17(2): 97-123

Wendorf, F., R. Schild, et. al. (eds.). 2001. *Settlement of the Egyptian Sahara.* Volume 1: *The Archaeology of Nabta Playa.* New York: Kluwer Academic

Wetterstrom, W. 1993. Foraging and farming in Egypt: the transition from hunting and gathering to horticulture in the Nile Valley. In Shaw, T., P. Sinclair, B. Andah, & A. Okpoko (eds.) *The Archaeology of Africa: Food, Metals and Towns*. London: Routledge

Wilkinson, T. 1999. *Early Dynastic Egypt*. London: Routledge

Williams, B. 1986. *Excavations between Abu Simbel and the Sudan Frontier. The A-Group Royal Cemetery at Qustul L.* Chicago: The Oriental Institute of the University of Chicago. Oriental Institute Nubian Expedition III

Preliminary Report on the 2000 Poznan Symposium

Juan José Castillos

International Symposium: "Cultural Markers in the Later Prehistory of Northeastern Africa and Our Recent Research"
Poznan, Poland, August 29 - September 2, 2000
Organizers: Prof. Lech Krzyzaniak and Prof. Michal Kobusiewicz
Ca. 50 attendees

This Symposium lasted four days, from August 29th to September 2nd, 2000, and the papers dealt mainly with the Western Desert, Predynastic Egypt and the Sudan. Due to airline problems, I arrived one day late for the Symposium, so I will not comment on most of the papers of the first day which I could not attend; for the few that are included I thank the speakers for giving me a written copy of their presentation. In one case a paper can be included because it was almost the same as the corresponding paper delivered at the Cairo Congress.

This Preliminary Report consists of my personal notes on the papers I attended and reflects my personal perceptions and interests. It is published here in anticipation of the official Proceedings[1], in the hope of thus bringing the topic of the Later Prehistory of Northeastern Africa to the attention of a larger audience.

K. Nelson, "Chronology and technology: The pottery of Nabta Playa (Egypt)"

The ceramic chronology at the sites of Nabta Playa spans the entire Neolithic sequence. From the onset of ceramic production in the desert to the latest sites predating the Predynastic, ceramic vessel form, construction and distribution are defined by changing ceramic technology and use. The first ceramics of North Africa are among the first pottery to appear in the world, second only to early pottery in Japan as part of the Jomon culture dating to 12,700 B.P. The earliest date for ceramics from Nabta is approximately 9,000 B.P. The pottery of Nabta Playa provides not only a chronology of types but indicates changes in subsistence and a changing relationship with the Nile Valley. Late Neolithic pottery is, for the first time in the Nabta sequence, similar to types found to the north and along the Egyptian Nile. The onset of the Late Neolithic, the arrival of herders, the differential use of cattle, seen in the changes of hearths and the first evidence of roasted cattle bone along with pottery similar to if not the same as Badarian pottery, provide evidence of this transition. With further study including the relationships of types, use wear analysis and further chemical studies, the function and distribution of pottery may reveal the dynamics of the people that once occupied Nabta Playa.

J. Linstädter, "Middle and Late Neolithic in the Wadi Bakht (Gilf Kebir)"

Assuming that the symposium title "Culture Markers" also means "Chronological Markers", the speaker contributed to the chronological discussion of the Neolithic Phase in the Eastern Sahara. As a result of several years of archaeological fieldwork in Northern Africa, he has noticed that if we do chronological work with just a single artefact type or a ceramic decoration, this feature – let's call it a marker – can be connected to different archaeological material coming from very different periods. To get a more precise image of an archaeological group or culture, we should add as many features as possible to our description, such as stone tool production sequences, mobility patterns, and other

[1] For the Poznan Archaeological Museum and its publications, please see this URL: http://www.muzarp.poznan.pl/muzeum/eindex.html

indications of land use strategies. To illustrate this idea, the speaker presented the outlines of recent work in the Gilf Kebir. Over the past 20 years the upper reaches of several wadis have been subject to archaeological research by scientists of the University of Cologne. By combining these single features we get a first impression of the Middle Neolithic inhabitants of our study area. During the rainy season the groups always came back to their places on the barrier dune, where the large amount of artefacts shows an extensive settlement. The plateau was only used on its edges for the acquisition of raw materials. Several indications suggest a hunter-gatherer economy. Late Neolithic people always had their camps on the Playa surface. These camps can be identified as remains of a single stay. Earlier camps might be covered by the Playa deposits. The Plateau was integrated in the subsistence activities, even far from its edges. The reason may have been a different precipitation regime or a change of subsistence strategies. The existence of domesticated lifestock is proved. The field work in Wadi Bakht region will continue this year [i.e., 2000/2001]. Besides an enlargement of the survey area on the plateau, the main aim will be the dating of the plateau sites and their assignment to one of the two main phases. The preliminary results have to be proved and the team's hypotheses have to be tested. So it needs to be checked to which period the stone structures belong, or whether the fauna and the tool kit indicate a hunter-gatherer economy in the Middle and a pastoral society in the Late Neolithic. And, if there was a change of subsistence strategy, how, when, and why it took place.

R. Kuper, "Through Abu Ballas to the back of the beyond: Pharaonic tracks in the Libyan Desert"

Around 6,000 B.C. there were many human settlements in the Western Desert and none in the Nile Valley. By 4,000 B.C. the situation had drastically changed, with settlements in various parts of the Nile Valley showing the beginning of agricultural activity, and none in the Western Desert. It would seem that by then the situation had become similar to what we find later in Pharaonic texts, that is, that life ceased outside the oases. What a Hungarian traveller (the hero of the recent film "The English Patient") had speculated in the first half of the XXth century, that is, the existence of a ancient commercial route going west from Ballat, through the oases and Gilf Kebir, has now been found to be correct as more settlements along this route have been located. Old Kingdom and later pottery was found, along with occasional inscriptions indicating that a certain Egyptian official had passed on his way "to meet the inhabitants of the desert". A relief was also found representing an ass. Curiously, the Old Kingdom pottery is a good example of the contemporary manufacture but the New Kingdom one is of bad quality, the actual term used by the speaker was "rubbish". The expedition took excellent aerial photographs by means of a simple device consisting of flying a kite with a digital camera hanging from it with a wire to trigger it from below. This system enabled them to photograph objects as small as 5 cm. long, which were clearly identifiable. In other places they found even older objects, and in some sites there were stone circles. Among the finds were a stone stela mentioning the god Amun and a satirical engraving representing a cat and a mouse made on the rock wall. Near Ballat the expedition found another engraving representing the god Seth.

Discussion: To the question of what the ancient Egyptians could want from such an inhospitable and remote area of the desert situated so far west, the speaker replied that it was probably not a commercial route but in fact a military one used during the New Kingdom in order to transfer troops without being detected by the enemy.

K. Kindermann, "New investigations into the Neolithic settlement system of Djara (Abu Muharik Plateau, Western Desert)"

An old caravan route connecting the Farafra Oasis with Assiut goes through the Djara area. One of the subjects of the speaker's research is the conglomeration of Neolithic sites there, which were first visited by a German explorer of the XIXth century, G. Rholfs, who went from Assiut to Farafra. Djara was rediscovered in 1989 by Carlo Bergmann who found Neolithic stone artefacts around the entrance to the cave and noticed the presence of rock engravings. The greater part of the Djara archaeological sites situated between the Bahariya and Kharga oases, falls into the Middle Neolithic (6,400 - 5,400 calC14 B.C.). The Abu Muharik Plateau, like other parts of the Western Desert during the Early and Middle Holocene, was not an environment in which it was easy to live and survive. Vegetation was sparse and concentrated around ephemeral lakes. In the Western Desert many Holocene sites were also discovered near playa sediments. The special attraction of the Djara region consists of the abundance of easily available raw material such as flint and the especially favourable environmental circumstances. After 5,400 B.C. the occupation seemed to have broken off. Possibly the drying trend at the end of the 6th millennium, which interrupted the relatively moist mid-Holocene, forced people out of this region as it did other such groups from other parts of the Western Desert, towards more fertile lands such as the Egyptian oases or the Nile Valley. Cultural markers (distinct key forms of prehistoric periods) at the site are arrowheads, side-blow flakes and bifacially retouched knives, like the "Gerzean flint knives" of Egypt.

J. J. Castillos, "Social development in Predynastic Egypt: a study of cemeteries in the Badari area"

Using a methodology borrowed and adapted from quantitative sociology, the speaker found out that a different evolution could be determined from the cemetery data for each of the major sites in northern Upper Egypt at Matmar, Mostagedda and Badari. The cemeteries at Matmar appear to have declined in social differentiation during the Protodynastic, a phenomenon not uncommon in other parts of Predynastic Upper Egypt and which seems to have affected out-of-the-way settlements, far from the political and economic centres at the time. At Mostagedda, the evolution appears as, what we might call, the most "normal", with a continuous increase in wealth and social differentiation. At Badari itself, which was, according to the speaker, the probable centre of the Badarian culture, the natural resistance of such a privileged location to accept the role of a provincial outpost under later cultures caused a less clear development. In social inequality, Badarian levels were only surpassed during the Protodynastic. The speaker also mentioned that in all the probable élite tombs that could be identified in these cemeteries at Badari itself, not one occupant was a woman, men apparently being privileged in this respect.

Discussion: Some of the scholars in the audience pointed out that in their excavations in Sudan and in Lower Egypt, they had found a very different picture, with some obviously élite tombs being occupied by women. The speaker replied that his findings did not imply that elsewhere the situation could not be different, and also that in many cases the occupant of large or rich tombs in the Badari area could not be identified, so his results were tentative. It remains strange, though, that if the general practice was as mentioned by some of the members of the audience, none of the cases in which the sex of the occupant could be determined agreed with findings elsewhere. From his previous research, the speaker could infer that, although the picture in this respect is fairly complex, men and women were on the whole treated in a roughly equal manner in the Predynastic cemeteries. But it would appear that women seem to have been favoured in the wealth of the tombs in early Predynastic times, while men saw their status improved as time went by.

S. Hendrickx, "The Badarian living site Mahgar Dendera II and the distribution of the Badarian in Upper Egypt"

This site was excavated 10 years ago by the speaker and B. Midant-Reynes. It was a rescue excavation because half of the site had already disappeared due to modern human activity. Now nothing remains of this site. Under a top layer of sand, the gravel was found to contain archaeological material, among which were graves without matting. Only large post holes belonging to the settlement could be identified because the smaller ones would not have gone deep enough into the gravel to be recognizable. The excavators found Badarian pots and sherds, hearths that could be dated to about 6,480 B.P. (4,300 calC14 B.C.), and milling stones that were permanently installed on the ground. Several areas had storage pits. The big post holes indicated the presence of large structures but these could not be precisely defined. Pottery was fairly rare in this site, lithics were much more frequent – the opposite of what has been found elsewhere. Some evidence of the so-called Tasian pottery, which also appears occasionally in Badarian contexts, was found. The lithics were studied by B. Midant-Reynes; the stone used was local, half of the tools were borers, but bifacial tools such as axes were also found, common in Naqada I and II contexts but also in Badarian ones. Agriculture does not seem to have been important in this site, for flints that had to do with it were comparatively rare. The lithics seem to have been meant for cutting wood and other perishable items. Faunal remains analysis indicated seasonal occupation of the site. The occupants would take their flocks to the Nile. A permanent settlement was not found, only this apparently seasonal one. From all the evidence the excavators could infer a sequence of land use during the whole year, before, during and after the annual flood. The occupants would fish in the Nile with nets, and slaughter young goats while the water was receding. This site offers a glimpse of life in Badarian times. The speaker pointed out that Tasian (Badarian related) items were also found in the desert west of Luxor, at Dakhla Oasis, and in the Eastern Desert from Matmar to Hierakonpolis.

Discussion: K. Kroeper objected to the use of C14 dates because in this period (around 5,000 B.P.) it leads to ranges of plus / minus 300 years. Only calC14 dates plus / minus 50 years could be useful. Some people have even quoted such C14 dates that would seem to agree with their pet theories, which is not an acceptable procedure. P. Vermeersch said that at this time period the calibration curve is quite flat which makes determinations difficult.

L. Watrin, "Five or six lost generations: what was going on in the Delta after Maadi and before Naqada"

Buto I has been dated by C14 to around 5,000 B.C. Maadi, contemporary with Naqada Ib to Naqada IIc,d, shows marked Palestinian influences, and was dated to 3,800 - 3,500 B.C. (F. Hassan). The semi-underground structures found there are related to EB I Palestinian ones. The rest of the Maadi material is related to Upper Egyptian artefacts (fish-tail blades, maceheads, etc.). Buto IIa seems to be contemporary with Naqada IIa and with the closing of Maadi. The current excavations at Tell el Farkha seem most promising because they are bringing up evidence of the material culture that would fill the gap between the end of Maadi and early Naqada penetration in the north.

M. Chlodnicki, "Recent research at Tell el Farkha"

The structures found here by the Italians in the late eighties, just north of the modern village, were of Old Kingdom date, with the earliest being Predynastic, contemporary with Buto. Magnetic surveys have indicated the underground presence of brick walls which stretch far beneath the modern village. In the Early Dynastic layers, structures (brick walls) orientated in a NE-SW direction were found as well as other round ones, probably ovens.

K. Cialowicz, "Tell el Farkha 2000: Excavations at the Western Kom"

The strata seem to confirm the replacement of Delta artefacts by Upper Egyptian ones. The latest layer of the Western Kom seems to date to the First Dynasty. The excavators found the remains of the rounded corner of a brick wall structure and some badly preserved remains of the floor. The wall was built in two ways, an inner layer consisting of yellow bricks and an outer one consisting of grey bricks. The orientation of the building was the normal one at Tell el Farkha, that is, NE-SW. The building seems to have had rectangular chambers divided by thin walls. This was a large Naqada II structure that at some point was destroyed and abandoned. Another structure found seems to have been a brewery and it would be the oldest such building found so far in the Delta. The building was erected using regularly placed D-shaped bricks (20 cm. by 10 cm.). This type of brick seems to have been used from the earliest periods in this site. Inscribed objects and seal impressions as well as fragments of statuettes were also found.

J. Kabacinski, "Lithic industry from the Predynastic and Early Dynastic settlement at Tell el Farkha (Nile Delta)"

The evidence indicates that the tools were used but not manufactured at the site since there is no evidence of such work. Three basic types of raw materials were found: semitransparent flint, opaque flint of the normal colour, and a variety of flint that is almost black. The manufacturing technique used was pressure. Two basic types can be found: denticulated blades to be inserted into handles and probably used to cut grass or other plants since they have the gloss that indicates such a use, and bi-truncated blades without gloss, the use of which is still unknown. About 70% of all flints were sickle blades. Other tools found are bi-truncated blades at the base with a pointed top, probably used as perforators (burins). Bifacial flints (mostly broken) and a fragment of a flint bracelet were also found. A small proportion of the tools seems to have been manufactured using techniques other than pressure, so it seems that there were basically two technological approaches present in the tools found here.

A. Maczynska, "Lower Egyptian culture at Tell el Farkha: Preliminary Report"

The excavators found some large pits about 3 m. in diameter, and a number of smaller pits, probably post holes. Several types of pottery were found; about 92% were rough ware of a brownish colour (brownish red or brownish grey), and 4% were red slip ware. Many of the pots were decorated with incised geometric, continuous or dotted, zigzag patterns.

M. Jucha, "Research on Predynastic and Early Dynastic pottery from Tell el Farkha"

Some of the pots found at the site were broken and were probably used to dry grain. Decorated ware consists of the so-called "water lines", typical of Naqada III. In phases IV and V the excavators also found W ware. Some of those cylindrical pots (Naqada IIIa2) had degenerated wavy handles made by pushing the clay while soft. Evidence from before the Early Dynastic strata shows an early occupation by people with a Lower Egyptian tradition contemporary with Naqada II.

B. Gabriel, "Cultural relics as Saharan landscape elements"

Stone formations in the desert, which an untrained eye would miss altogether, were mostly hearths used by Neolithic herders. Some geomorphologists have defined them as natural formations, but since the times of Caton-Thompson's work at Kharga Oasis and Fred Wendorf's work in the Western Desert, they have been identified as archaeological features. Many of these formations are being destroyed by traffic through the desert routes. A number of clusters of pits (about 3 m. wide and 0.5 m. deep) can be seen in the Western Desert (for example, at the latitude of Malawi or at Dakhla) and they cannot be explained as natural formations. They have different shapes (oval or more or less rectangular) and have been explained by some as Palaeolithic mining pits for the procurement of different types of raw materials. Isolated stone boulders in the desert seem to have been the result of human activity as well. Even waste products of today may become the relics of tomorrow, as testified by the remains of Second World War campaigns scattered in the desert or even by discarded trucks by the road.

B. Barich, E. Garcea, C. Conati Barbaro, C. Giraudi, "The Ras el Wadi sequence in the Jebel Gharbi and the Late Pleistocene cultures of Northern Libya"

A systematic surface collection permitted assembling collections of Aterian points and Levallois tools. In an area of about 100 square metres a high density of objects could be recovered including Epipalaeolithic tools. The high percentage of bladelets probably indicates plant collection activities. Two major sources of flint were located in this area. It seems that the Aterian was later here than in other Saharan sites, followed by Epipalaeolithic and other later stages.

B. Keding, "Environmental change and cultural development in the Middle Wadi Howar (Northwestern Sudan)"

The first detectable activity at the site is dated around 4,000 calC14 B.C. The land had previously been too swampy for human occupation. Some authors have overemphasized either environmental elements, social elements, or the context (contacts with other people) as determining change, but the speaker tried to integrate these different approaches into a comprehensive one. A climatic deterioration can be detected until aridity set in permanently in about 4,000 B.C. A large number of lithics related to plant gathering and plant processing activities have been found, but also big game hunting took place. No evidence of animal domestication could be found. So far, the evidence indicates that the environment at the time here was a savanna type. The population consisted of hunter-gatherers and the transition to animal domestication took place between the 4th and the 2nd millennium calC14 B.C. It seems that domestication reached the site from two sources: from more westerly Saharan sites and from the Nile Valley. As this process advanced, the settlements became smaller as people became pastoralists. Finds of cattle bones and almost complete pots indicate the existence of ritual ceremonies having to do with cattle, the source of this people's main livelihood. We can observe the second change in the main activity of this people, from cattle pastoralism to the keeping of small livestock. At about 1,000 calC14 B.C., cattle, sheep, and goats are being kept, a diversification more suitable for an environmental change towards even greater aridity. The climatic change cannot be considered the main stimulus for the change into pastoralism here because the climatic change was not drastic enough, but the increasing aridity must have encouraged change. The transition seems rather to have been triggered by socio-cultural changes due to decreasing usable land, changes in group size, and a tendency to sedentarization or semi-sedentarization.

Discussion: To a suggestion that change could have been due to the intrusion of new people rather than to people changing their ways, the speaker replied that in her opinion that was not the case because she could not find a clear break in the material remains such as the pottery. L. Krzyzaniak asked if the cattle bones were related to human burials. The speaker replied that this was not the case so far, but since modern ethnographic practice (Dinka) is to bury people and cattle bones simultaneously in different places, a similar practice might have taken place here, although no evidence has been found yet. P. Vermeersch asked if there was any indication of warfare, or were these people peaceful? The speaker replied that having found so few burials she can not possibly say one way or the other, but she feels there was at the time enough space for everyone and therefore warfare might not have been necessary. J. J. Castillos remarked that since too many interpreters resorted too quickly to war and conquest to explain change in human cultures in the past, it is only reasonable that nowadays archaeologists are reticent to explain change that way unless there is compelling evidence.

Ph. van Peer, "Survey of Palaeolithic sites on Sai Island (Northern Sudan)"

The speaker found many sites of quartz lithic implements, many of them stratified. He confined himself to one special site among those found: an area of 30,000 square metres with three principal layers, one of Middle Palaeolithic date mixed to some extent with Neolithic tools, and two other layers. Over the underlying Nubian sandstone he found a layer with stone boulders associated with Acheulian tools. The expedition tried, sometimes successfully, to reassemble the original nucleus by re-grouping the tools, but with quartz this is a difficult task. Then they found a break represented by a layer with boulders associated with re-worked Acheulian tools and over this, another layer with stones which the speaker called grinding stones, one of the sides of which was smooth (polished). On top of that there was a layer of black silt and then a layer of Middle Palaeolithic occupation. The raw material (quartz) was brought to the site as nuclei that were processed in situ. The sites seem to have been normal habitation sites like those found in later Nubian phases elsewhere. On top of this there was a layer which the speaker called "Aterian" but that was very deteriorated. Many boulders over this top layer may have helped prevent further erosion of this site. He pointed out that in his opinion the origin of the Aterian should be found in Sub-Saharan Africa.

M. Lange, "Finds of the A-Group from the Eastern Sahara and the chronological development of the ceramic in the Laqiya region (Northwestern Sudan)"

The site under discussion could be dated to 3,200 to 3,100 calC14 B.C. A domestic cow skull dated to 3,000 calC14 B.C. shows that these people were pastoralists. The skull was found at a certain distance from other remains and was not associated with them. The speaker described the ceramics from this site that were decorated with several incised patterns and that later deteriorated when the site was resettled after temporary abandonment. A large sandstone palette was found, similar to some of those bird-shaped ones found at Naqada.

B. Gratien, "Palaeogeomorphology and human occupation in the northern Wadi el Khouri (Sudan)"

A cemetery covering an area of about 100 metres by 100 metres was found, with Neolithic tombs that were in very bad condition because they had been disturbed. Wind erosion was very strong, and it was estimated that about 60 to 80 cm. of upper layer had been lost due to it, affecting a Kerma, pre-Kerma and early Kerma settlement. In a layer datable to about 2,000 B.C. (Middle Kerma), the expedition found the first structures of irregular rectangular shape with granaries surrounded by post holes belonging to a fence around it all. They also found rectangular houses with a hearth, a 4 m. wide structure that probably was a sanctuary or shrine, the remains of large farms with houses that were rebuilt many times, a large courtyard, kitchen outside the house, etc. The site was abandoned in about

1,500 B.C. judging by the pottery sequence. Among the discoveries were also large Classical Kerma structures, built with strong brick walls that reveal that the builders were acquainted with contemporary Egyptian building techniques. The upper level dates to the 18th Dynasty. They found houses of Kerma type, kilns, etc. The material is late Kerma and Egyptian. Scarabs of Tuthmosis III and typical contemporary Egyptian pottery were also found.

I. Takamiya, "Egyptian pottery in A-Group cemeteries, Nubia: Towards an understanding of pottery production and distribution in Predynastic and Early Dynastic Egypt"

The speaker undertook quantitative analysis of the Naqadian pottery distribution patterns as compared to A-Group Nubian pottery, mainly from funerary contexts. She selected about 30 A-Group cemeteries suitable for her study, which were divided into three groups: Naqada II, Naqada II-III and Naqada III. She used Hierakonpolis as the Naqada assemblage most suitable for this study because the periods are well represented there. She found the pottery distribution in contemporary A-Group and Hierakonpolis to be mostly in agreement in the case of Hard Orange Ware (HOW), abundance in the latter site indicating capacity to supply. The distribution centre for A-Group people was probably situated at Elephantine because this site dates back to Naqada II when HOW started to be produced in quantities suitable for export. For Naqada II, the study showed a linear system of exchange based on simple transfer of products. In Naqada III, there were two systems of distribution, one for wavy handled vases and another for wine jars.

Discussion: One objection was raised by H.-Å. Nordström, namely that in this study only intact tombs should have been used because it was a common practice to rob Egyptian pots for later re-use, a phenomenon that would distort the picture. The speaker admitted this was possible.

M. Honegger, "Neolithic and Pre-Kerma occupation at Kerma"

The work of the speaker was confined to the area around the third cataract, covering a time span between the 5th and the end of the 4th millennium B.C. Most of the sites were in bad condition due to wind erosion. The excavation now comprises more than 10,000 square metres and hundreds of pits measuring about 2 m. in diameter, containing jars or potsherds, grinding stones, and, rarely, cattle remains. Their function was storage of foodstuff or liquids. There are also many post holes belonging to structures that measured from 1 to 7 m. in diameter. A very common distance was 4 m., which most probably indicates houses. According to contemporary practice, branches were intertwined with the posts to enclose those structures which were probably workshops or the houses of important people. Two rectangular buildings were also identified, one of them apparently important because it was rebuilt three times with exactly the same design. The second had thicker post holes. Post holes belonging to palisades to divide dwelling areas were also found. The total area of this site is of about two hectares. The western part was eroded by the wind until the lower Neolithic layers were exposed on the surface. Other areas were destroyed by Middle Kerma tombs or other structures. These Pre-Kerma agglomerations appear to be more complex than those of modern times. The pottery had pointed bottoms (sometimes flat) and wide mouths; no Egyptian imports were found here. A Neolithic settlement dating back to 4,500 B.C. was surveyed, and post holes around oval structures were found. The design does not seem as coherent as the Kerma one. The remains of caprines and other domestic animals were found. The pots were globular in shape and had rippled decoration. Fluctuations in the course of the Nile seem to indicate that at the time it flowed closer (towards the east) to the structures described here. The radiocarbon dates agree with the expected dates if charcoal is used, but for mummified tissue and bone remains the dates appear to be more recent because of some sort of pollution that the speaker has not been able to explain yet.

L. Chaix and J. Hansen, "The 'Bent Horned' cattle from Kerma (Sudan): A cultural marker?"

During Kerma times it is possible to observe a progressive decrease in the number of cattle kept by these communities, until in Late Kerma cattle are comparatively rare while caprines and sheep increase in numbers to fill the gap. The tombs found in a tumulus consist of the burial chamber with a wooden bed and caprine and sheep bones. The central burial depression is surrounded by buried bucrania (cattle skulls). Large numbers, even thousands, of bucrania are found in the tombs of important people buried in the immediate vicinity of the people. These bucrania were buried in Early Kerma with the nasal bones, in Middle Kerma with the lower part of the skull cut in half, and in Late Kerma only the horns and the top of the skull are present. The skulls were buried facing the dead human being. Some cases of deformed cattle horns were found in which they had been deformed by mechanical means while the animal was alive, starting with young calves. These constitute about 15% of the total number of bucrania. The idea was to make the horns parallel to each other. Female animals seemed to predominate with about 70% of the total. This practice has been observed in most of North Africa, apparently arising from the central Sahara. The first examples of this practice appear in Tassili (7th millennium B.C.) and decrease in time. A similar practice can still be seen among contemporary people living near the Ethiopian border, in which case the horns of the animal are tied with ropes.

Discussion: R. Kuper asked the speaker if there were other types of deformed horns (both falling down, one deformed and the other normal, etc.), as found in the Sahara rock art. The speaker said that such odd deformities were present in his sites. He was also asked whether the horns were ornamented as can also be seen in the rock art, and he replied that no examples of this could be found among his bucrania. M. Kobusiewicz asked after the reason for the progressive decrease in the number of cattle. The speaker replied that it was most probably a consequence of the aridification process as time went by and cattle became a precious commodity. P. Vermeersch pointed out that a similar trend towards sheep and goats can also be observed in the Neolithic of Western Asia. K. Kroeper asked whether the cattle for the burials were killed at one time and then buried, or whether they were collected over a long period of time and buried with the dead. The speaker replied that both could be possible, and that he is only sure that they were buried at the same time. F. Wendorf pointed out that cattle and sheep/goats require different and separate management, and probably the aridification made it no longer feasible to manage both on an equal level. As a result, only small groups continued keeping cattle, while most kept only sheep and goats. He also asked the speaker what he had done with the numerous bucrania he had found. The speaker replied that he kept some for further study, the rest he reburied in the sand.

J. Kabacinski, "Stone Age lithic industries in the Letti Basin"

In this area, Middle Palaeolithic sites were found at the edge of the desert, in natural elevations of the ground, and the later Neolithic remains were found in the valley. The Middle Palaeolithic tools exhibit a very simple technology. Very few were endscrapers, and segments accounted for the plurality (30%). Some burins were found. In later Neolithic settlements the same simple lithic techniques were observed (direct percussion, etc.). Some of the pots found resemble those of Late Kadero (beakers). Some of the pots had incised dotted decoration in zigzag that goes right through the pot, so they were obviously not meant to contain liquids. More than 70% of the lithics associated with these objects were retouched denticulates, very few were endscrapers. Most of these lithics were made of chert or agate and very few of quartz.

E. Garcea, "A review of the El Melek group in the Dongola reach (Sudan)"

The sites studied were Late Neolithic and can be compared to the other Neolithic sites on the other bank of the Nile. Near the river, the sites were smaller, with fewer chipped lithics but with more abundant pottery and grinding stones. In the lithic work the cores were not deeply exploited, a few chips were struck away and the cores were discarded. The stonework was generally rough and careless. Quartz, agate, and chert were used. Their percentages vary, quartz predominated only in one site. In most sites notched denticulates and perforators were the most numerous tools in the industry. The poor condition of the pottery that was discovered is due not only to the erosion (they were lying exposed on the surface) but also to poor firing. The simple tool kit of these people indicates pastoralists in the Late Neolithic. These Late Neolithic sites can be compared with Pre-Kerma sites elsewhere because of the predominance of denticulates and perforators and because of the poor quality of the pottery.

P. Osypinski, "Palaeolithic sites from the SDRS"

This study involved the Dongola area of the Sudan, where Lower Palaeolithic tools were found in the slopes of the natural elevations of the terrain. Upper Palaeolithic tools with a Levallois technique, and Late Palaeolithic artefacts, were also present. The raw materials were quartz, agate, etc.

D. Welsby, "Survey in the Fourth Cataract region (Sudan)"

This survey was carried out in the area above the Fourth Cataract. It was a rescue operation because of the planned construction of a dam that today appears to be a doubtful project. The human occupations attested range from the Middle Palaeolithic to the Medieval Period in about 120 sites found in only 6 km. of the team's 40 km. concession. The dam may yet be built in 2006, so it is almost too late now for any long term projects in this area.

L. Krzyzaniak, "Excavation at Kadero: A summary of results"

This site is situated about 20 km. from Khartoum, and the excavation team has already found 191 Neolithic burials in the centre of the mound, where they had expected to find a settlement. The tombs could be dated to the 5th millennium B.C. (4,850-4,250 B.C.). In the midden they found thousands of grinding stones, animal bones, and plant remains. Most of the tombs found had no objects as funerary items. However, some of the tombs occupied by women were richly furnished with pots, ivory bracelets and anklets, bone harpoons, and Red Sea shells. The male burials appeared less rich: men were only buried with maceheads. Early Kadero pottery consists of globular pots with wide mouths, while the Late Kadero pots are tall decorated beakers. No rippled pottery was found at the site, which is unusual for the Neolithic. Thus far, and in spite of several flotation and other attempts, no evidence of sorghum has been found. The speaker thinks that the occupants used wild sorghum as there are many grinding stones at the site, and that they were definitely pastoralists. Erosion was severe at Kadero and it is estimated that what we have now is probably 1 m. less than the original surface level. There were also some very rich tombs of subadults and even of babies, which – if we go along with current anthropological thinking – would indicate the presence of an élite in this pastoralist cattle-breeding community. The richest graves were clustered on one side of the cemetery. The speaker pointed out that there may still be another cluster of rich tombs. He showed a graphic representation of the tombs in this cemetery divided in three classes: very few class I (more than 5 objects) and class II (between 1 and 4 objects) tombs, and a large majority of class III graves without any items.

Discussion: J. J. Castillos inquired whether the tombs were all found intact, to which the speaker replied that this was indeed the case for the rich burials, but that he cannot be certain about the poor ones, although at least he could not find any robbers' pits. J. J. Castillos also asked whether all these tombs were roughly contemporary (same archaeological horizon) or whether they belonged to several periods, to which the speaker replied that he cannot say since he does not yet have C14 dates for them, but they seem to be contemporary. J. J. Castillos further asked about the size of the graves, and why their volume was not determined to use as another guideline to evaluate them. The reply was that the size was hard to define for the poorer graves. K. Kroeper, who also worked at this site, later clarified that because no intrusions of one grave into another were detected it appears to be the cemetery of one group at a given time.

M. Kaczmarek, "Biological consequences of environmental change in the Post Pleistocene Nubia"

Referring to the Kadero cemetery, the speaker gave the following numbers: a total of 229 graves, of which 205 were Neolithic, 23 Meroitic (100 B.C. - 400 A.D.) and 1 Christian (550 - 1400 A.D.), formed as oval pits. Of the 205 Neolithic burials, 48 belonged to immature individuals (0 to 14 years, 31.8% of the total of identified human remains), 56 to males, 39 to females (ratio 1.43:1), 56 to persons of unknown sex, and 6 were unidentified. The mean age at the time of death was 30.8 years for men (standard deviation = 2.43) and for women 27.7 years (standard deviation = 1.70). Dental studies would place the Kadero population between Mesolithics and agriculturalists, although from other evidence the speaker would place the Kadero people close to intensive agriculturalists and far from Mesolithic.

Discussion: J. J. Castillos pointed out that the percentage of subadults in this Neolithic cemetery is similar to the one he found for the Early Predynastic and the Badarian in the Nile Valley. He asked whether no palaeopathological studies had been carried out on the human remains in order to determine possible illnesses or congenital malformations. The speaker replied that those studies will be carried out in the future.

M. Winiarska-Kabacinska, "Neolithic gouges from Kadero: what were they used for?"

The speaker explained that microwear analysis of such artefacts, abundant at Kadero (more than 100 have already been found), revealed that they were used for cutting or working with soft or hard materials, so they were in fact multi-task tools.

General meeting

After these presentations and discussions, a final general meeting took place in which several decisions were made:

- M. Kobusiewicz wrote a letter to Prof. J. Desmond Clark, eminent prehistorian, founder and staunch supporter of these regular meetings, wishing him well and expressing our regret at not being able to have him with us. This letter was signed by all present and sent to Prof. Desmond Clark.
- In the name of all present, H.-Å. Nordström thanked the organizers of this Symposium for their hospitality, for the excellent accommodations provided and for the efficient day to day running of this event.
- It was announced that in order not to overlap with other related conferences, our next Symposium in Poznan will take place on August 25-30, 2003.

- F. Wendorf encouraged all those who would be willing and able to carry out rescue archaeological work in areas of the Sudan threatened for various reasons, so as to recover and register the information before it is too late.
- J. J. Castillos suggested (and it was unanimously accepted) that the next volume with the Proceedings of this Symposium be dedicated to Prof. J. Desmond Clark.

Finally, F. Wendorf gave an overview of the meeting, mentioning the papers that had impressed him most and summarizing the main themes that had marked this event. Following this, each participating member of the audience gave a short description of his or her future work.

The writer wishes to thank the Director of the Archaeological Museum of Poznan, Prof. Lech Krzyzaniak, for his hospitality in arranging for a brief stay at the Museum Guest Rooms until he could fly back home.

Juan José Castillos
juancast@yahoo.com
Director of the Uruguayan Institute of Egyptology

Socio-Political Hierarchy of First Dynasty Sites: A Ranking of East Delta Cemeteries Based on Grave Architecture

Joris van Wetering & G. J. Tassie[1]

Introduction

This article is an updated version of the second part of a paper originally presented at the *8th International Congress of Egyptologists*, Cairo, 28th March - 3rd April 2000. The first part of the paper is published in the proceedings of this Congress (Tassie & van Wetering 2002). The focus of the present article is on examining social complexity in the East Delta through analysing mortuary architecture during the Protodynastic Period (3,200 – 3,050 B.C.E.) and First Dynasty (3,050 – 2,890 B.C.E.)[2]. It tries to redress the balance of research into this phenomenon, which has in the past mainly focused on sites in the Nile Valley, such as Abydos, Naqada, and Hierakonpolis. This analysis of mortuary architecture presents a modelling of regional socio-political hierarchy of sites during the period of state formation.

The East Delta, especially the north-eastern part, is from an archaeological point of view a reasonably well-investigated region (*Fig. 1a/b*). Based on archaeological work at numerous sites, several surveys and geoarchaeological investigations, a regional picture has emerged of two main branches of the Nile, the old Pelusiac branch and the old Tanitic branch, and their tributaries (Bietak 1975: 122; van Wesemael & Dirksz 1986), and of a settlement pattern that is orientated upon these river ways (van den Brink 1993). In the Nile Valley, sites were primarily located on the desert edge near the floodplain, while later on sites were also located within the floodplain, on the levees (ancient river banks). However, in the Delta, most sites were located on the levees of the multiple waterways and on the geziras (natural sand-hills) in the floodplain[3]. The settlements were situated on the lower slopes of the geziras, where they were close to the fertile land of the floodplain. The cemeteries were located higher up the geziras to avoid the graves being flooded during the annual inundation (van den Brink 1992: 56; Kroeper, in van den Brink 1992: 69). The settlements and the cemeteries changed through time, increasing in size or being located in a slightly different location at the site (Kroeper, in van den Brink 1992: 69). Unfortunately, the location of most ancient settlements and cemeteries is usually also the location of modern villages, agriculture usage (e.g. orchards), or cemeteries.

Socio-Political Differentiation & Cemetery Evidence

A recent study by E. C. M. van den Brink (1993) indicated that 'in the spatial distribution of sites . . . a rather linear and egalitarian pattern' existed in the northern East Delta during the Early Dynastic Period, and that 'on the level of social structuring intrasite and intersite differentiations do seem to occur' (van den Brink 1993: 298–301). For this nearest-neighbour analysis, the location of

[1] The authors wish to thank: Prof. F. A. Hassan for his advice and support; Dr. Zahi Hawass, Prof. Gaballa A. Gaballa, Prof. Dr. Abdel-Helim el-Nour el-Din, and the SCA (Canal Zone) for their support; the members of the UCL-SCA joint mission to Kafr Hassan Dawood for their comments, advice, and support, especially J. M. Rowland for checking the English of this paper as well as her invaluable advice and comments; and Dr. S. Hendrickx, Dr. E. C. M. van den Brink, and Drs. W. van Haarlem for their comments on earlier drafts. The authors are staff members of the UCL-SCA joint Mission to Kafr Hassan Dawood and trustees of the Egyptian Cultural Heritage Organisation (ECHO) [http://www.e-c-h-o.org].

[2] The focus of part I of this study is grave good analysis for East Delta sites during this period, principally Kafr Hassan Dawood, Minshat Abu Omar, and Tell Ibrahim Awad, although grave architecture is also alluded to; see Tassie & van Wetering, in press.

[3] For settlement location and patterning in both the Nile Delta and Valley, see: Bietak 1979; Hassan 1993; Hoffman 1980.

sites and their distance to each other was the main focus (van den Brink 1993: 290–297). A comparative analysis of the East Delta cemeteries of Kafr Hassan Dawood, Minshat Abu Omar, and Tell Ibrahim Awad showed striking differences in grave architecture that might point to social differentiation on a regional level (Tassie & van Wetering, in press). For example at Tell Ibrahim Awad elaborate mud-brick tombs and mastaba tombs point to a higher status of individuals interred there than at Minshat Abu Omar, where no mastaba tombs are reported, although the presence of mud-brick tombs suggest a higher status than at Kafr Hassan Dawood, where no mud-brick architecture has been found (Tassie & van Wetering, in press). The apparent lack of mud-brick constructions in the Kafr Hassan Dawood cemetery, as well as the comparatively less prestigious nature of the grave good assemblage, points to a lower status of this cemetery, while the elaborate grave type at Tell Ibrahim Awad points to a higher status for this cemetery (Tassie & van Wetering, in press).

The first stage of the present study was to conduct a survey of all the literature on the surveyed and excavated sites in the East Delta. Because the criteria for this study are mortuary architecture and grave goods, sites which have only been surveyed have largely been omitted from this research. Twelve sites that have been or are currently being excavated were selected, based on the amount, quality, or availability of the investigated material.

To establish whether a regional hierarchy of cemetery sites existed during the First Dynasty in the East Delta, grave types based on grave architecture from the cemetery sites mentioned below are compared[4]. It needs to be stressed that of all these cemeteries only Kafr Hassan Dawood and Minshat Abu Omar have so far been extensively investigated. Therefore, this analysis, for the time being, can only be preliminary and further investigations at the East Delta cemeteries will either substantiate and complement these findings, or refute them.

The following grave types are used in this study[5]:

Type 1 — **Inhumations in ceramic vessels or coffins**: Interment in either a ceramic pot or coffin with grave goods inside or beside the receptacle.

Type 2 — **Simple pit graves**: Interment in a simple pit, dug in the sedimentary matrix, which can be oval or rectangular [Type 2a]. The body can also be laid on or covered by a (reed) mat [Type 2b].

Type 3 — **Mud-lined graves**: Interment in an oval pit of which the tapering walls are lined with mud [type 3a] or a rectangular pit with straight walls lined with mud [Type 3b].

Type 4 — **Mud-brick tombs**: Interment, generally in rectangular pits, with a single chamber [type 4a] or with multiple chambers divided by mud or mud-brick walls [Type 4b].

Type 5 — **Mastaba tombs**: This type is primarily characterised by its superstructure and by thick walls, constituting more than a single course of mud-bricks. Large simple mastaba tombs [Type 5a] consist of a substructure made up of a burial chamber with storerooms on either side, with an uncomplicated superstructure. Very large elaborate mastaba tombs [Type 5b] consist of a burial chamber with several storerooms around it (substructure) and a superstructure with 'palace niche-façade' and sometimes other decorations. The structure can occasionally be enclosed by a wall or stand on a platform. Type 5b tombs have not been uncovered in the East Delta.

Type 6 — **Royal tomb complexes**: This type only occurs at the Royal Cemetery of the First Dynasty at Abydos. It has a unique architectural development, although aspects of it do

[4] In this discussion a grave is distinguished from a tomb, a grave being a pit dug in the ground with a body inhumed, often marked by a mound or other kind of marker, and a tomb being a large, especially underground, structure for the burial of the dead (demarcated by mud or mud-brick walls).

[5] This typology is based on published material from the cemeteries used in this study and cannot be considered to be comprehensive since other unpublished grave types are known from the Delta (E. C. M. van den Brink, *pers. comm.* 2000).

show up in non-royal grave architecture (Dreyer 1999: 109-144; van Wetering, in press).

Sites of the East Delta

About 46 sites with remains dating to the late Predynastic, Protodynastic and/or Early Dynastic periods are known, either through surveys, individual finds (usually without archaeological context), or excavations. Most of these sites have been identified through (surface) surveys, such as Tell Daba'a[6], Tell Gezira el-Faras, Gezira el-Faras, Kom Om Sir[7], Tell el-Diba, Umm el-Zaiyat, Gezira Sangaha, Tell Abu Dawad, Tell ed-Dab'a, Tell Gandiya, Mendes / Tell el-Ru'ba, Tell el-Ain, and Tell Gherier (van den Brink 1998). Some of these survey sites have been or are being excavated, like Mendes, Tell

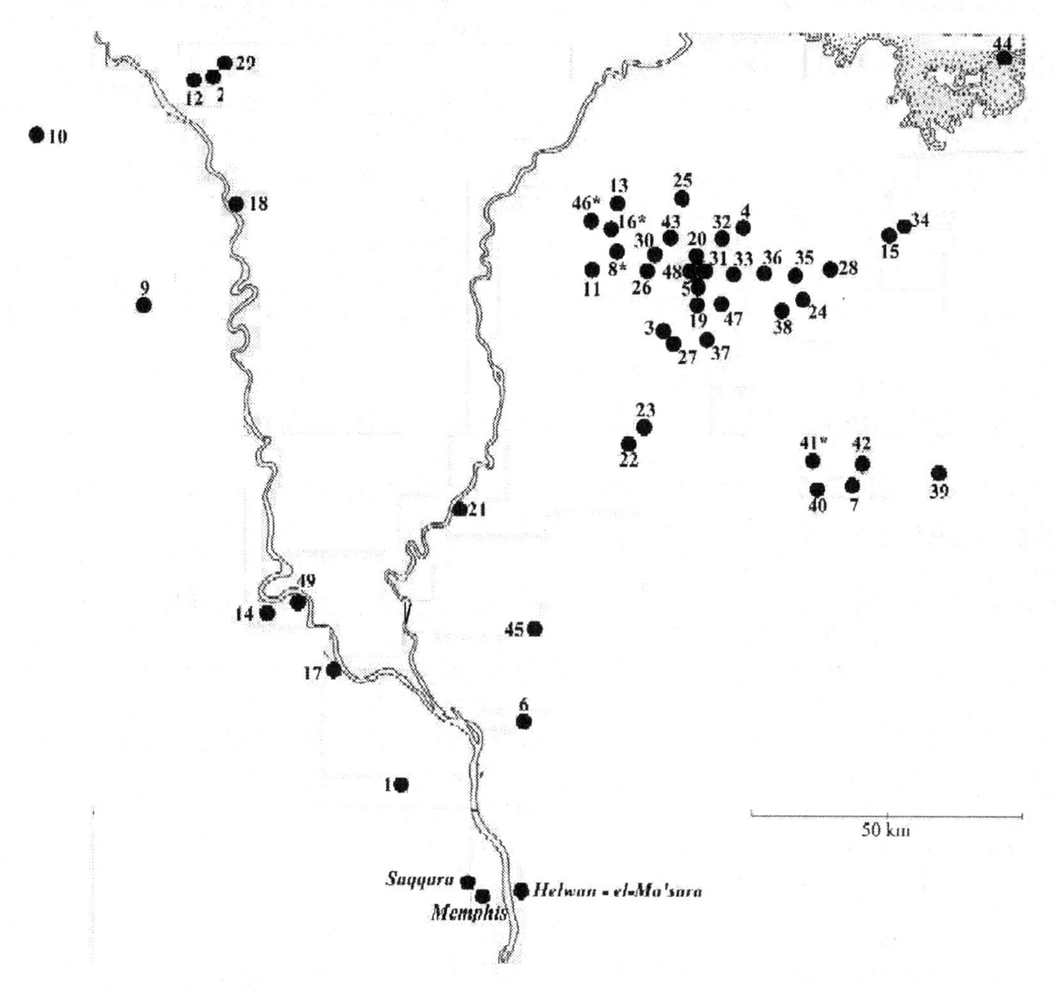

Figure 1a – Sites of the East Delta.[8]

[6] *Tell* is the Arabic word for mound, and is used to describe a mound of accumulated debris from a long lived settlement, usually superimposed phases of mud-brick buildings. Many Egyptian toponyms have incorporated the word in their name, though it is often inaccurate, such as Tell El-Amarna, which is not a *tell* at all but a single occupation site.

[7] *Kôm* is colloquial Egyptian Arabic for Classical Arabic *kawm*, 'pile', 'heap', i. e. it denotes whatever is piled on top and rises higher. In Egyptian toponyms, the word *kom* is used to describe an accumulated mound of debris from ancient settlement remains; it is a synonym for *tell.*

[8] Based on: Hendrickx & Van den Brink 2002; Hendrickx 1994: map 2.

el-Farkha, and et-Tell el-Iswid (south), although primarily settlement remains have been uncovered there (Wilkinson 1999: 364-365).

Other sites are known either through sporadic finds that have been made by locals or by old ‘excavations’ or through the antiquities trade, but have never been properly investigated, such as Bilbeis, Ghita, Zawamil, Minayar, Shibin el-Qanatir, Tell el-Yahudiya, Abu Zabal, Gebel el-Asfar, Wadi Gafra, Tell Nishabe, Tell er-Rataba, Tell el-Mashkuta, Tell Tennis, el-Amid, el-Ghassana, Tell el-Ginn[9], Tell Atrib, Tell Fag’i, el-Khudariya, il Khulga’an[10], and Tell el-Zragy (Hendrickx & van den Brink 2002: 346-402).

1 – **Abu Rawash** – ED
2 – **Ezbet el-Qerdahi** – MA
3 – **Ezbet el-Tell / Kufur Nigm** – PD / PRT / ED
4 – **Gezira el-Faras** – PRT / ED
5 – **Gezira el-Sangaha** – PRT / ED
6 – **Heliopolis** - MA
7 – **Kafr Hassan Dawood** – PRT / ED
8 – **il-Khulga’an / el-Khilgan** * – MA / PRT / ED
9 – **Kom el-Hisn** – ED
10 – **Kom el-Kanater** – MA
11 – **Kom Om Sir** – PD / PRT / ED
12 – **Konayiset es-Saradusi** – MA
13 – **Mendes / Tell el-Ru’ba** – MA / PD / PRT / ED
14 – **Merimde Beni salame** – ME / MA / ED
15 - **Minshat Abu Omar** – PD / PRT / ED
16 - **Minshat Ezzat** * – PRT / ED
17 – **el-Qatta** – PRT / ED
18 – **Sa el-Hagar / Sais** – ME / MA
19 – **Tell Abu Dawud** – PD / PRT / ED
20 – **Tell el-Ain** – PD / PRT / ED
21 – **Tell Atrib** – ED
22 – **Tell Basta** – PRT / ED
23 – **Tell Beni Amir** – PD / PRT / ED
24 – **Tell ed-Dab‘a** – ED
25 – **Tell Dab’a / Tell Qanan** – PRT / ED
26 – **Tell el-Diba** – ED
27 – **Tell Fag’i** – ED
28 – **Tell Fara‘on / el-Huseiniya** – MA / ED
29 – **Tell el-Fara‘in / Buto** – MA / PD / PRT / ED
30 – **Tell el-Farkha** – MA / PD / PRT / ED
31 – **Tell Gandiya** – ED
32 – **Tell Gezira el-Faras** – PRT / ED
33 – **Tell Gherier** – PD / PRT / ED
34 – **Tell el-Ginn** – PD / PRT / ED
35 – **Tell Ibrahim Awad** – MA / PD / PRT / ED
36 – **et-Tell el-Iswid (south)** – MA / PD / PRT / ED
37 – **Tell el-Khasna** – PD / PRT / ED
38 – **Tell el-Masha’la** – MA / ED
39 – **Tell el-Mashkuta** – (?) ED
40 – **Tell el-Nishabe** – PRT and/or ED
41 – **Tell el-Niweiri** * – MA
42 – **Tell er-Rataba** – PRT / ED
43 – **Tell es-Samara** – PD / PRT / ED
44 – **Tell Tennis** – PRT and/or ED
45 – **Tell el-Yahudiya** – PRT and / or ED
46 – **Tell el-Zragy** * – ED
47 – **Tilul Mohammed Abu Hassan** – ED
48 – **Umm el-Zaiyat** – ED
49 – **Wardan** – ED

ME = **Merimde complex** MA = **Maadi complex** PD = **Predynastic Period**
PRT = **Protodynastic Period** ED = **Early Dynastic Period** * = Location unclear

Figure 1b – Sites of the East Delta (Key).

[9] The material from Tell el-Ginn (Ismailia Museum) is assumed to have come from a cemetery but there is no confirmation of this information and Krzyzaniak (1989: 271) states that this site needs to be re-investigated to substantiate this claim.

[10] Recently (2002), a French mission has started excavations at this site, and the preliminary report (Midant-Reynes, in press) indicates the existence of a Protodynastic (– ? Early Dynastic / First Dynasty) cemetery.

Several sites have cemeteries, which either have been fully or partially investigated, or are still under investigation. Kafr Hassan Dawood is one of the early cemeteries still under investigation. It is the largest cemetery in the region and, since most of it has been excavated, it provides an interesting comparison to other cemeteries in the region (Hassan *et al.* 2000: 37-39; Hassan *et al.*, in press): Minshat Abu Omar, Tell Ibrahim Awad, Tell el-Farkha, Tell Beni Amir, Ezbet el-Tell / Kufur Nigm, Mendes / Tell el-Ru'ba, Tell el-Fara'on / el-Huseiniya, Tell Basta, Tell es-Samara, and Minshat Ezzat.

The map of the East Delta (*Fig. 1a/b*) during the Predynastic, Protodynastic, and Early Dynastic periods can be deceptive because it shows a great number of sites/places, but only a small number of these have actually been extensively investigated. The sites of the southern part of the East Delta are only known through finds or surveys and need to be investigated before the area becomes inaccessible for excavations due to expansion of modern occupation, pollution, salinisation, and the rising water-table. Although several excavations have taken place in the northern part of the East Delta, many of the survey sites need to be investigated further to get a more detailed understanding of their cultural development. Additionally, the sites that have been partially excavated need to be more fully investigated to correlate the evidence from the settlement with that of the cemetery and vice versa.

Period	Dates	Naqada	Maadi
8th Dynasty	2,185 – 2,150		
6th Dynasty	2,360 – 2,185		
5th Dynasty	2,502 – 2,360		
4th Dynasty	2,612 – 2,502		
Old Kingdom	**2,612 – 2,150 BCE**		
3rd Dynasty	2,686 – 2,612		
2nd Dynasty	2,890 – 2,686	Naqada IIID	
1st Dynasty	3,050 – 2,890	Naqada IIIC$_1$ – IIID	
Early Dynastic Period	**3,050 – 2,612 BCE**		
Protodynastic B	3,150 – 3,050	Naqada IIIB	
Protodynastic A	3,200 – 3,150	Naqada IIIA	
Protodynastic Period	**3,200 – 3,050 BCE**		
Late Predynastic B	3,300 – 3,200	Naqada IID$_1$ – IID$_2$	
Late Predynastic A	3,400 – 3,300	Naqada IIC	Maadi
Middle Predynastic	3,500 – 3,400	Naqada IC – IIA – IIB	Maadi
Early Predynastic B	3,800 – 3,500	Naqada IA – IB	Maadi
Early Predynastic A	5,500 – 3,800		
Predynastic Period	**5,500 – 3,200 BCE**		

Figure 2 – Chronology of the East Delta cemetery sites.[11]

Cemetery Sites of the East Delta

The 12 sites listed below form the core of this research. These sites have been excavated and published. Some, such as Kafr Hassan Dawood, Minshat Ezzat, Tell el-Farkha, and Tell el-Masha'la, are still in the process of being excavated. All these sites have provided insights into the funerary practices of the East Delta and furnish information for a tentative site hierarchy based on grave architecture, although again the majority of the evidence comes from the northern East Delta.

[11] Chronology sources: Adams & Ciałowicz 1997: 5; Hendrickx 1996: 64.

The site of **Kafr Hassan Dawood** is located on the southern bank, about halfway up the Wadi Tumilat[12]. From 1989 to 1995 the local SCA excavated 921 burials dating to the Protodynastic and Early Dynastic periods (Naqada IIIA-D), and to Saite – Ptolemaic periods. The early and later graves are found interspersed with each other within the western cemetery[13]. Since 1995, the SCA-UCL joint mission has concentrated its work on the earlier segment, excavating burials dating to the Protodynastic and Early Dynastic periods, increasing the total number of graves found to 1057. Geophysical investigations may have identified the contemporary settlement, located at the edge of the floodplain, slightly northeast of the early burials in the western cemetery. Unfortunately it seems to be directly underneath a year-round drainage lake and partially in an area that is flooded by rising ground water during the winter months. The results of a test-pit in this area indicated that the settlement is about 3 to 4 metres below the present ground level and at a level where the ground water fluctuates. This would have damaged the settlement remains greatly. The early cemetery consists of 745 burials, although the original number might be closer to 1300 but unfortunately modern funerary use in the northern part of the cemetery will prevent the complete investigation of the cemetery (Hassan *et al.*, in press). The cemetery has been extensively investigated on all sides, except the northern one. The earliest segment of the cemetery is expected here, close to the floodplain and to the presumed location of the settlement (Hassan *et al.*, in press). It would seem likely that this earliest segment dates back to the late Predynastic Period (Naqada IIC-D), based on comparison with Minshat Abu Omar (Hassan *et al.*, in press). Preliminary analysis shows that the central area of the cemetery contains elite burials (Hassan *et al.*, in press). The investigated graves date from the Protodynastic Period / Naqada IIIA to the Early Dynastic (First Dynasty) Period / Naqada IIID (Hassan *et al.* 2000: 37-39). The majority of the graves are Type 2 graves, with a few Type 1 (ceramic coffin) and Type 3 graves (Tassie & van Wetering, in press). The majority of the community seems to have used small graves (Type 2) while the elite of the community used large graves (also Type 2) during the Protodynastic period and large rectangular graves (Type 3) during the First Dynasty (Tassie & van Wetering, in press). The majority of the graves are oval shaped, while the rest are rectangular in shape, with nine coffin burials. The superstructures were all made by creating either a mud cap or a sand mound over the pit, while the elite burials had an elaborate version of this layout. None of the graves had a roof made of wooden beams, mats, and mud, as has been found at other places (Hassan *et al.*, in press).

Minshat Abu Omar is located in the north-eastern part of the East Delta[14]. It consists of two tells. At Tell A, a large cemetery, dating to the late Predynastic, Protodynastic and Early Dynastic periods and the Saite – Roman periods, has been found. At Tell B, settlements, dating to the Saite – Roman periods, have been found (Kroeper & Wildung 1994: XI-XII). To the south-east of the cemetery (Tell A) augering (drill-cores) has exposed the existence of the earlier settlements, one tentatively dated to the early Predynastic (sherds resembling those of the Merimde cultural complex and the Maadian cultural complex were recovered – Kroeper & Wildung 1994: XIII), and the other contemporary with

[12] For general site information about Kafr Hassan Dawood, see: Hassan *et al.* 2000; Hassan, Tassie, Tucker, Rowland & van Wetering, in press; Tassie & van Wetering, in press; Rowland & Hassan 2002; Tucker 2002. Cemetery plan with probable location of the settlement: Tassie & van Wetering, in press: fig. 1.

[13] There are at least four, possibly five, cemeteries at Kafr Hassan Dawood: a modern Islamic cemetery (1) and a modern Coptic cemetery (2) lying just to the north and slightly overlapping the ancient western cemetery. This main cemetery actually consists of two cemeteries: the early cemetery with burials from the Protodynastic to Early Dynastic periods (3) and a late cemetery with burials from the Late Period and Ptolemaic Period (4). The late cemetery partially overlies the southern, western, and central part of the early cemetery but starts much further east. As these later graves were dug to a greater depth, the result was that the two sets of graves were interspersed. The late cemetery may in fact be two distinct cemeteries – a Late Period cemetery and a Ptolemaic cemetery – but determining this will need further excavation between the main western cemetery and the eastern sector of the site 100 m. east of the main western cemetery, where Late Period to Ptolemaic burials have been found in several other trenches in this area. As well as more extensive excavation, further spatio-temporal analysis would have to be undertaken. However, this is not the main focus of the current research design.

[14] Minshat Abu Omar site map and cemetery plan: Kroeper & Wildung 1994: plan 3.

the cemetery (Krzyzaniak 1989; 1992). Evidence for a Saite – Roman period settlement has also been found. However, at present none of the settlements have been investigated. The early cemetery, although not completely excavated, has yielded 422 graves dating to the late Predynastic, Protodynastic, and Early Dynastic periods (Kroeper & Wildung 1994: XII). These graves have been divided into four chronological groups, MAO I (Naqada IIC-D) to MAO IV (Naqada IIIC-D / First and early Second Dynasty), to provide a relative date for each burial (Kroeper 1988: 18; 1996: 81; 1999: 529). The cemetery is located on the eastern slope of the gezira. The oldest graves (Naqada IIC-D) are found in the southern and eastern parts of the cemetery, while the latest graves, dating to the First and Second dynasties, are located in the extreme northern and western parts of the cemetery (Kroeper 1994: 29). Most of the burials were undisturbed, although some, including the Early Dynastic elite tombs, were robbed in antiquity, and there are even indicators that this happened not long after the burial (Kroeper 1988: 12; 1996: 70). The majority of the graves are of Type 2, with some Type 1 inhumations, several Type 3 graves and a number of mud-brick tombs (Type 4) with burial chamber and one or more storerooms (Kroeper 1999: 529-531). The majority of the Minshat Abu Omar community used Type 2 and 3 graves, whereas the elite of this community utilised elaborate Type 3 graves during the Protodynastic Period and Type 4 tombs during the First Dynasty and the beginning of the Second Dynasty (Kroeper 1992: 140; 1996: 79; Kroeper & Wildung 1994: 116-122). It would seem that there were superstructures or markings over the pit. That some graves have an empty space around them seems to indicate that the superstructure extended beyond the grave pit (Kroeper & Wildung 1985: 25; Kroeper 1992: 144). The cemetery has been almost completely investigated while the nearby settlements are only known through augering. No excavation is currently taking place.

Tell Ibrahim Awad, situated near the modern village of Umm Agram, is one of the most interesting sites from early Egypt. Few sites provide such a full picture: multiple settlements, a long sequence of temples (superimposed upon each other) and cemeteries (van Haarlem 2001: 13-16). The original size of the site (van Haarlem 2001: fig 1) is about 450 m. x 360 m., as indicated by borings, but only the southern part (250 m. x 125 m.) rises above the present surrounding floodplain (van den Brink 1988: 76). Extensive land reclamation and the plantation of an orange orchard have strongly affected the site (van den Brink 1988: 76). This orchard has effectively divided the site in two parts, area A and B; unfortunately due to erosion the stratigraphical relationship between the corresponding layers of both areas is often not clear. Like Minshat Abu Omar and Tell Beni Amir, this site was located near a waterway as indicated by the nearby levees, and like Tell Beni Amir, it is favourably located between the two main Nile branches of the East Delta (van den Brink 1988: 77). The settlement will be discussed in detail below. The cemetery situated north-west of the Temple Area (van Haarlem 2001: map) has only been partially investigated and hardly published (Tassie & van Wetering, in press)[15]. Although the nature of the cemetery is difficult to establish, it might have been an elite cemetery[16]. The cemetery with First Dynasty tombs, north – north-east of the Temple Area (van Haarlem 2001: map), is hardly known. The two cemeteries might form one large cemetery. At least two mud-brick tombs of Type 4 have been investigated in the First Dynasty cemetery north-west of the Temple Area, together with a large tomb of Type 5a and a transitional tomb of Type 4a-5a, all dating to the First Dynasty and early Second Dynasty (van den Brink 1992: 50-51; van Haarlem 1998: 509; 1997: 145; Tassie & van Wetering, in press). The Type 5a tomb dates to the middle of the First Dynasty (van den Brink 1992: 51; van Haarlem 1996: 7), while the transitional Type 4-5a tomb dates to the early First Dynasty (van den Brink 1988: 77). The transitional tomb has mud-brick walls erected

[15] No plan of this cemetery nor one of the north – north-east cemetery is available, but the cemetery locations as well as the Temple Area and the Settlement Area with the large building (Area B near First Dynasty cemetery) are marked on the site map: van Haarlem 2001: fig. 1.

[16] According to W. van Haarlem (*pers. comm.* 2000), it seems to be a cemetery for the local elite. The high status of the burials in the cemetery and the absence of Types 2 and 3 graves support this opinion (van den Brink 1992: 50-51; van Haarlem 1993: 37-41; 1996: 7-13).

at intervals, maybe to support the roofing of wooden poles and the reed matting. It has four compartments, a central one with two annexes to the north-east and one to the south-west, that were roofed separately by wooden poles (9-10 cm. diameter) and by several layers of reed mats (van den Brink 1988: 77-78; 1992: 51; van Haarlem 1993: 39-40). The compartments were separated by 1 m. high mud-brick walls (Brink 1988: 77). Unfortunately, the plan of the tomb (Brink 1988: fig. 11) does not show the configuration of the mud-brick walls and this plan does not show the third storeroom that was discovered later (van Haarlem 1993: 37). Excavation is currently under way in area A and will, in due time, be resumed in area B (W. van Haarlem, *pers. comm.* 1998). It is likely that graves older than the First Dynasty are present here, since surface indicators, namely a wide spread of (Degenerated) Wavy Handled Jar fragments, show that this cemetery had been in use from an earlier period of time (van den Brink 1988: 78).

Tell Beni Amir is located near the modern villages of Beni Amir and Ezbet el-Sheikh el-Saiyid Abu Hashem[17]. It forms one large tell that is intersected by the Beni Amir Canal (Abd el Moneim 1996b: 253; Abd el-Hagg Ragab 1992: 207). A large number of tombs, dating to the late Predynastic – Early Dynastic periods, were excavated but the exact amount is unclear. According to Abd el-Hagg Ragab (1992: 208), about 250 graves (Early Dynastic and Saite – Roman periods) were found and the majority of these date to the early period, but according to Abd el-Moneim (1996a: 241; 1996b: 258) only 36 tombs were found that date to the late Predynastic – Early Dynastic periods. Unfortunately, no plan or map of the cemetery, located in the lower stratum (B) in the western part of the Western (Main) Tell, is available. The cemetery information is equivocal. It is unclear how many graves were uncovered and whether the cemeteries were completely investigated (Krzyzaniak 1989: 277). The excavated graves consisted of Types 2, 3, 4, and 5a (Abd el-Moneim 1996a: 241-249). During the First Dynasty, the elite of the community used large mud-brick tombs (Type 4), and simple mastaba tombs (Type 5a). The large mud-brick tombs (including one simple mastaba tomb and one very large mud-brick tomb) point to the high status of those interred here (Abd el-Moneim 1996a: 248-249). No traces of the contemporary settlement have been found. This indicates that the settlement was not located on the present tell but nearby, either under the agricultural fields, or under the modern village on the eastern side of the canal, or under the estates on the western side of the canal (Abd el-Moneim 1996b: 259). Tell Beni Amir is located at the crossroads of major routes, at a point where the two important Nile branches of the East Delta are closest (Abd el-Moneim 1996b: 260). It also lies at the western end of the Wadi Tumilat, the route to the Sinai and the copper and turquoise mines located there (Abd el-Monein 1996b: 260). Excavation is not taking place here anymore, and the site is now covered with modern occupation (Krzyzaniak 1989: 277).

The site of **Ezbet al-Tell / Kufur Nigm** is located between Tell Ibrahim Awad and Tell Beni Amir[18]. It consists of a large tell that has been dissected by canals and roads into three smaller koms (Krzyzaniak 1989: 277; Bakr 1994: 9). Surface finds indicated that an Early Dynastic to Old Kingdom settlement was located on all koms (Krzyzaniak 1989: 277). The local SCA had investigated the site in the early 1960s and then again in 1978, and the Old Kingdom settlement had been partially excavated as well as a Protodynastic to Early Dynastic cemetery (Krzyzaniak 1989: 277). The cemetery is located in the northern half of the Northern Kom, and consists of about 400 graves[19] (Krzyzaniak 1989: 277; Bakr 1988: figs. 1 and 2). According to Kroeper (1988b: 18) objects in the Zagazig Museum are

[17] Tell Beni Amir site map: Abd el-Moneim 1996b: 255 – no plan of the cemetery is available.

[18] Ezbet al-Tell / Kufur Nigm site map: Bakr 1993: fig. 1 – no plan of the cemetery is available.

[19] About 300 graves were excavated by the Egyptian Antiquities Organisation in the 1960s and 1970s at a cemetery dating to the Protodynastic period and the Early Dynastic period. Zagazig University investigated about 100 graves in the 1980s and early 1990s. It is not clear whether the two groups of graves form part of the same cemetery (Krzyzaniak 1989: 277, 280).

similar to the ones found at Minshat Abu Omar during Phase IV – the latest phase[20]. This might indicate that the occupation at Ezbet el-Tell / Kufur Nigm continued into the Second Dynasty, like that at Minshat Abu Omar. The graves can be identified as Types 1, 2, 4 and possibly 3, dating to the late Protodynastic Period and the First Dynasty (Bakr 1993: 10-12; Krzyzaniak 1989: 277). The majority of the burials consisted of simple graves (Types 2 and 3). Some of these graves had ceramic coffins in them, with the grave goods beside the coffin but inside the grave-pit, whereas others had no grave goods in them at all (Bakr 1993: 12-13). Some of the ceramic coffins used as the burial receptacle were not within a tomb construction, but placed tight inside a grave-pit with the grave goods inside the coffin (Bakr 1993: 12-13). The number of pottery coffins at this site is striking, and points to highly localised burial customs. From the excavation reports, most of the coffins seem to be rectangular; the number and utilisation of these ceramic coffins has not been encountered anywhere else in the East Delta[21]. Several inhumations of young children or foetuses in ceramic vessels were also found (Bakr 1993: 12-13). Several First Dynasty mud-brick tombs (Type 4), either consisting of a single chamber or of a burial chamber and a storeroom divided by compact-mud walls or mud-brick walls, have also been investigated (Bakr 1993: 12-13). The two-chambered tombs of mud-brick are the most elaborate grave type and might have served as elite graves, although this is difficult to substantiate, as there are no published records pertaining to the grave good assemblage of these tombs. It is unclear whether the cemetery has been completely excavated or future excavation is intended.

Tell Basta is located near Tell Beni Amir[22]. A large number of graves have been excavated here but unfortunately, the precise number is unclear. The seemingly earlier graves were excavated in an area of the Middle Kingdom palace while the 'Protodynastic' tomb was excavated to the south-east of this area, about 200 metres away (see map in el-Sawi 1979). Underneath the foundation layer of the Middle Kingdom palace and the underlying storerooms, an unknown number of graves (Type 2) were uncovered by the Zagazig University Mission (Bakr *et al.* 1992: 156; Kroeper 1988: 18-19; Abd el-Moneim 1996a: 241-242). These graves seem to belong to the earliest occupants of the site, but a date for this occupation is unknown. Only one of the excavated graves had a grave good, a single pot (Bakr *et al.* 1992: 20). According to Abd el-Moneim (1996a: 241-242), several simple graves consisting of a pit in the sand just large enough for the body and a few grave goods were found here (Bakr 1982: 156). Probably part of the same cemetery is the large tomb found in 1970 by el-Sawi (1979: 63). This mud-brick tomb, burial 137, consists of a shaft with two chambers and measures about 4 m. by 2 m. with a depth of about 2 m. Several grave goods were found in the northern chamber. They seem to be associated with a child's burial, although other human remains were also found. The grave good assemblage includes three diorite bowls, a squat porphyry jar, a cylindrical ceramic jar, and a copper mirror (el-Sawi 1979: figs. 105-109). In the southern chamber, a single skeleton was found with a diorite bowl (el-Sawi 1979: 63). This tomb is dated to the Protodynastic and beginning of the Early Dynastic Period (el-Sawi 1970: 63), although, based on the architecture, a late Early Dynastic date, probably Second or even Third Dynasty, is more likely. It therefore falls outside the scope of this study. The architecture of the graves found at Tell Basta is very difficult to analyse, but at least two grave types were present.

The site of **Tell Fara'on / el-Huseiniya** is located near Minshat Abu Omar and Tell Ibrahim Awad[23]. In the cemetery a large number of graves (Types 2 and 4) that date to the Protodynastic and

---

[20] 24 objects from the excavations are displayed in the Zagazig Museum and have partly been published in Krzyzaniak 1989, while a large amount of objects are displayed in the Archaeological Museum of the University of Zagazig. The authors visited this museum in 2000 and the objects there confirm the dating.

[21] The use and distribution of ceramic coffins in the East Delta cemeteries, especially Ezbet el-Tell / Kufur Nigm, will be discussed in Tassie, Rowland & van Wetering, in prep.

[22] Tell Basta site map: el-Sawi 1979: 4 – no plan of the cemetery is available. The location of the 'Protodynastic' mud-brick tomb is marked, the Type 2 graves were found under/near the Middle Kingdom Palace (Bakr et al. 1992: 20).

[23] Tell Fara'on / el-Huseiniya site map: Mustafa 1988: fig. 1 – no plan of the cemetery is available.

Early Dynastic periods were found, but it is unclear how many were excavated and whether the entire cemetery was fully investigated (Mustafa 1988: 75-76; Kroeper 1988: 18; Krzyzaniak 1989: 271; Abd el-Moneim 1996a: 242). Excavation is not currently taking place.

The site of **Minshat Ezzat** is located 10 miles to the south of the modern city of Mansura near the town of el-Simbillawein and south-west of the site of Mendes / Tell el-Ru'ba. It was discovered in 1998 by the local SCA, which has since been excavating the extensive cemetery, covering an area of 50 acres, as well as investigating the settlement remains (el-Baghdadi & el-Said Nur, in press). Up to a 100 graves of Types 2, 3, and 4 dating to the Protodynastic and Early Dynastic periods, as well as later periods, have been uncovered[24]. Recent investigations indicate the existence of elaborate burials (possibly Type 4) with a large number of grave goods (S. el-Baghdadi, *pers. comm.* 2002). The cemetery seems to date to the Protodynastic and Early Dynastic periods, but earlier graves, possibly from the Maadian cultural complex might be located here, under an orange orchard.

The local SCA is currently excavating the cemetery at **Tell es-Samara** where graves have been found dating to the Protodynastic and Early Dynastic periods[25]. Up to 75 graves of Types 2, 3, and 4 have been found (el-Baghdadi & el-Said Nur, in press).

A single First Dynasty mud-brick tomb (Type 4) was uncovered by the local SCA at **Tell el-Masha'la** in 1988–89 (Abd el-Hagg Ragab 1992: 210-213). The tomb is indicative of a cemetery but no other tombs were investigated by the SCA. The recently started re-investigation (Canadian Mission) might shed light on the size and status of this cemetery and the associated settlement.

The site of ancient **Mendes** comprises two tells: **Tell el-Ru'ba** with the early material and Tell Timai with later (Saite – Roman) material (Wilkinson 1999: 341). The location of the site on a Nile branch and near the sea might have given it an important place in the communication and economic transport systems through river routes with the rest of Egypt and sea routes to the Near East and North Africa (Brewer & Wenke 1992: 193; Wilkinson 1999: 341; Wenke 1999: 499). The settlement information will be discussed below. During excavations in the 1960s, Early Dynastic graves were uncovered within the temple enclosure. Due to the scarcity of grave goods, the graves were dated to the First Dynasty and Second Dynasty on the basis of pottery in the stratum the graves were found in (Wilkinson 1999: 341-342). The architecture of the graves found is almost impossible to reconstruct so an analysis is not feasible at the moment, although at least Type 2 graves occur (Abd el-Moneim 1996a: 241-242).

Recently, a number of graves have been uncovered at **Tell el-Farkha**, a site where the settlement has also been extensively investigated. The settlement, situated on the Central and Western Koms, is discussed below; the cemetery is located on the Eastern Kom. The six to ten graves (Types 2 and 4) are all dated to the Early Dynastic Period (Debowska, in press; Jucha, in press).

Ranking Cemetery Sites

'The most abundant evidence of social ranking in centralised societies comes from burial, and from the accompanying grave goods' (Renfrew & Bahn 2000: 209-210). Therefore in this discussion aspects of funerary evidence are considered as a means of identifying social stratification (Rowland, in press [a]; in press [b]). To date, the evidence from settlements (next section) is still too sketchy to utilise in a ranking of sites, although the evidence from the settlements that have been partially excavated is providing very useful information on the intrasite social dynamics of Delta sites.

---

[24] Members of the Kafr Hassan Dawood team visited the site in December 1998 and again in October 1999. The information used here is based on our observations, as well as on: el-Baghdadi 1999: 9-11; el-Zahlawi 2000; el-Baghdadi & el-Said Nur, in press.

[25] Members of the Kafr Hassan Dawood team visited the site in October 1999 and the information used here is based on our observations.

The absence of some grave types (specifically Type 4 and/or Type 5a) in certain East Delta cemeteries seems to indicate that certain communities were unable or did not wish to invest that kind of expenditure into the tombs of the elite of their community. The most elaborate grave constructions at Kafr Hassan Dawood are those of Type 3b, while those at nearby Tell Beni Amir (approximately 40 km. away) are Type 5a. Although contact between these two communities can be assumed, either through trade or other forms of intersite activity, this contact did not encourage the adoption of mud-brick as a building material for graves by the elite of the Kafr Hassan Dawood community. A similar situation occurs at Minshat Abu Omar, where no Type 5a constructions have been found, although this grave type is present at the nearby site of Tell Ibrahim Awad. So, certain communities did not incorporate certain grave types within their funerary practice, which indicates social differences between the East Delta communities, distinguishing those sites with high status graves (utilising mud-brick architecture) from other, possibly more rural, sites (without mud-brick architecture). The use or absence of mud-brick within a settlement seems to be indicative of how urbanised the site was (Spencer 1976) and as such its utilisation (or lack thereof) within the cemetery can also be regarded as significant. If it is assumed that the absence of Type 4 and/or Type 5a tombs at a cemetery indicates a lower social rank than a cemetery where these types are present, then the following tentative ranking system can be put forward[26]:

- **High-ranking cemeteries**, based on the presence of Type 5a tombs: Tell Ibrahim Awad – Tell Beni Amir.
- **Middle-ranking cemeteries**, based on the presence of Type 4 tombs and the absence of Type 5a tombs: Minshat Abu Omar – Ezbet el-Tell / Kufur Nigm – Tell al-Fara'on / el-Huseiniya – Tell Basta – Tell el-Farkha – Tell es-Samara – Tell el-Masha'la – Minshat Ezzat – possibly Mendes / Tell el-Ru'ba.
- **Lower-ranking cemeteries**, based on the absence of both Type 4 and 5a tombs: Kafr Hassan Dawood.

Corroboration from Settlement Evidence

At several Delta sites, settlement remains have been investigated but the information is still insufficient to provide a comprehensive picture[27]. Hardly any of these settlements have been extensively investigated (i.e. horizontally and vertically). Sites where both the settlement and cemetery have been extensively investigated are rare in Egypt, but the potential for this is present at Minshat Abu Omar and Kafr Hassan Dawood, where the locations of the contemporary settlements are known through augering. However, due to difficulties in accessing the settlement and the costs involved in using pumping equipment to excavate them, the settlements at both sites have not been fully investigated. Below we present an overview of the settlement evidence from the Delta sites Tell Ibrahim Awad, Tell el-Fara'in / Buto, Tell el-Farkha, Mendes / Tell el-Ru'ba, and Tell el-Iswid (south), followed by a comparison of settlement evidence, including the architectural design of local temple constructions.

At **Tell Ibrahim Awad** a multiple-layered settlement (Tell A and B) with a sequence of superimposed temples (area A) from the Predynastic Period to the Middle Kingdom, frequently with associated cemeteries, gives a rare insight into the temporal development of an East Delta community.

---

[26] Due to the scarcity of published and on-going as well as new excavations at East Delta sites, this is a very preliminary attempt whereby the ranking of sites like Tell el-Farkha, Tell el-Masha'la, Minshat Ezzat, and Tell Samara are subject to change.

[27] For a more detailed description and discussion of (east) Delta settlement evidence, see Tassie, Rowland & van Wetering, in prep.

The settlement remains are impressive with several settlement layers and contemporary temple constructions. The earliest settlement layers date back to the Maadian cultural complex (van den Brink 1992: 53-54) although the temple constructions only date back to the early Protodynastic Period (Eigner 2001: 17).

Eight superimposed shrines have been investigated at Tell Ibrahim Awad for the period from the beginning of the Protodynastic to the end of the Early Dynastic Period[28], and four more temple layers are present in the period up to the Middle Kingdom (Eigner 2001: 17). These shrines consist of small mud-brick (mostly single-chambered) constructions around a brick platform and seem to be an integral part of the settlement (Eigner 2001: 25-36). The shrine/temple sequence at Tell Ibrahim is not only important for understanding the development of temple architecture in ancient Egypt, but also for the social function of temples and for dating comparisons. The Protodynastic A shrine construction seems to be the oldest confirmed Pre-Formal shrine in Egypt. In the earliest shrine construction of the Protodynastic A Period, ceramic cultic vessels have been found that seem to be unique in Egypt (van Haarlem 1999: 193). The large number of votive deposits found in the shrine constructions of the Early Dynastic Period (First Dynasty and Third Dynasty) and underneath the Old Kingdom temple constructions is unique for Lower Egypt (van Haarlem 1998: 510-513). At shrines/temples such as Tell Ibrahim Awad, the religious festivals may have been a focal point for a larger regional community.

The settlement of the late Protodynastic and Early Dynastic (First Dynasty) periods also hints at the importance of the place with a large building in which the names of several Protodynastic and Early Dynastic kings (*serekhs*) were found (van den Brink 1992: 52-53). The mud-brick walls are about 1.2 m. thick, and, although the size of the building could not be determined, one wall could be traced over a distance of 15.5 m. before it made a corner (van den Brink 1992: 52). Although the function of this building could not be ascertained, the discovery of ceramic import goods from the Southern Levant and the *serekhs* might point to an administrative function (van den Brink 1992: 51-52).

The current excavations at **Tell el-Farkha** are providing and will continue to provide further valuable information on northern Delta interaction. The settlements are situated on three koms and are dated from the Predynastic (Maadian complex contemporary with Naqada IIC) to the Old Kingdom – Fourth Dynasty. Important for this study are Phases 4 / IIIA-B – Protodynastic Period, 5 / IIIB-C – early First Dynasty, and 6 / IIIC-D – late First Dynasty to late Second Dynasty (Chlodnicki & Ciałowicz 2002: 89-90).

Architectural remains have been recovered from all phases. Although the settlement has been extensively investigated, the latest phase of occupation on the Western Kom is Phase 5, so either the area was disturbed by ancient and/or modern activities or abandoned during the First Dynasty (Chlodnicki & Ciałowicz 2002: 94).

On the Central Kom there also seems to be less activity from the Phase 5 / First Dynasty onwards, with hardly any evidence for Phases 6-7 (Chlodnicki & Ciałowicz 2002: 100; Chlodnicki, in press).

Interesting evidence for both socio-political and economic complexity comes from layers preceding the First Dynasty. At the Western Kom, a monumental structure (with at least two occupation phases) consisting of several long and narrow rooms, walls 2.5 m. thick, and with dimensions of 17 m. in length by 12 m. in width, has been uncovered in Phases 2 / Naqada IID and 3 / IID-IIIA (Chlodnicki & Ciałowicz 2001: 87-89; 2002: 89; Ciałowicz, in press). Numerous storage vessels and possibly imported vessels from the Southern Levant have been found associated with this structure, the so-called 'Naqadian residence' (Chlodnicki & Ciałowicz 2002: 90-91; Ciałowicz, in press). Phase 2 also provides information about a domestic brewery, one of the oldest of its kind found to date (Chlodnicki & Ciałowicz 2002: 92). Between Phases 3 and 4, there is evidence of wide-scale burning and flooding at the Western Kom (Ciałowicz, in press). Another large building consisting of several rooms and with

[28] A shrine is a small-scale single- or two-chambered structure, while a temple is a large-scale multiple-chambered structure mostly within a walled enclosure.

a size of 25 m. in length and 15 m. in width is located in Phase 4 / IIIIA-B (Ciałowicz, in press). Here too, a large number of storage vessels and even status objects (such as palettes) have been uncovered at the Western Kom (Ciałowicz, in press). Part of this complex seems to have been a shrine, as indicated by the deposit of figurines of faience, clay, and stone, including figurines of baboons (Ciałowicz, in press; 2003). Similar objects have been found associated with the temple constructions at Tell Ibrahim Awad (van Haarlem 1998: 509-514), although at this site the shrine seems to be an independent building, while at Tell el-Farkha it is part of a complex (Ciałowicz, in press; 2003).

Preliminary analysis shows that the site decreased in size (abandonment of the Western Kom – Chlodnicki & Ciałowicz 1999: 70) and maybe also in socio-political and economic complexity during the Early Dynastic Period. There seems to be evidence for natural disaster(s), like flooding and an earthquake (Ciałowicz, in press).

The settlement evidence at **et-Tell el-Iswid (south)** is too limited (two test-trenches of 4 m. x 5 m.) to provide extensive insights, but the 10 settlement layers dating from the Predynastic (Maadian cultural complex) Period to Early Dynastic Period do give insight into the stratigraphical and chronological development of the community (van den Brink 1989: 59, 77-79). Most interesting is the introduction of mud-brick architecture at the site during Phase B (strata VII-VIII, at the end of the Protodynastic Period / beginning of the Early Dynastic Period) within a domestic context (van den Brink 1989: 64). This utilisation of mud-brick seems to have important socio-political implications (Spencer 1976), but it is too early to understand these implications fully. No cemetery has been found associated with this site, although four simple pit graves, non-contemporary with each other, were found within the various settlement layers.

The settlement remains of the Protodynastic – Early Dynastic periods excavated at **Mendes / Tell el-Ru'ba** are at present too limited to provide detailed insights into the community (Brewer & Wenke 1992: 273-275). Indirect evidence, surface finds, and information from older investigations point to extensive Early Dynastic occupation (Brewer & Wenke 1992: 268). The later Old Kingdom remains point to the important role of this settlement in the Delta during the Old Kingdom (Brewer & Wenke 1992: 275-279). It will be interesting to see the early development, but most of the excavated pre-Old Kingdom remains, within the Temple Enclosure, were heavily disturbed and came from key-hole trenches (Brewer & Wenke 1992: 273-275). The evidence points to domestic architecture, with no indications of a temple or administrative structures (Brewer & Wenke 1992: 273-275). Interestingly, but not surprisingly, hardly any imported goods have been found in a domestic context, except for 'a few small sherds of possibly Syro-Palestinian origins' (Brewer & Wenke 1992: 272). The location of this community, near the sea and on the river (Brewer & Wenke 1992: 271), suggests favourable conditions for (inter-)regional interaction.

One of the most important sites from the Delta with elaborate settlement remains is **Tell el-Fara'in / Buto**. At this site excavations have uncovered settlement layers from the Predynastic Period (Maadian cultural complex – Layers I-II) to the Old Kingdom (Layer VI) and beyond (Late Period – Layer VII). It falls outside the scope of this study to give a complete overview of this site. However, Layers IV-V-VI (Protodynastic Period to early Old Kingdom) from the Sechmawy area are included for comparison with the East Delta settlements (Ziermann 2002).

The Sechmawy area probably represents the central area of the ancient settlement. It has been identified as a royal residence or administrative centre (Ziermann 2002: 498-499), although, since the extent of the ancient settlement is not yet known, this is still very difficult to substantiate. The excavated area consists of numerous walls and rooms forming an intricate complex that went through several building phases. The predecessor of the complex only emerges in the later phases of Layer IV (Protodynastic – early First Dynasty) (Ziermann 2002: 467, 498-499). During the different phases of Layer V (early First Dynasty – end of the Second Dynasty), a large complex is apparent, with different sized rooms and an enclosure wall to separate it from the rest of the settlement (Ziermann 2002: 472-

487, 498-499, figs. 7-8-9). In Layer VI (end of the Second Dynasty – Fourth Dynasty), the complex continues with new buildings being added to it (Ziermann 2002: 463, 498-499).

In area Aa a shrine has been identified for Layer V with possible similar structures in the preceding Layer IV and the succeeding Layer VI. The identification is based on the building construction and architectural design as well as the presence of 'ritual' sand in the presumed sanctuary (Ziermann 2002: 464, 473-474), although no votive objects seem to have been found to confirm the identification.

Although the settlement evidence from the East Delta is still too sketchy to testify how the communities functioned, the evidence from Tell el-Fara'in / Buto, Tell Ibrahim Awad, and Tell el-Farkha does provide some useful insights.

The constructions at Tell el-Fara'in / Buto and Tell el-Farkha point to distinct differentiation within the settlement, with possible enclosed complexes within the settlement (former site) and structures with thick walls (latter site). The large building at Tell Ibrahim Awad has only been subject to cursory investigation, and no plans exist. Only part of the construction is known, which makes it more difficult to analyse and compare with structures at other sites.

Importantly, all three sites have cultic structures or shrines. In a study of early temples and shrines, B. Kemp (1989) devised a model to classify the religious structures of ancient Egypt. He divided them into Pre-Formal and Early, Mature, and Late Formal styles (Kemp 1989: 66). The Pre-Formal style is characterised by an open architecture (accessible courtyards) and mud-brick architecture. This style can be traced from the Predynastic Period to the end of the Old Kingdom, at which time all known Pre-Formal temples and shrines had been rebuilt in the Early Formal style. The Early Formal style is still characterised by mud-brick architecture as the primary building material, although stone slabs were used for decorative purposes and the entire structure was 'closed' as opposed to the 'open' Pre-Formal feature. This means the actual sanctuary was hidden, only accessible to authorised personnel who formed an agency controlling movement and knowledge, a feature that points to a royal/state sponsored policy (Kemp 1989: 65-83; Wilkinson 1999: 272, 305-307)[29], whereas the 'open' Pre-Formal shrines seem not to have been under the patronage of the state.

All three sites have small shrines from the Late Protodynastic B Period to the beginning of the Early Dynastic Period. At **Tell Ibrahim Awad,** the shrines of Phases 3 to 6d / Third Dynasty to Protodynastic A fall within the Pre-Formal style, all small-scale shrines of mud-brick situated within the settlement, although it seems each were free-standing structures[30]. At **Tell el-Farkha**, the one-chambered shrine appears to be part of a larger complex with several developmental stages (Ciałowicz 2003). This shrine is identified by the numerous votive objects, similar to the ones found at Tell Ibrahim Awad (Ciałowicz 2003). At **Tell el-Fara'in / Buto** a clearer picture has emerged as the shrine and the surrounding settlement area have been extensively investigated. In area Aa of Layer V, a shrine was identified (Ziermann 2002: figs. 7-8 – *Heiligtum/Kultraum*), based on the building construction and architectural design as well as the occurrence of sand in the main room (Ziermann 2002: 473-474). This layer, with several occupation phases, is dated from the early First Dynasty to the end of the Second Dynasty. The shrine is part of a large complex that has been identified as a royal residence or administrative centre (Ziermann 2002: 498-499). If this identification is correct, and as stated above this can be substantiated, the shrine would have royal or at least administrative connections and as such one would expect an Early Formal style shrine. However, the two-chambered construction very much resembles the Pre-Formal shrines of other sites where no state/royal connection is attested. On the

[29] The Mature Formal and Late Formal styles are not relevant to this study as these styles date to the time of the New Kingdom and onwards.

[30] Unfortunately, hardly any information is available on the interaction of the temple within the settlement as only the temple / shrine constructions have been excavated. In some cases, hardly more than a few parts of the wall of a shrine (phase 3 – Eigner 2001: fig. 7) have been recovered.

other hand, the possible utilisation of stone slabs at this shrine (Ziermann 2002: 474 - note 64) points to the Early Formal style.

Several Pre-Formal shrines from the Protodynastic – Early Dynastic periods have been discovered in Upper Egypt. At **Badari**, a small mud-brick two-chambered construction Loc3003/3 is identified as the Protodynastic B – early First Dynasty shrine (Wilkinson 1999: 315; Holmes 1999: 163). An important sequence of the Pre-Formal shrines of the local goddess Satet has been uncovered on the island of **Elephantine** (First Cataract) with layers dating to the Protodynastic B Period, the early First Dynasty, late First Dynasty/early Second Dynasty, and beyond (Kaiser 1999: 283-4). These superimposed shrines are constructed between the huge granite boulders on the island. It is therefore of an atypical architectural design, very much influenced by local circumstances. Nevertheless, a basic two-chambered room design is clear for the early structures (Kaiser 1999: 283, Seidlmayer 1996: 11-112) [31]. An example of absence of royal patronage is demonstrated at the shrine of early First Dynasty. At some point during the early part of the First Dynasty a state-sponsored fortress was built right in front of the Satet shrine. This fortress, rather than encompassing the shrine within its protective walls, destroyed part of the shrine's forecourt and resulted in re-construction of the layout of the shrine, particularly the entrance, which had to be moved to an inconvenient position, thus showing disrespect for the local shrine (Seidlmayer 1996: 111-112; Ziermann 1993: 140).[32]

A number of seemingly Early Formal shrines and temples are known: the Netjery-khet shrine at **Heliopolis**; the Horus temple on Thoth Hill at **Thebes**[33]; the Hathor sanctuary at **Gebelein**; the Nekhbet sanctuary at **Nekheb (Elkab)**; and the Horus sanctuary at **Nekhen (Hierakonpolis)**[34].

The question of which sites constituted the most important nodes in the regional settlement pattern, and possibly functioned as regional administrative centres in the northern East Delta, cannot yet be fully answered. The cemetery ranking indicates that Tell Ibrahim might have had such a role, but to date, the settlement evidence is too limited to corroborate this.

The on-going research at Tell el-Farkha might shed light on this question; it will be of interest to compare the cemetery at Tell el-Farkha with the other cemeteries (especially the one at Tell Ibrahim Awad). The same comment applies to Tell el-Fara'in / Buto, although no trace of any early cemeteries has yet been located.

Another site where this question could be answered is Tell Beni Amir where the cemetery ranking indicates a similar situation to that found at Tell Ibrahim Awad. The location of Tell Beni Amir between the northern East Delta and the Memphite region would support a regional administrative function, but settlement information is completely lacking and the tell is now covered by modern occupation.

Ancient River Ways & Site Location

An important feature that emerges from this settlement pattern of ranked cemetery sites in the East Delta is the location of communities along the river system. As stated above, most settlements are located on higher ground, geziras or levees, of ancient Nile branches. For the East Delta, the river system consists of the old Pelusiac branch and the old Tanitic branch, which nowadays are roughly

[31] See maps of the Satet shrines of the early First Dynasty and Second Dynasty in Kaiser 1999: 283 - fig. 30.

[32] The majority of pottery found in the early town is of Egyptian Naqada ware, with a small admix of A-Group pottery, so it may be better to see it as lying in the zone of influence of both cultures with no clear cut cultural differentiation.

[33] Based on the present evidence, this is the earliest occurrence of the later classical design of pylon with sanctuary (Vörös & Pudleiner 1998: 335-340).

[34] Horus temple area of Nekhen mound. The so-called 'Temple' / HK29A lacks the usual objects associated with contemporary temples at Tell Ibrahim Awad and Elephantine. The more likely function of this structure (HK29A ceremonial structure) is as a palace instead of a temple, especially since the recently discovered evidence from the reign of Narmer suggests this identification (Friedman 2002/2003; 2003).

represented by the Bahr Faqus and the Bahr Muweis, respectively (van Wesemael & Dirksz 1986: 8)[35]. These two river ways and their tributaries were the main arteries for communication, and the communities of the East Delta had their homes along these rivers[36]. According to Bietak (1979: 102), the sites were not only situated on geziras at or near the main branches but also at river junctions and the junctions of rivers with overland routes (Bietak 1979: 102). Smaller towns and villages were situated around these centres (Bietak 1979).

Both Tell Ibrahim Awad and Tell Beni Amir, the two sites with high-ranking cemeteries, were situated at strategic locations for communication and transport. Tell Beni Amir was located at the point where the old Tanitic branch and the old Pelusiac branch divided, and was the road junction for overland routes to the northern part (Tell Ibrahim Awad and the routes to the Southern Levant), the southern part (Memphite region), and the eastern part of the East Delta (Wadi Tumilat and the Sinai). Tell Ibrahim Awad was located downstream of Tell Beni Amir, near the old Pelusiac branch; the ancient levees of this river way (or a tributary) have been located nearby (van den Brink 1988: 77). This place was well-suited as a collection point for long-distance trade (on the overland route to the Southern Levant, cf. Bard 2000: 68), and to facilitate riverine transport downstream to Tell Beni Amir, which was located about halfway between the Memphite region and the northern East Delta. The focus of central administration on economic resources made Tell Ibrahim Awad and Tell Beni Amir ideal locations for regional administration and distribution centres (Wilkinson 1999: 113-114, 125). All the middle-ranking cemetery sites were located on one of the major branches or their subsidiaries, facilitating contact with the high-ranking cemetery sites. Tell Basta was located near Tell Beni Amir, probably on the old Pelusiac branch. Minshat Abu Omar (old Pelusiac branch), Tell el-Masha'la (old Tanitic branch), and el-Fara'on / el-Huseiniya (old Pelusiac branch) were near Tell Ibrahim Awad. Ezbet el-Tell / Kufur Nigm was about halfway between Tell Ibrahim Awad and Tell Beni Amir, on the old Tanitic branch. The cluster of Tell el-Farkha, Tell es-Samara, and Minshat Ezzat was also situated on the old Tanitic branch or its tributaries. Mendes / Tell el-Ru'ba seems to have been located on a different waterway, the old Mendesiac branch, but it seems very likely this waterway was connected to the main waterways of the East Delta through tributaries. A pattern of high-ranking cemeteries surrounded by middle-ranking cemeteries with low-ranking cemeteries at the periphery (Kafr Hassan Dawood in the Wadi Tumilat) can thus be identified, attesting to the early socio-political landscape of the region[37].

Socio-Political Hierarchy of First Dynasty Sites

A distinction is drawn between an administrative hierarchy, as indicated by the number of tiers of state administrators evident in the settlement system, and a settlement hierarchy, as indicated by the number of tiers of settlements (Flannery 1998: 16). This results in a four-tier site settlement hierarchy (based on cemetery evidence) and a two-tier administrative hierarchy for the Memphite region and the East Delta (*Fig. 3*), since state administrative institutions can only be attested at high ranking East Delta sites and within the Memphite region. That Egypt was an early state by the beginning of the Early Dynastic Period is generally accepted (Baines & Yoffee 1998: 216-218; Bard 2000: 67) and, according to Flannery (1998: 15-16), early states were characterised by incorporating at least four levels of settlements in their socio-political hierarchy: city, town, large village, and small village. This

---

[35] See van den Brink 1987: fig 1; 1993: fig. 6, for ancient river ways and associated sites of the Early Dynastic and Old Kingdom periods in the northern East Delta.

[36] The geographical situation of the East Delta, the ancient river ways, and the distribution of sites will be discussed in detail in Tassie, Rowland & van Wetering, in prep.

[37] The placing of the sites on the ancient, now defunct, Nile branches is based on van Wesemael & Dirksz 1986: 8 and van den Brink 1987: fig 1; 1993: fig. 6.

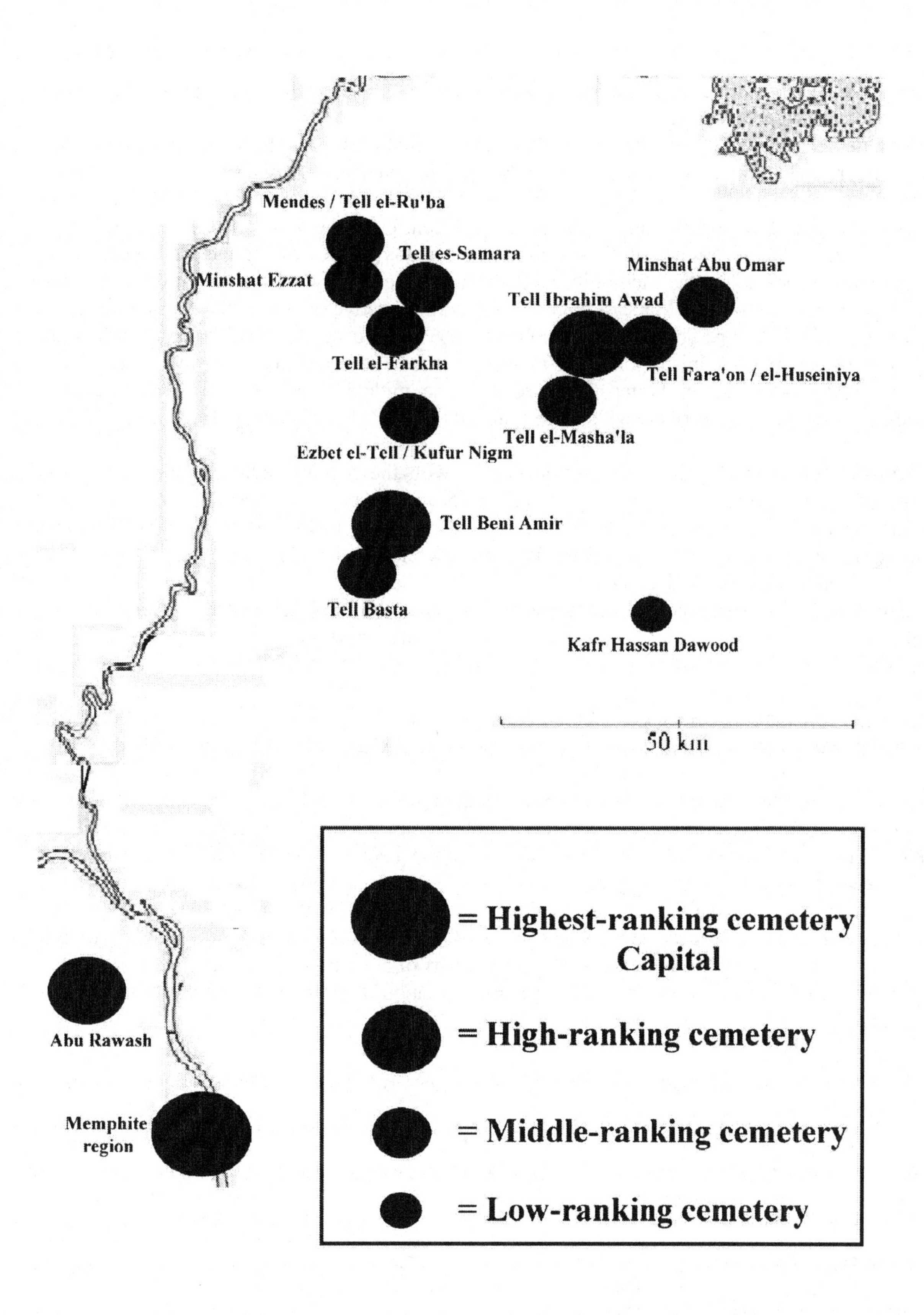

Figure 3 – Preliminary Socio-political Landscape of the East Delta.

is a hierarchical system based on settlement evidence whilst the hierarchical system of the present analysis is based on cemetery sites.

The first tier comprises the 'capital' (and associated cemeteries) and the royal cemeteries (van Wetering, in press)[38]. The exact relationship between Memphis and Abydos is still not entirely understood. Memphis was the administrative centre and This-Abydos was the royal cemetery of the First Dynasty kings, and possibly also a regional sub-centre of the Early Dynastic state[39] (van Wetering, in press). The occurrence of Type 6 tombs at Abydos is self-evident and the occurrence of Type 5a-b tombs (mastabas) in the Memphite region is also expected. The sheer size of the cemeteries at North Saqqara and Helwan–el-Ma'sara points to the administrative nature of the Memphite region (van Wetering, in press). Type 5b tombs (elaborate mastabas) are located at North Saqqara, where the highest officials of the administration, presumably members of the royal family, were buried (Wilkinson 1999: 360-361; van Wetering, in press). The middle class cemetery was located at Helwan–el-Ma'sara, across the river, with more than 10,000 burials (Wilkinson 1999: 114-115; van Wetering, in press).

The regional high-ranking cemeteries would equate with the regional administrative centres or towns, the second tier. The occurrence of mastaba (Type 5a) tombs could be a result of the presence of state administrators in these regional administrative centres. These towns would probably, within a pre-existing set of economic and political regulations, exercise limited influence within their local sphere of dominance (Hassan 1993: 564-567).

The regional middle-ranking cemeteries would then equate with the local centres or large villages, the third tier. Although indications of interregional trade are present at Minshat Abu Omar, the evidence is still ambiguous[40]. However, it may be possible to substantiate this when Tell el-Farkha and the probable settlement at Tell el-Masha'la are fully excavated.

The regional low-ranking cemeteries would equate with the small farming villages, the fourth tier, many of which would have existed (Bard 2000: 69). However this again is difficult to verify due to the absence of settlement information. At Kafr Hassan Dawood this is suggested by the absence of elaborate grave types, the lack of mud-brick architecture, and the generally less 'prestigious' nature of the grave good assemblages, as well as the few Southern Levantine import objects (Tassie & van Wetering, in press). The grave good assemblages and the dental wear patterns of the cemetery population seem to indicate an economy based on fishing, hunting and pastoralism supplemented with small-scale cultivation of cereals and other crops (Tucker, in press), with the pottery suggesting that the elite were enjoying beer, bread, and probably wine[41] (Hassan *et al.*, in press). The site's location in the Wadi Tumilat could possibly indicate that it formed part of a separate site hierarchy, not linked to that of the East Delta. However, there is not enough information from other sites in the Wadi Tumilat to substantiate this.

---

[38] It has to be realised that the word 'capital' has a meaning in our vocabulary that should not be applied directly to its use in an archaeological context.

[39] This has to be substantiated by the settlement evidence from This and Abydos (if these are different places), but this kind of evidence is sadly lacking.

[40] According to Wilkinson (1999: 363), the 'very existence' of Minshat Abu Omar is based on 'trade with Palestine'. This seems to be a very premature statement as only the funerary aspect of this community has been investigated. Nothing is known about the settlement and its social differentiation (Kroeper 1999). About 20 import pieces (because of grave robbing, the 'true' amount of imports – ceramic and non-ceramic – cannot be established) from the Southern Levant (Kroeper 1989) in a cemetery with more than 3500 objects (Kroeper 1996: 82) seems to argue against identifying this community as a trading centre. This kind of statement is too easily made on the occurrence of artefacts of which the exact context and nature is not yet completely understood.

[41] One has to be careful not to interpret all the so-called 'wine jars' as actual containers for wine; other liquids or goods might also have been transported by these 'wine jars', such as oil, although residues of wine have been found in several 'wine jars' (Joffe 1998; Murray 2000).

The fourth tier communities would probably have interacted with the third and second tier communities, but direct interaction between the first tier, the central administration, and fourth tier communities would have been sporadic, if it existed at all.

The interaction described above, i.e. Egypt's internal interactions and the external interaction of Egypt with the Southern Levant, are developments that can be interpreted via Core-Periphery Interaction and World Systems Theory, respectively, interpretative models currently in use in archaeology[42].

Imports & *Serekhs*

Most known sites in the Delta are characterised by Southern Levantine import pieces and seem to have been places that could tap into the trade network with the Southern Levant (Wilkinson 1999: 364-365). The almost total absence of imported goods at Kafr Hassan Dawood is therefore striking and has interesting implications for the general view we have of the area. It would seem unlikely that every site with imported goods was a trading station. It is far more likely that some well-located places along the route were trading posts but that this was not their primary function. The most important functions of all the East Delta communities were hunting, fishing, agriculture, and the keeping of live-stock, a community's basic subsistence strategies. The exchange of luxury and imported goods must have been secondary. Hence the appearance of one or a few imported goods does not make a trading station and we need to balance such finds with the fact that these kinds of goods are more likely to show up in the archaeological record since imported goods are more likely to be 'treasured' and taken into the grave than 'regular' goods that have to do with day-to-day work (Joffe 1998). Also, the discovery of a large Egyptian settlement with elaborate fortifications near Gaza (Protodynastic B to early First Dynasty) places the interaction between Egypt and the Southern Levant in a whole new, possibly aggressive / military, perspective (de Miroschedji & Sadek 2002: 34-43), so such imported goods need not necessarily be the outcome of friendly reciprocal trade.

Similar caution should be taken with the occurrence of *serekhs*[43] in the archaeological record. Their implications are not yet known, but their distribution is relatively large. Several sites in the East Delta have storage jars with royal names on them. A closer look at these *serekhs* shows that there are differences between these royal names, even between names of the same king. This will be seen, for example, if one compares the *serekh* of Narmer found at Kufur Nigm (Bakr 1988: pl. 1a) with a similar one found at the same place (Bakr 1988: pl. 1a), or one found at Minshat Abu Omar, grave 1590 (Kroeper 1988: pl. 13), or if one compares the *serekhs* found at such sites with the ones found in the royal tombs at Abydos; there are big differences in the style and the way they are executed.[44] So it cannot be said that there is uniformity between the known *serekhs*, since some are beautifully executed while others are not, and this does not seem to point to a single place of origin (a royal workshop). If at least some of the storage jars with *serekhs* were not issued from the royal administration, it might be that the others were issued by local or regional administrations (in the East Delta), but that they were still treasured enough to take into the grave, or perhaps the local or regional administration issued a jar with a *serekh* for the burials of local elites, not necessary in the same place. The implications of the occurrence of a *serekh* at a site is not fully understood, so using it to substantiate claims about contacts with the central administration must be done very carefully (van Wetering, in prep.). Therefore, the

[42] The application of World Systems Theory and Core-Periphery Interaction to early Egypt will be examined in Tassie, Rowland & van Wetering, in prep.

[43] *Serekh*: Ancient Egyptian word for the rectangular device, represented a section of the royal palace-façade, which served as a frame enclosing the king's Horus name; the *serekh* is usually surmounted by the figure of a falcon (Wilkinson 1999: 374).

[44] For an extensive study of the incised *serekh*-signs of the Protodynastic and Early Dynastic periods, see van den Brink 1996 and 2001.

claim by van den Brink (1988: 78) that the occupant of a Type 5a mastaba tomb at Tell Ibrahim Awad 'obviously had access to the royal workshops' seems unfounded, because investigations at Minshat Abu Omar and Kafr Hassan Dawood have shown that local elites were able to procure or produce, and control the circulation of, these kind of goods. It would seem unlikely that they all had direct access to the royal workshops. It is more likely that certain items with the royal name / *serekh* on them were occasionally sent by the administration in the course of regular economic interaction to the local elites, and that these objects were deemed so valuable that they were taken to the grave.

Concluding Remarks

The absence of grave types (specifically Type 4 and/or Type 5a) in certain East Delta cemeteries seems to indicate the existence of a regional three-tier hierarchy and a national four-tier hierarchy of cemetery sites during the First Dynasty. These results complement the nearest-neighbour study by van den Brink (1993), based on several sites surveyed in the late 1980s, which, together with surveyed sites of the Italian survey and of the American survey, has made the northern East Delta one of the best-surveyed areas in Egypt (van den Brink 1988; 1998; Chlodnicki *et al.* 1991: 5-33; Brewer *et al.* 1996: 29-42). However, a surface survey does not give a full picture of the site and cannot indicate its character and stratigraphy. A community's economic, political, and social functions can only be identified by extensive excavation and considerable artefactual evidence (Brewer & Wenke 1992: 273). The call by the SCA to concentrate archaeological work in the Delta must not only lead to surface surveys but also to a co-ordinated effort to excavate a viable sample of Delta sites. Ideally, both surface and sub-surface surveys should take place all over the Delta, and, based on this information, a regional plan of excavation work should be established within a larger Delta cultural heritage management project. Also, it is of great importance to be able to compare the evidence from the northern East Delta with the rest of the Delta. At present all the extensive surveys have taken place in the northern East Delta and most of the archaeological work is also concentrated in this area. Hardly any work is taking place on the early communities of the other parts of the Delta, except for the excavations at Tell el-Fara'in / Buto, Sais, and Kom el-Hisn, all located in the West Delta.

As this paper demonstrates, settlement evidence is still too incomplete to utilise it reliably in a regional analysis of socio-political dynamics. Apart from a few well-investigated cemetery sites, the mortuary information is also too limited to construct a map of the socio-political landscape of the East Delta[45]. Although our knowledge of cultural dynamics in the Delta has expanded enormously over the last 25 years, it is still very difficult to detect regional developments. However, from analysing the available evidence, social complexity seems to be implied during the First Dynasty, through ranked cemetery sites. The results of this preliminary analysis, which is only one of the steps in a regional evaluation, are provisional and need to be substantiated or refuted by future investigations, since many of the sites considered here are either only partially investigated or are still under investigation. Extensive evidence from settlements and cemeteries needs to be collected to illuminate the processes of early state formation that took place in Egypt during the Predynastic and Early Dynastic periods, especially on a regional level. Therefore, further research in the East Delta, especially into the function of the various sites, is needed to increase our knowledge of this very important region where the archaeological remains are constantly endangered by salinisation, modern agricultural practices, land reclamation involving flooding of vast areas of land, pollution, a high water-table, *sebakhin*[46], demolition, building and development projects, and general urban sprawl caused by increasing population growth.

---

[45] An attempt to model the socio-political landscape and its economic interaction of the East Delta in particular and early Egypt in general will be undertaken in Tassie, Rowland & van Wetering, in prep.

[46] Diggers for ancient mud-brick and mud-brick wash, to use as fertiliser or in manufacturing new mud-bricks.

Joris van Wetering & G. J. Tassie
Egyptian Cultural Heritage Organisation
jflvwetering@e-c-h-o.org
gtassie@e-c-h-o.org

References Cited

Adams, B. & K.M. Ciałowicz 1997. Protodynastic Egypt. Shire Egyptology 25, Buckinghamshire [UK].

Abd el-Hagg Ragab, M. 1992. A report on the excavations of the Egyptian Antiquities Organisation at Beni Amir and Gezira el-Masha'la in the Eastern Delta; in E.C.M. van den Brink (ed.) The Nile Delta in Transition: 4th - 3rd Millennium B.C. Tel Aviv (Israel): 207-214.

Abd el-Monein, M. 1996a. Late Predynastic – Early Dynastic cemetery of Beni Amir (Eastern Delta); in L. Krzyzaniak, M. Kobusiewicz & K. Kroeper (eds.) Interregional Contacts in the Later Prehistory of North-eastern Africa. Poznan (Poland): 241-251.

Abd el-Monein, M. 1996b. Late Predynastic – Early Dynastic mound of Beni Amir (Eastern Delta); in L. Krzyzaniak, M. Kobusiewicz & K. Kroeper (eds.) Interregional Contacts in the Later Prehistory of North-eastern Africa. Poznan [Poland]: 253-275.

el-Baghdadi, S.G. 1999. La Palette décorée de Minshat Ezzat (Delta); in *Archéo-Nil* 9: 9-11.

el-Baghdadi, S.G. & N.M. el-Said Nur in press. The Late Predynastic – Early Dynastic cemeteries of Minshat Ezzat and Tell el-Samarah (el-Dakahliy Governorate, northeastern Delta). To be published in the Proceedings of the *Origins of the State. Predynastic and Early Dynastic Egypt* symposium, held at Krakow Aug.-Sept. 2002. Krakow [Poland]. Abstract of paper (2002).

Baines, J. & N. Yoffee 1998. Order, Legitimacy, and Wealth in Ancient Egypt and Mesopotamia; in G. Feinman & J. Marcus (eds.) Archaic States. Santa Fe [USA]: 199-260.

Bakr, M. 1988. The new excavations at Ezbet el-Tell, Kufur Nigm; the first season 1984; in E.C.M. van den Brink (ed.) The Archaeology of the Nile Delta: Problems and Priorities. Amsterdam [Netherlands]: 49-62.

Bakr, M. 1993. Excavations at Kufur Nigm; in C. Berger, G. Clerc & N. Grimal (eds.) *Hommages à Jean Leclant. Vol. 4: Varia*, Bibliothèque d'Étude 106-4: 9-17, Cairo [Egypt].

Bakr, M. *et al.* 1992. Tell Basta I. Tombs and Burial Customs at Bubastis. The Area of the So-called Western Cemetery. Cairo [Egypt].

Bard, K. 2000. The Emergence of the Egyptian State (c. 3200-2686 B.C.); in I. Shaw (ed.) The Oxford History of Ancient Egypt. Oxford [UK]: 61-87.

Bietak, M. 1979. Urban archaeology and the town problem, in K. Weeks (ed.) Egyptology and the Social Sciences. Cairo: 93-144.

Bietak, M. 1975. Tell el Dab'a II. Der Fundort im Rahmen einer archäologisch-geographischen Untersuchung über das ägyptische Ostdelta. Vienna [Austria].

Brewer, D., J. Isaacson & D. Haag 1996. Mendes regional archaeological survey and remote sensing analysis; in *Sahara* 8: 29-42.

van den Brink, E.C.M. 1988. The Amsterdam University Survey Expedition to the North-eastern Delta; in E.C.M. van den Brink (ed.) The Archaeology of the Delta: Problems and Priorities. Amsterdam [the Netherlands]: 65-114.

van den Brink, E.C.M. 1989. A transitional late Predynastic – Early Dynastic settlement site in the North-eastern Delta, Egypt; in *Mitteilungen des Deutschen Archäologischen Instituts, Abteilung Kairo* 45: 55-108.

van den Brink, E.C.M. 1992. Preliminary report on the excavations at Tell Ibrahim Awad, seasons 1988-1990; in E.C.M. van den Brink (ed.) The Nile Delta in Transition: 4th – 3rd Millennium B.C. Tel Aviv (Israel): 43-68 (69 discussion).

van den Brink, E.C.M. 1993. Settlement patterns in the North-eastern Nile Delta during the fourth – second millennia B.C; in L. Krzyzaniak & M. Kobusiewicz & J. Alexander (eds.) Environmental Change and Human Culture in the Nile Basin and Northern Africa until the Second Millennium B.C. Poznan (Poland): 279-304.

van den Brink, E.C.M. 1996. The incised *serekh*-signs of dynasties 0-1, Part I: Complete jars; in J. Spencer (ed.) Aspects of Early Egypt. London [UK]: 133-151.

van den Brink, E.C.M. 1998. Regional, Diachronic Investigations into Settlement Patterns in the Eastern Nile Delta, Egypt. Ph.D. dissertation – University of Tel Aviv [Israel].

van den Brink, E.C.M. 2001. The pottery-incised *serekh*-signs of dynasties 0-1, Part II: Fragments and additional complete vessels; in *Archeo-Nil* 11: 23-100.

Chlodnicki, M. in press. Excavations at the Central Kom of Tell el-Farkha, 1999-2002. To be published in the Proceedings of the *Origins of the State. Predynastic and Early Dynastic Egypt* symposium, held at Krakow Aug.-Sept. 2002. Krakow [Poland]. Abstract of paper (2002).

Chlodnicki, M., R. Fattovich & S. Salvatori 1991. Italian excavations in the Nile Delta: fresh data and new hypotheses on the 4th Millennium cultural development of Egyptian prehistory; in *Rivista di Archeologia* 15: 5-33.

Chlodnicki, M. & Ciałowicz, K.M. 1999. Tell el-Farkha explorations 1998; in *Polish Archaeology in the Mediterranean* X: 63-70.

Chlodnicki, M. & Ciałowicz, K.M. 2001. Tell el-Farkha (Ghazala) interim report, 2000; in *Polish Archaeology in the Mediterranean* XII: 85-97.

Chlodnicki, M. & Ciałowicz, K.M. 2002. Tell el-Farkha seasons 1998-1999. Preliminary report; in *Mitteilungen des Deutschen Archäologischen Instituts, Abteilung Kairo* 58: 89-117.

Ciałowicz, K.M. 2003. From the Residence to Early Temple: the case of Tell el-Farkha. Abstract of workshop *Early Temples*, held at Dutch-Flemish Institute Cairo / NVIC 9-10 January 2003. Cairo [Egypt].

Ciałowicz, K.M. in press. Tell el-Farkha 2001-2002. Excavations at the Western Kom. To be published in the Proceedings of the *Origins of the State. Predynastic and Early Dynastic Egypt* symposium, held at Krakow Aug.-Sept. 2002. Krakow [Poland]. Abstract of paper (2002).

Debowska, J. in press. Recent discoveries in the necropolis of Tell el-Farkha. To be published in the Proceedings of the *Origins of the State. Predynastic and Early Dynastic Egypt* symposium, held at Krakow Aug.-Sept. 2002. Krakow [Poland]. Abstract of paper (2002).

Dreyer, G. 1999. Abydos, Umm el-Qa'ab; in K. Bard (ed.) Encyclopaedia of the Archaeology of Ancient Egypt. London [UK]: 109-114.

Eigner, D. 2001. Tell Ibrahim Awad: Divine Residence from Dynasty 0 until Dynasty 11; in *Ägypten und Levante* 10: 13-16.

Flannery, K. 1998. The ground plans of Archaic states; in G. Feinman & J. Marcus (eds.) Archaic States. Santa Fe [UK]: 15-57.

Friedman, R. 2002/2003. Excavating Narmer's Temple (HK29A – Season 2002-2003); at the following URL: http://www.archaeology.org/interactive/hierakonpolis/field/index.html.

Friedman, R. 2003. The Ceremonial Complex (HK29A) at Hierakonpolis. Abstract of workshop *Early Temples*, held at Dutch-Flemish Institute Cairo / NVIC 9-10 January 2003. Cairo [Egypt].

van Haarlem, W. 1993. Additions and corrections to the publications of a First Dynasty tomb from Tell Ibrahim Awad (Eastern Nile Delta); in *Göttinger Miszellen* 133: 37-52.

van Haarlem, W. 1996. A Tomb of the First Dynasty at Tell Ibrahim Awad; in *Oudheidkundige Mededelingen uit het Rijksmuseum van Oudheden* 76: 7-33.

van Haarlem, W. 1997. Imitations in pottery of stone vessels in a protodynastic tomb from Tell Ibrahim Awad; in *Archéo-Nil* 7: 145-150.

van Haarlem, W. 1998. The excavations at Tell Ibrahim Awad (Eastern Nile Delta): recent results; in J. Eyre (ed.) Proceedings of the Seventh International Congress of Egyptologists, Cambridge 1995. Leuven (Belgium): 509-514.

van Haarlem, W. 2001. An introduction to the site of Tell Ibrahim Awad; in *Ägypten und Levante* 10: 13-16.

van Haarlem, W. 1999. A Predynastic Triple vessel from Tell Ibrahim Awad; in *Göttinger Miszellen* 173: 193-195.

Hassan, F.A. 1993. Town and village in Ancient Egypt: ecology, society and urbanization; in T. Shaw *et al.* (eds.) The Archaeology of Africa. Food, Metals and Towns. London [UK]: 551-569.

Hassan, F.A. *et al.* 2000. Kafr Hassan Dawood: A late Predynastic to Early Dynastic site in the East Delta, Egypt; in *Egyptian Archaeology* 16: 37-9.

Hassan, F.A., G.J. Tassie, T.L. Tucker, J.M. Rowland & J. van Wetering in press. Social dynamics at the late Predynastic to Early Dynastic site of Kafr Hassan Dawood, East Delta, Egypt; in *Archeo-Nil* 12.

Hendrickx, S. 1994. Analytical Bibliography of the Prehistory and the Early-Dynastic Period of Egypt and Northern Sudan. Leuven [Belgium].

Hendrickx, S. 1996. The relative chronology of the Naqada Culture: Problems and Possibilities; in J. Spencer (ed.) Aspects of Early Egypt. London [UK]: 36-69.

Hendrickx, S & van den Brink, E. C. M. 2002. Inventory of Predynastic and Early Dynastic cemetery and Settlement sites in the Nile Valley, in E. C. M. van den Brink & T. E. Levy (eds.) Egypt and the Levant: Interrelations from the 4th Through the Early 3rd Millennium B.C.E. London [UK]: 346-402.

Hoffman, M. A. 1980. Egypt Before the Pharaohs. London [UK].

Holmes, D. 1999. el-Badari district, Predynastic sites; in K. Bard (ed.) Encyclopaedia of the Archaeology of Ancient Egypt. London [UK]: 161-164

Joffe, A.H. 1998. Alcohol and social complexity in ancient Western Asia, *Current Anthropology* 39 (3): 297-322.

Jucha, M. in press. Tell el-Farkha 2001-2002. The pottery from the Tombs; to be published in the Proceedings of the *Origins of the State. Predynastic and Early Dynastic Egypt* symposium, held at Krakow Aug.-Sept. 2002. Abstract of paper (2002).

Kaiser, W. 1999. Elephantine; in K. Bard (ed.) Encyclopaedia of the Archaeology of Ancient Egypt. London (UK): 283-289.

Kemp, B. 1989. Ancient Egypt: Anatomy of a Civilisation. London [UK].

Kroeper, K. 1988. The excavations of the Munich East Delta Expedition in Minshat Abu Omar; in E.C.M. van den Brink (ed.) The Archaeology of the Delta. Problems and Priorities. Amsterdam [the Netherlands]: 11-46.

Kroeper, K. 1989. Palestinian Ceramic Imports in Pre- and Protohistoric Egypt; in P. de Miroschedji (ed.) L'Urbanisation de la Palestine à l'age du Bronze ancien. Oxford [UK], BAR 527: 407-422.

Kroeper, K. 1992. Tombs of the Elite in Minshat Abu Omar; in E.C.M. van den Brink (ed.) The Nile Delta in Transition: 4th – 3rd Millennium B.C. Tel Aviv (Israel): 127-150.

Kroeper, K. 1994. Minshat Abu Omar. Pot burials occurring in the Dynastic cemetery; in *Bulletin de Liaison du groupe international d'Étude de la Céramique Égyptienne* 18: 19-32.

Kroeper, K. 1996. Minshat Abu Omar. Burials with palettes; in J. Spencer (ed.) Aspects of Early Egypt. London (UK): 70-92.

Kroeper, K. 1999. Minshat Abu Omar; in K. Bard (ed.) Encyclopaedia of the Archaeology of Ancient Egypt. London (UK): 529-531.

Kroeper, K. & D. Wildung 1985. Minshat Abu Omar. Münchner Ostdelta Expedition. Vorbericht 1978-1984. München [Germany].

Kroeper, K. & D. Wildung 1994. Minshat Abu Omar. Ein vor- und fruhgeschichtlicher Friedhof im Nildeltas I. Graber 1-114. Mainz [Germany].

Kroeper, K. & D. Wildung 2000. Minshat Abu Omar. Ein vor- und fruhgeschichtlicher Friedhof im Nildeltas II. Graber 115-210. Mainz [Germany].

Krzyzaniak, L. 1989. Recent archaeological evidence on the earliest settlement in the Eastern Nile Delta; in L. Krzyzaniak & M. Kobusiewicz (eds.) Late Prehistory of the Nile Basin and the Sahara. Poznan (Poland): 267-285.

Krzyzaniak, L. 1992. New data on the late Prehistoric settlement at Minshat Abu Omar, Eastern Nile Delta; in L. Krzyzaniak & M. Kobusiewicz (eds.) Environmental Change and Human Culture in the Nile Basin and Northern Africa until the Second Millennium B.C. Poznan [Poland]: 321-325.

Leclant, J. & G. Clerc 1991. Fouilles et travaux en Égypte et au Soudan, 1989-1990. Tell Ibrahim Awad; in *Orientalia* 60: 170-172.

Leclant, J. & G. Clerc 1992. Fouilles et travaux en Égypte et au Soudan, 1990-1991. Tell Ibrahim Awad; in *Orientalia* 61: 226-227.

Midant-Reynes, B. in press. Kom el-Khilgan. To be published in the Proceedings of the *Origins of the State. Predynastic and Early Dynastic Egypt* symposium, held at Krakow Aug.-Sept. 2002. Krakow [Poland]. Abstract of paper (2002).

de Miroschedji, P. & M. Sadek 2002. Gaza et l'Egypte de époque Prédynastique à l'Ancien Empire. Premiers résultats des fouilles de Tell es-Sakan; in *Bulletin de la Société Française d'Égyptologie* 152: 28-52.

Murray, M. A. with N. Boulton, & C. Heron 2000. Viticulture and wine production; in P. Nicholson & I. Shaw Ancient Egyptian Materials and Technology. Cambridge [UK]: 577-608.

Mustafa, I. 1988. Some objects dating from the Archaic Period found at Tell Fara'on – Imet; in *Göttinger Miszellen* 102: 73-84.

Renfrew, C. & P. Bahn 2000. Archaeology. Theories, Methods and Practice. (3rd edition) London [UK].

Rowland, J.M. in press [a]. The transition to state society in Egypt: problems and possibilities of applying mortuary evidence; in A. Cooke & F. Simpson (eds.) Proceedings of Current Research in Egyptology II, Liverpool 2001. Oxford [UK].

Rowland, J.M. in press [b]. Application of Mortuary Data to the Problem of Social Transformation in the Delta from the Protodynastic to the Early Dynastic period. To be published in the Proceedings of the *Origins of the State. Predynastic and Early Dynastic Egypt* symposium, held at Krakow Aug.-Sept. 2002. Krakow [Poland].

Rowland, J.M. & F.A. Hassan in press. The computerised database and potential for a Geographical Information System at Kafr Hassan Dawood; in Z. Hawass and L. Pinch Brock (eds.) Egyptology at the Dawn of the Twenty-First Century. Proceedings of the Eight International Congress of Egyptologists, Cairo 2000. Vol. 1: Archaeology. Cairo [Egypt].

el-Sawi, A. 1979. Excavations at Tell Basta. Report on seasons 1967-1971 and catalogue of finds. Prague [Czech republic].

Seidlmayer, S. 1996. Town and state in the early Old Kingdom: a view from Elephantine; in J. Spencer (ed.) Aspects of Early Egypt. London [UK]: 108-127.

Spencer, A. J. 1976. Brick Architecture in Ancient Egypt. Warminster [UK].

Tassie, G.J. & J. van Wetering in press. Early Cemeteries of the East Delta; Kafr Hassan Dawood, Minshat Abu Omar and Tell Ibrahim Awad; in Z. Hawass and L. Pinch Brock (eds.) Egyptology at the Dawn of the Twenty-First Century. Proceedings of the Eight International Congress of Egyptologists, Cairo 2000. Vol. 1: Archaeology. Cairo [Egypt].

Tassie, G.J., J.M. Rowland & J. van Wetering in prep. The Social-Political Landscape & Economic Interaction of Egypt during the late 4th and early 3rd Millennium B.C.E.: State Formation from a Delta Perspective.

Tucker, T.L. in press. Biocultural Investigations at Kafr Hassan Dawood; in Z. Hawass and L. Pinch Brock (eds.) Egyptology at the Dawn of the Twenty-First Century. Proceedings of the Eight International Congress of Egyptologists, Cairo 2000.Cairo [Egypt].

Vörös, G. & R. Pudleiner 1998. Preliminary report of the excavations at Thoth Hill, Thebes: The Pre-11th Dynasty Temple and the Western Building (Season 1996-1997); in *Mitteilungen des Deutschen Archäologischen Instituts, Abteilung Kairo* 54: 335-340.

Wenke, R. & D. Brewer 1996. The Archaic – Old Kingdom Delta: the evidence from Mendes and Kom el-Hisn; in M. Bietak (ed.) House and Palace in Ancient Egypt. Vienna [Austria]: 265-285.

van Wesemael, B. & P. Dirksz 1986. The Relation between the Natural Landscape and the Spatial Distribution of Archaeological Remains. A Geo-Archaeological Survey in an Area North of Faqus, Eastern Nile Delta, Egypt. Reports of the Laboratory of Physical Geography and Soil Sciences 24, University of Amsterdam [Netherlands].

van Wetering, J. in press. Royal cemetery of the Early Dynastic Period at Saqqara and the Second Dynasty royal tombs. To be published in the Proceedings of the *Origins of the State. Predynastic and Early Dynastic Egypt* symposium, held at Krakow Aug.-Sept. 2002. Krakow [Poland].

van Wetering, J. in prep. The Egyptian 'Presence' in the Sinaï and the Southern Levant, 4.000 – 2.500 B.C.E. Archaeological Perspectives on Egypt's 'Foreign' Contacts. Doctorandus dissertation – Faculty of Archaeology, University of Leiden (the Netherlands).

Wilkinson, T. 1999. Early Dynastic Egypt. London [UK].

el-Zahlawi, N. 2000. Tombs in an Old Kingdom Necropolis, Work still going on to reveals more objects discovered in Monshaet Ezzat; at: http://guardians.net/sca/monshaet_ezat.html

Ziermann, M. 1993. Elephantine 15, Befestigungsanlagen und Stadentwicklung in der Frühzeit und im frühen Alten Reich. Mainz [Germany].

Ziermann, M. 2002. Tell el-Fara'in – Buto. Bericht über die Arbeiten am Gebäudekomplex der Schicht V und die Vorarbeiten auf dem Nordhügel (site A). *Mitteilungen des Deutschen Archäologischen Instituts, Abteilung Kairo* 58: 461-499.

A Spatial Analysis of Deposits in Kom el-Hisn

Anthony Cagle

1.0 Introduction

Egypt has always represented something of a problem for those scholars studying the evolution of complex societies. Rather than large, deeply stratified tell sites such as those found in Mesopotamia, Egyptian sites tend to be fairly shallow, often below the present water-table, and buried by ongoing sediment deposition and later habitations. This is especially true in the Delta where moist conditions and intense agricultural activity make preservation of accessible sites difficult. A relative lack of large mortuary and temple sites with stone architecture has further hindered extensive excavations in the Delta. Consequently, until fairly recently, most of what was known about Predynastic and Old Kingdom demographic and settlement distributions has come from epigraphic sources related to temple and mortuary centers, notably the pyramid and temple complexes around Saqqara and the Giza plateau among others (e.g., Badawy 1967; Kanawati 1977, 1980; Strudwick 1985).

In the absence of extensive archaeological data on a range of settlements, scholars have relied upon epigraphic sources and limited settlement excavations to build a picture of a largely rural Egypt with little of the highly urbanized character of Mesopotamian city-states. Useful though it may be, epigraphic information is inherently biased and does little to provide a detailed picture of the socioeconomic relationships that existed among the rural settlements that constituted the bulk of the population, and between these settlements and the central government. For example, it is unclear whether and to what degree individual settlements produced all necessary goods and services locally and interacted with the central authority through taxes and tribute only (i.e. they were functionally redundant), or whether they were part of a larger regional production and exchange system (i.e. they were functionally interdependent).

The picture of a largely rural Egypt with the population distributed among numerous smaller agricultural villages and related to the central authority primarily through taxes and other appropriative measures is often contrasted with the large, fortified urban city-states of Mesopotamia (e.g. Trigger 1993). This idea of Egypt as a "civilization without cities" (Wilson 1960) has persisted even though a clearer picture of a range of variation in both site size and function has steadily appeared (e.g., Bietak 1979a, 1979b; Emery, et al. 1979; Hoffman, et al. 1986; Kemp 1977; van den Brink 1987, 1988, 1992a, 1992b; Wegner 1998). Indeed, much of the difficulty scholars have had in placing Egypt within a larger context of cultural evolution reflects deeper theoretical issues involving the definition of "cultural complexity" and the methodologies used to describe and compare complex civilizations in a meaningful way (e.g., Sjoberg 1960; Wenke 1997). Clearly, what is needed is a diverse set of well-excavated settlement sites of different sizes from a variety of locations and micro-environments. One such site, Kom el-Hisn, was chosen for detailed excavations by Robert Wenke and Richard Redding in 1984. The work continued for two more seasons in 1986 and 1988. Kom el-Hisn is a large site with abundant, largely undisturbed settlement remains from the Old through Middle Kingdoms, in addition to extensive cemeteries of various dynastic dates (mostly post-Old Kingdom) and a New Kingdom temple complex (no longer extant). The site has seen occasional archaeological and philological work since 1884 but only the most recent work has concentrated on the Old and Middle Kingdom architecture and associated deposits (but see Kirby 1998).

Several publications have appeared describing this most recent work at Kom el-Hisn (Wenke and Brewer 1992; Wenke, et al. 1988; Wenke and Redding 1985, 1986; Buck 1990; Moens and Wetterstrom 1988; Redding 1992; Sterling and Wenke 1997). These and earlier reports indicate that, at least from the Old Kingdom and perhaps earlier, Kom el-Hisn may have functioned as a regional capital, was heavily involved in agricultural activity, maintained herds of sheep, goats, and pigs, and was possibly involved in the specialized production and export of cattle. More recently, I undertook a detailed spatial analysis of the site with a special emphasis on the depositional history and the distribution of various artifact types with respect to each other and to architectural features (Cagle 2001). The present paper summarizes my own analyses and integrates them into previous work at Kom el-Hisn and some of the larger issues involving Old Kingdom settlement patterns.

2.0 Kom el-Hisn: Setting

The site of Kom el-Hisn lies approximately 100 km. northwest of Cairo and about 10 km. southeast of the village of Dilingat near the western margin of the Delta. The Rosetta branch of the Nile flows approximately 14 km. to the south-southeast and several canals lie within a few kilometers of the site. The (now extinct) Canopic branch of the Nile may have been nearby in Old Kingdom times, and another smaller distributary, the Alexandria branch, is shown by Toussoun (1922) flowing to the west of Kom el-Hisn along the western margin of the Delta and emptying into Lake Maryut. The migration of waterways over time makes it difficult to determine the exact locations of the distributaries at the time of Kom el-Hisn's occupation in the Old Kingdom without extensive coring and dating of the deposits, but it seems clear that at least these two Nile branches and possibly other large canals were within a few kilometers of Kom el-Hisn.

The major geological deposits in the area are a series of Quaternary sands and gravels overlain by Holocene fluvial sediments ("Nile mud"). Coring of the area in 1984 and 1986 revealed that the Quaternary deposits, known as *gezira,* extend in a general north-south direction for about 700 meters in length and 400 meters in width and at a maximum of 7 meters above sea level (Buck 1990). The *gezira* consists of well-sorted medium sand with a mean grain size of 1.5-2.5 phi and may be several meters thick. Van den Brink (1987) found these *gezira* deposits to be very important in the location of settlements, positing that habitations were placed there to avoid high Nile floods while remaining as close as possible to agricultural lands.

In antiquity, Kom el-Hisn was located in the third nome of Lower Egypt, Ament or Imenti (*ʾImnty*), also the location of the "Estate of the Cattle", *ḥwt-iḥwt* (Helck 1974: 154). The "Estate of the Cattle" is one of the oldest state foundations in Egypt (Helck 1974: 154-155; Zibelius 1978: 149-151) and is known from a seal in the tomb of Merneith at Abydos from the 1st Dynasty (Petrie 1900, pl. 20) and several impressions on jar lids from Tomb #6 at Abu Roash (Montet 1946). The nome-sign of *ʾImnty* consists of the Horus falcon sitting atop a standard, demonstrating the association of the nome with Horus (Montet 1957: 57).

The goddess Hathor, by her name (*Ḥwt-Ḥr*, "House of Horus") connected to Horus, likely originally as his mother (Bonnett 1952: 277), was also associated with this area. She merged at an early date with the local cow goddess Sekhat-Hor (*Sḫ3t-Ḥr*) who was wet-nurse of Horus and patron of cattle (Bonnett 1952: 402). Hathor is also frequently combined with the goddess Sekhmet (*Sḫmt*), creating a dual goddess, Sekhmet-Hathor, based on the fact that both were, by the New Kingdom, equated with the "Solar Eye". In the Destruction of Mankind myth, part of a larger work known as "The Book of the Cow of Heaven" (Lichteim 1976: 197; Watterston 1999: 42), the gods advise Re to punish rebellious mankind: "Let your Eye go and smite them for you. [...] May it go down as Hathor!". After a day of slaying mankind, Re receives the returning Eye with the words: "I shall have power [= *sḫm*] over them [i.e. mankind] as king by diminishing them", which leads to the conclusion: "Thus the Powerful One [= *sḫmt*] (Sekhmet) came into being." However, when the Eye performs her task too well, Re decides to save mankind by secretly pouring a red brew, made by maidservants, over the fields. The Eye drinks, gets drunk, and forgets about destroying mankind. And thus the pacified Eye is received by Re with the words: "Welcome in peace, O gracious one [= *im3(y)t*]!". The episode concludes, referring to the maidservants: "Thus beautiful women [= *nfrwt*] came into being in the town of Imu [= *imꜣw*]". (All quotes are from Lichtheim 1976: 198-199, with additions in square brackets.) There are several word-plays in the tale, one which connects the gracious (*imꜣ*) goddess Sekhmet-Hathor with the town of Imu (*imꜣw*).

Imu or Imau, which Kom el-Hisn has often been equated with, is known to be the capital of *Ỉmnty* and is normally written with a tree glyph, either alone or in groups of two or three:

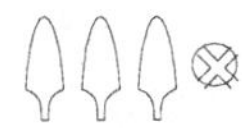

or with a slightly different spelling, as in Zibelius (1978: 35):

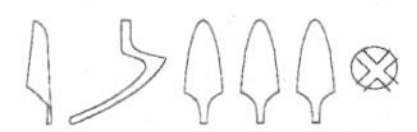

The tree glyph (*imꜣ* or *iꜣm*, Gardiner's code M1) was, according to Buhl (1947: 86), the masculine form of the word for the date palm, and Griffith refers to Kom el-Hisn as the "(city of) palm trees" (1888: 82; also Bonnett 1952: 83, 85, 279). This designation is generally no longer accepted (e.g. Faulkner 1996: 20), though the specific type of tree the glyph represents is not known (for several suggestions, see Hannig 1995: 69). The tree was the symbol of a local tree goddess, who became identified with Hathor (Bonnett 1952: 380). Also in other places, Hathor as goddess of heaven merged with local tree goddesses; for example, in Memphis she was referred to as "The Mistress of the Southern Sycamore" (*nbt nht rśt*) (Buhl 1947: 86; Bonnett 1952: 85).

3.0 Kom el-Hisn: Previous Work at the Site

Several researchers have visited the area over the years and provided comment on Kom el-Hisn, both on its general archaeology and its status as Imu. Petrie, while working at Naukratis in 1884-1885, located an offering tablet, supposedly from Kom Firin, mentioning the "Mistress of Imu, Sekhmet", *nbt imꜣ sḫmt* (Petrie 1886: 94-95; pl. XXXVI, #2). Petrie also noted a similar inscription to the "Mistress of Imu" on a Ramesside statue at Kom el-Hisn. In view of these observations, Petrie was unable to determine whether Kom Firin or Kom el-Hisn is the Imu being referenced.

More extensive work was carried out by Griffith in December 1885 and published as an appendix in Part II of the Naukratis monographs (Griffith 1888: 77-84). At that time, the enclosure walls and the foundation of one pylon of the Ramesside temple enclosure were still visible, along with four inscribed statues (labeled I-IV in the 1888 volume). Griffith's map of the enclosure (pl. XXIV) showed it to be approximately 127 yards east – west by 70 yards north – south, and that it was four yards thick, rested on "rubbish", and was built of mud bricks of nine by eighteen inches. Griffith's estimate of the span of the entrance defined by the pylons is approximately ten yards. Two of Griffith's statues are still extant at Kom el-Hisn (Coulson and Leonard Jr. 1981). His statue I (Coulson and Leonard's statue A) shows Ramesses II seated with a goddess, presumably Hathor, with inscriptions on the back and side referring to, respectively, "Sekhmet, beloved mistress of Imu" (*Sḫmt nbt imꜣ mry*) and "Sekhmet-Hathor, beloved mistress of Imu" (*Sḫmt-Ḥt-Ḥr nbt imꜣ mr*). Griffith's statue II (Coulson and Leonard's statue C) is the lower portion of a quartzite statue and refers to the "Beloved mistress of Imu" as Sekhmet and Hathor separately.

Griffith did no excavating at Kom el-Hisn and initially thought that Kom el-Hisn was, in fact, Imu. However, he seems to have had second thoughts. The main part of the appendix Griffith wrote for the Naukratis Volume II was written in November of 1887. In a later addendum, written in May of 1888, he posits that Kom el-Hisn was, instead, a residence of Ramesses II and that "the nome capital [i.e. *Ỉmꜣ*] is to be looked for elsewhere" (Griffith 1888: 82-83). Petrie regarded Kom Firin as a larger site but was uncertain which (if either) was the nome capital (Petrie 1886: 94-95).

Georges Daressy next visited the site, copying several inscriptions from at least two statues (Daressy 1903). The inscriptions provided by Daressy (no drawings or plates of the statues themselves were provided by him) are similar to those in the other monuments but do not mention either Sekhmet-Hathor or Imu. One of those described by Daressy may be Statue B of Coulson and Leonard, a "badly weathered statue of Ramesses II advancing, left leg forward, on a high base" (Coulson and Leonard 1981: 82). The dimensions, material (sandstone), and inscriptions on the back of the statue match those described by Daressy (1903: 282-283). However, Coulson and Leonard (1981: 82) note some discrepancies between the inscriptions published by Daressy and those observed on Statue B so it is possible that the statue Daressy described is not Statue B of Coulson and Leonard or that there is some error in Daressy's copy.

The other major monument at Kom el-Hisn is the tomb of Khesu-wer (*Ḫsw-wr*) which was studied in some detail by C. C. Edgar in 1910 (Edgar 1909-1915) and later by Silverman (1988). The tomb lies in the southwest portion of the site near a modern village and is constructed of limestone blocks with (in Edgar's time) traces of mud brick walls surrounding the structure. The inscriptions on the walls of the tomb were drawn but never published in translated form. *Ḫsw-wr* was described as being "Overseer of the beautiful women" (*ỉmỉ-rꜣ nfrwt*) (who were, presumably, servants of the Hathor cult; cf. the *nfrwt* in the Destruction of Mankind myth above), and a priest (*ḥm-nṯr*) of Hathor; he had charge of the temple (*mr ḥwt*). Edgar argued that the temple where he held this office was Kom el-Hisn because of the Ramesside monuments found there by Griffith and Daressy which all mention Imu as being associated with Sekhmet-Hathor. Edgar dated the tomb to the reign of Amenemmes (Amenemhat) III based on the character of the religious texts (Edgar 1909-1915: 61) and the form of a basalt head found in the tomb (Plate XXXII in the original) which Edgar argued is typical of those found during the Middle Kingdom reigns of Amenemmes III or Sesostris III (both 12th Dynasty). Still, as noted by Edgar, and by Coulson and Leonard (1981: 83) after him, the head was obviously intrusive and the dating of the tomb is best determined through a more detailed analysis of the inscriptions (see also Helck and Otto 1980).

The next detailed work at Kom el-Hisn was carried out by Hamada, el-Amir, and Farid in the late 1940s (Hamada and el-Amir 1947; Hamada and Farid 1947, 1948, 1950). Their primary interest was the abundant graves, most of which were at the north end of the main midden area. Most of the graves are dated by seals or scarabs to the New Kingdom. For example, the following are illustrated in Hamada and Farid (1950): an oval seal with the name of Thutmosis III (Plate VII, #16); a scarab with the name of Thutmosis III within a cartouche (Plate VII, #17); and a scarab with the name of Amenophis III within a cartouche (Plate VII, #19; all 18th Dynasty objects are described on page 371 of that volume). Hamada and Farid exposed a small area in the main midden to the northeast of the excavations on which the current research is based. These graves were contained within the existing architecture, much like the burials excavated in this general area in 1986 and 1988. Unfortunately, Hamada and Farid were unable to provide secure dates for these graves.

The last work at Kom el-Hisn before the current project was a brief visit to the site by Coulson and Leonard in connection with their work at Naukratis. In their published work (Coulson and Leonard 1981; Silverman 1988) they described in more detail the three remaining Ramesside statues at the site (now all situated near the rest house) and correlated them with those described by Griffith and Daressy. By that time almost all the large walls surrounding the site described by Griffith had disappeared, along with the pylons. In their brief visit, Coulson and Leonard were able to clarify some of the glyphs described by Griffith and collect a small sample of sherds.

The excavations on which the present research is based were conducted over three seasons, 1984, 1986, and 1988. The main occupation areas are located in the southern portion of the site. The majority of the site is covered by a layer of coarse salt-encrusted sediment with abundant ceramics (hereafter: the Upper Pottery Layer or UPL) which is thought to be a lag deposit resulting from *sebakh* digging. Mud brick architecture is found throughout the main midden within a few centimeters of the surface.

Preliminary analyses of the material excavated in 1984 and 1986 (Buck 1990; Moens and Wetterstrom 1988; Redding 1991, 1992, ND; Wenke and Redding 1985, 1986, 1987) indicate the following. First, the majority of cultural remains are of Old Kingdom date and are largely undisturbed by later occupations. A small area of Middle Kingdom remains is located in the southwestern end of the site and two burials thought to be of First Intermediate date are also intrusive into the Old Kingdom architecture. The date of the occupations is based primarily on ceramic types. Most forms are known from other Old Kingdom occupations at Giza, Saqqara, and other Lower Egyptian sites, and specifically represent wares common in Dynasties 5-6. No evidence of ceramic production, such as kilns or slag, has been found. Radiocarbon dates (Wenke et al. 1988) also correct to Early Dynastic or Old Kingdom times. Epigraphic material came mostly from the Middle Kingdom portions of the site, but several fragments of Old Kingdom date were also found.

Second, the structure of the site indicates mostly domestic architecture extending over at least 900 m^2. No architectural use of stone was found. A large enclosure wall seems to ring at least a portion of the site, evidence for which appears near the modern village in the southwestern end of the site. Wenke et al. (1988) suggest that an administrative or religious sector was located in this southwestern area, now covered by the modern village. Structurally this would make Kom el-Hisn similar to other Old Kingdom centers where these functions were performed in a restricted, and possibly walled, area of the town. There seem to be at least two major building episodes represented for the Old Kingdom, and probably three. Earlier structures were apparently leveled with rubble before rebuilding rather than utilized in later constructions.

The artifacts analyzed thus far indicate basic domestic activities. Typical "bread mold" ceramics, fire-blackened (cooking?) jars, and various shallow bowls suggest food preparation and storage. The lithic component consisted of chipped and ground stone tools in almost equal parts. The majority of debitage and finished tools were blades and sickle blades, respectively. The character of the blades and the almost complete absence of cores or manufacturing debris indicate that most lithic reduction took place off-site and that either blade blanks or prepared cores were imported into the site. About 15% of the ground stone tools represent grinding implements (manos or metates) and the remainder, with the exception of some flakes of limestone, granite, and alabaster, seem to be reworked pieces of these grinding implements.

Taken together, the vertebrate remains indicate a mixed diet of domestic (sheep/goats, pigs, cattle) and wild (fish, hartebeest) animals. Mammal bone dominates the faunal assemblage numerically (identified specimens), by weight, and by meat yield as part of the diet, but all other major vertebrate classes are present except for amphibians (Redding ND). The majority of the mammalian remains are from sheep/goats (*Ovis/Capra* or ovicaprids) and pigs (*Sus scrofa*). Age and sex distributions, while tentative, indicate that pigs were eaten before maturity while ovicaprids had a bimodal distribution with some individuals killed as juveniles and others in late maturity. The remainder of the vertebrate remains primarily consist of animals present in the surrounding area, such as catfish (*Clarias*), Nile perch (*Lates*), various species of bird (e.g. teal), game animals such as hartebeest (*Alcelaphus*), and a unique example of a soft-shelled turtle (*Trionyx triunguis*). Curiously, few remains of domestic waterfowl – geese and ducks – which are characteristic of Egyptian villages of all periods, have been found.

Cattle bones (*Bos taurus*) are rare in the Kom el-Hisn assemblages and this fact, together with evidence from the botanical remains, suggests that at least one function of Kom el-Hisn was as a specialized cattle producer. The absence of cattle bones implies that cattle, if present, were not eaten on site but exported elsewhere, perhaps to religious or administrative centers. In addition, Moens and Wetterstrom (1988) note that the distributions of four general classes of plant remains – cereal straw, field weeds, reeds and sedges, and fodder plants – differ from those at other habitation sites in Egypt. Fodder plants (e.g., clover, vetch) compose almost 27% of the remains thus far analyzed and this suggests that the animals were fed in pens on specially raised crops and only allowed to browse in the open seasonally if at all. Little evidence of sheep or goat dung has been found and it is thus argued that the majority of these remains came from cattle dung used as fuel.

Much of the evidence indicates that Kom el-Hisn may have had strong ties to the central government. The absence of production facilities for basic goods and the similarity of ceramic types to those found in administrative and mortuary contexts at Giza, together with the evidence of specialized cattle production, indicate a smaller degree of self-sufficiency than is generally supposed for Old Kingdom Egypt. Whether Kom el-Hisn was initially settled as a pious foundation or a directly administered arm of the central authority, or simply developed that way, is unclear.

4.0 Stratigraphic Analysis

The goals of my own project were to clarify the stratigraphy present at the site, examine patterns of artifact association in detail, and develop a functional model of the excavated portion of the site. To do so I relied mainly on data collected during the 1986 and 1988 seasons (the 1984 season consisted only of mapping and excavating small test pits). Data from the 1986 season came from several 2-meter units placed according to a stratified random sample design. These units were designated by the coordinates of the southwest corner of the square in meters south and east of the origin of the grid system (to the northwest of the site). In addition, a shallow trench called the "block area" was also excavated in contiguous 2-meter units covering 72 m^2. Based on preliminary analyses of the 1986 data, three areas were chosen for wider excavation in 1988. The architecture in these areas lies close to the surface and was exposed by scraping off the upper 10-20 cm. of the Upper Pottery Layer. Structures identified as such were then excavated as individual rooms and were numbered sequentially (*Fig. 1a*). Several of the rooms excavated in 1988 were also partially excavated in 1986 as part of the block area. In these cases, any of the 1986 deposits that correlated with 1988 room structures were added to the 1988 room deposit sequences.

In both seasons, excavation was carried out following natural stratigraphy. The basic unit of excavation was the "sedimentary unit" or SU. The volume of each SU was determined by the number of baskets of a predetermined volume that were removed. All material (except where noted by the excavator) was dry screened through approximately 1/4" mesh and all artifacts were removed and bagged separately at the screens. Faunal material (bone and shell) was identified by species or element if possible and weighed. Lithics were weighed in aggregate and then analyzed typologically. Ceramics were separated first by diagnostic type (rims, bases, etc.) and body sherds, the latter being further sorted by fabric type (Nile A, B, and C) and weighed. Diagnostic sherds were washed and sorted according to a typology developed specifically for Kom el-Hisn. The typology is somewhat paradigmatic and is based on overall morphology, fabric type, and surface treatment, and broadly conforms to the whole-vessel categories derived from the standard corpus of Old Kingdom pottery types (Brunton 1927; Mond and Meyers 1937; Reisner 1931, 1932, 1942, 1955). I further collapsed the 27 types thus defined into a set of 12 types representing more or less equivalent functional groups (*sensu* Dunnell 1978, 1978b, 1983).

There are basically six sets of deposits in the excavated portions of Kom el-Hisn, five of which are of Old Kingdom date (Levels 1-5). The topmost layer, Level 0, is composed primarily of Upper Pottery Layer deposits and those dated to post-Old Kingdom times. Apart from the UPL, these later deposits were generally restricted to the southeastern portion of the excavated area and are of First Intermediate or Middle Kingdom date.

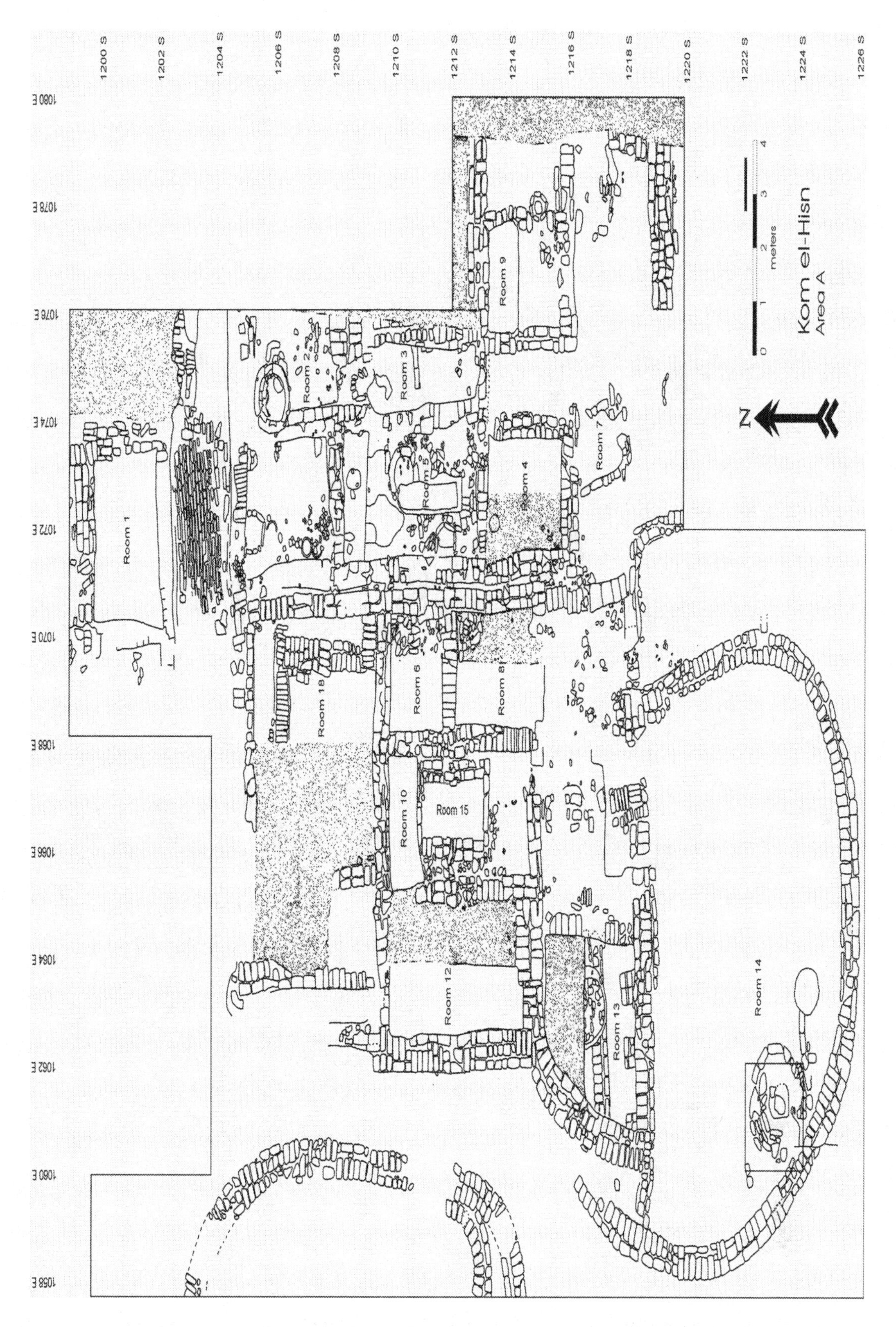

Figure 1a. Plan map of the architecture of Kom el-Hisn (Area A).

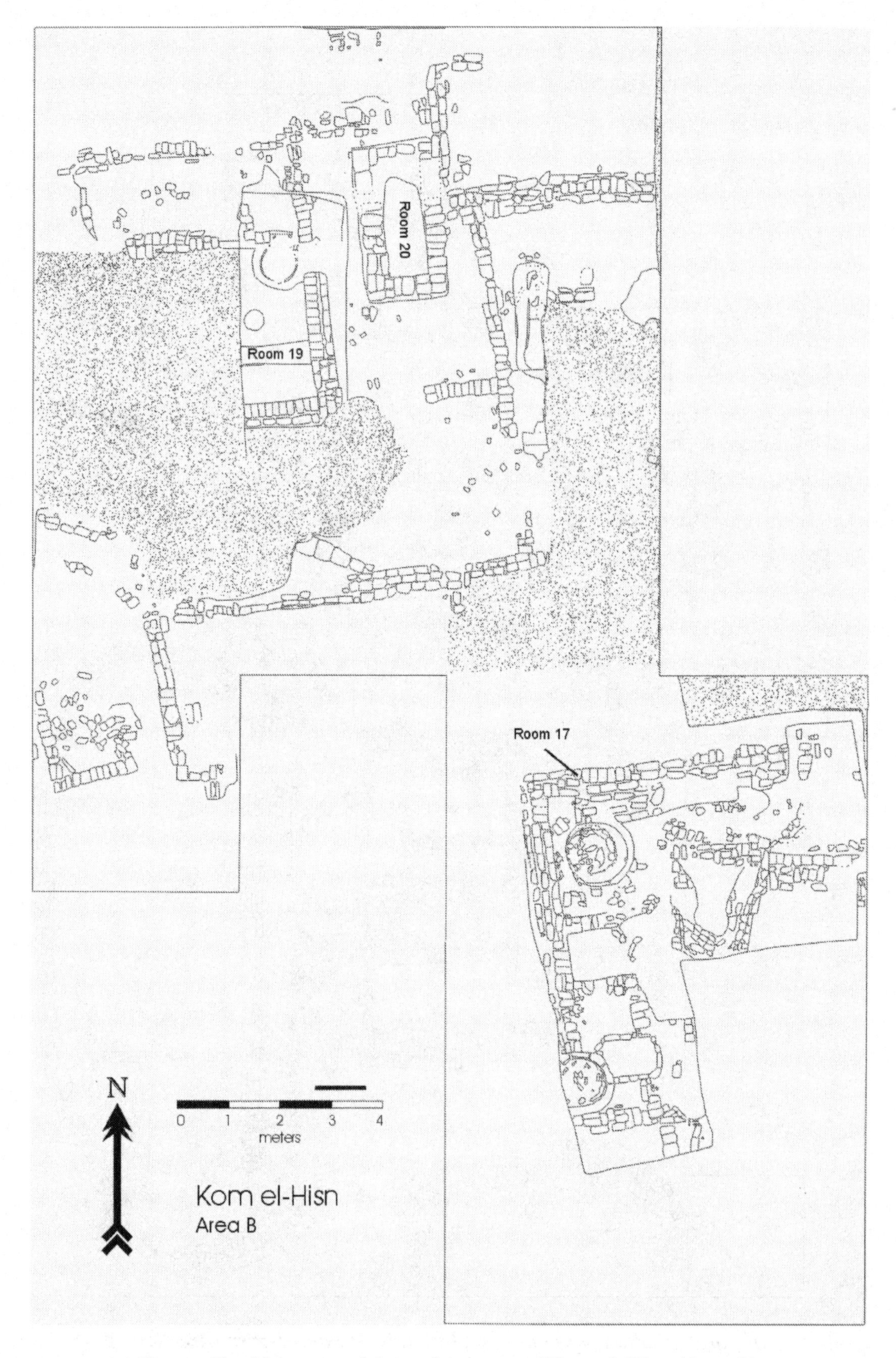

Figure 1b. Plan map of the architecture of Kom el-Hisn (Area B).

Below the Level 0 deposits are a series of dumps and other deposits associated with an Old Kingdom occupation which presumably had its major focus in other areas of the site. All of these deposits contained typical Old Kingdom ceramics and few, if any, later types that cannot be attributed to stratigraphic mixing. In addition, they are generally separated from underlying Old Kingdom deposits by earlier unconformable deposits, such as collapsed and partially dissolved mud brick walls.

Level 2 represents several burials found within this area of the site. A total of seven burials were found, three adult and four infants or children. All were associated with earlier Old Kingdom (Level 3) structures. An Old Kingdom date for these burials is indicated by the presence of deposits with typical Old Kingdom ceramics directly overlying them. In both Room 5 and Room 15, for example, typical dump deposits were found directly overlying the burials, neither containing later ceramic types. Consequently, since the burials are directly associated with the existing architecture and are stratigraphically below other Old Kingdom deposits, I concluded that these are Old Kingdom burials deposited in this area after the main occupations of Level 3, but before additional Old Kingdom remains were deposited in the same area, these later deposits being mostly dumped refuse (Level 1).

The majority of deposits are from Level 3, the main set of occupations dealt with in this study. They present a variety of deposit types such as floors, wall collapses, dumps, and pits, the latter representing a mixed bag of structures from small, excavated pits to structures composed of mud bricks. Most of the architecture shown in *Fig. 1a* is associated with this level. Dumping was primarily restricted to particular areas outside of the occupation structure (the basins of units 1192/1035 and 1204/1060) with some possible post-abandonment dumping within a few rooms and in some pit structures. Storage pits, presumably for grain, were restricted to three adjacent rooms, namely Room 2, Room 5, and the room directly north of Room 5 (not analyzed here). Room 17 (*Fig. 1b*) also contained pit structures with fire-hardened interiors. Thus, a rough sort of pattern can immediately be seen with certain areas devoted to food storage, others to food preparation, and dumping restricted to nearby topographic basins.

At least two occupation levels exist below Level 3. Most of these deposits are Level 4, with only a single wall collapse deposit in Room 18 and a small fluvial deposit in 1235/1056 assigned to Level 5. The remaining Level 4 deposits are largely restricted to five units: 1166/1066, 1235/1056, and Rooms 4, 17, and 18, with minor occurrences in other units (usually some wall collapse underlying Level 3 deposits). Units 1235/1056 and 1166/1066 contain substantial exposures with associated architecture and occupation floors. Other exposures with floors and substantial walls occur in Rooms 4, 17, and 18. All of these occur under Level 3 architectural units. There does not appear to be any relationship between the architecture of Levels 3 and 4; that is, the later Level 3 structures were not aligned with respect to earlier structures. Rather, it seems that the underlying structures were leveled and the later structures built according to a different plan altogether. Since little of the underlying architectural plan is exposed this is necessarily tentative, yet what is visible does not seem to correspond to anything in later levels.

To summarize, Level 3 represents the major set of Old Kingdom occupations at Kom el-Hisn. These were preceded by at least two earlier occupations (Levels 4 and 5) whose architectural plans seem unrelated to later structures. After the Level 3 structures were abandoned, some minor dumping occurred on the floors of these rooms, but most were allowed to collapse and to begin to decompose. During this period several adults and infants were interred within the structures as Level 2 deposits. Infants were generally laid next to walls and then covered with sediment, while adults had brick tomb structures built for them within structures defined by older Level 3 walls. Later, after the tomb structures had begun to collapse, more dumping occurred on top of these deposits, presumably from other Old Kingdom occupations in a nearby portion of the site. Thus, after the Level 3 structures were abandoned, the area was left to decay for some time, after which it was used as a cemetery. This cemetery was then abandoned and was used for refuse dumping, still during Old Kingdom times.

Seriation studies and radiocarbon dates are insufficient at this time to independently verify the above occupational history. All of the radiocarbon dates obtained thus far are from the 1984 season whose deposits are not well integrated into this scheme yet, though several carbon samples from the 1986 and 1988 seasons are available but not yet submitted for analysis. Attribute-level seriations are being investigated for their applicability to these deposits (see Sterling and Wenke 1997) and show some promise of establishing tighter chronological control.

5.0 Spatial Analysis

The goal of this analysis was to investigate patterns of co-occurrence among various classes of artifacts – ceramics, lithics, bone, etc. – and how they are distributed across the site. For the most part, these analyses were restricted to Level 3 deposits as these were the most numerous and spatially distributed; except where indicated, the following discussion refers only to Level 3 deposits.

Several statistical methods were used to examine the spatial patterning of artifacts and their co-occurrence. The basic unit of analysis was the deposit and the frequency of different artifact types within each deposit. All deposits in Level 3 were used, except those for which only one or two cases were present (e.g. burials) and those derived from mud brick walls, since the artifacts contained within these wall-derived deposits are not thought to be functionally relevant. This left various instances of refuse dumps (both *in situ* and redeposited), floors, and pit deposits for a total of 39 deposits distributed among 17 rooms and excavation units. Both parametric and non-parametric methods were used depending on sample sizes and the number of deposits available for the analysis (floral data was only available for a few deposits).

The architectural plan shown in *Fig. 1a* indicates perhaps two distinct areas of occupation within this complex of rooms, separated by a large wall running north – south dividing Rooms 6, 8, and 18 to the west from Rooms 4, 5, 7, and unit 1220/1072 to the east. The series of rooms to the west of this wall constitute a coherent larger structure consisting of Rooms 6, 8, 10, 12, 13, and 18. The only entrance to this block of rooms apparent from the visible architecture is along the south wall near the empty space south of Room 8. The main room in this block was not defined initially as a single entity but is L-shaped, with the northern arm excavated as Rooms 6 and 8. This large L-shaped room offers entrance to Room 13 and the structure containing Room 10 (hereafter, this larger structure containing Rooms 10 and 15 will be referred to as Room 10). Access to Room 12 is gained only via Room 10 and perhaps through a possible doorway in the northwest corner of Room 12. Thus, this block consists of four main rooms, with the wall between Rooms 6 and 8 being built sometime during Level 3 to split this northern wing of the L-shaped room in two. Room 15 is a later (but still Old Kingdom) tomb built within the confines of Room 10.

The eastern block of rooms is less coherent overall, but some structure is seen in the northern half, comprised of Rooms 2 and 3/5 and the structure immediately to the north of Room 5. The jumbled nature of the visible mud bricks makes interpretation of the overall structure difficult. There do not appear to be any connections between these structures and those to the west. Entrance to Room 5 may have been gained through what appears to be a doorway in the center of the south wall and through the southwest corner of Room 2. Rooms 3 and 5 comprise a larger structure in which the individual rooms were divided by a short wall. Room 3 also may have been further subdivided into two very small sections by the divider projecting from the east wall of that room. Entrance to the room to the west of Room 2 (unexcavated in depth) seems to have been gained only via Room 2. The "pavement" to the north of this room may be the north wall of that room having fallen over largely intact. The structures to the south of these four rooms (Rooms 4 and 7 and unit 1220/1072) do not have sufficient visible architecture to relate them to the overall structure.

All of the northern four rooms in this eastern area have specially constructed storage pits associated with them. Altogether, this block of four rooms contains at least six pits constructed of mud brick: three in Room 5, one in Room 2, and two visible in the unexcavated room. All of the pit structures in Room 5 are associated with Level 3 floors. Although there are two distinct floors associated with this room, both appear to be closely associated with the room walls with little time between them. Artifacts in these rooms are not very informative. Ceramics are generally sparse and not distinctive either in the floors or the pits themselves. Lithics are similarly sparse, as are plant remains, which is due to there being no burning in any of these rooms and thus little chance for organics to be preserved. Interestingly, two of the pits (one from Room 5 and that from Room 2) contained a large number of fish cranial elements (Cagle 1991) indicating the use of some pits as dumps for, apparently, fish heads after the main occupation was complete.

The structures to the south have no readily discernible function. The Level 3 deposits of Room 4 contained no detectable floor, which may have been obscured by the activities associated with the later infant burial in this room. Room 7 is a poorly defined structure with one curved wall and a shallow occupation surface. The artifactual remains from this room provide little insight into its function. The adjacent unit, 1220/1072, contained no architecture and seems to have been a shallow basin into which material from Room 7 flowed.

Room 9 is enigmatic and different from every other room excavated thus far. The floor deposit associated with the room consists of a series of square clay features approximately 45-50 cm. on a side and spaced approximately 20 cm. from the room walls and from each other. This may be only a portion of a much larger structure reminiscent of the columned porticoes observed fronting the residence blocks at, for example, Abydos (Wegner 1998: 15, Figure 8). If this is the case, then the structures in the main excavated areas may be the support structures for as-yet-unexcavated habitation units to the east.

Turning back to the western set of rooms, Room 12 contained a distinctive set of artifactual and faunal remains. It contained numerous ground stone fragments, most of which were of a single type of stone (sandstone) and represent both upper (mano) and lower (metate) parts of grinding apparatus. All of these were found in the southern portion of the room along with many of the faunal remains. Evidence that at least some of the grinding stone fragments are in their use locations is indicated by specimen #8 from this room, a thick, loaf-shaped fragment of a metate found resting on a clay setting. This room is also relatively rich in faunal remains containing nine different species, a large number of which are fish. A large portion of a *Trionyx triunguis*, the soft-shelled turtle, was found with the carapace largely intact in one corner of Room 12 and was probably butchered within the confines of this room. The ceramics from this room are predominantly small to medium bowls, notably flared bowls (Type F, Reisner's type D-XXXIX), bowls with a molded interior rim (Type I, Reisner's C-LXIII), and small, plain-rimmed bowls (Type O, probably Reisner's C-LXI or C-LXII). Two jar forms were also present, Type B (Reisner's types A-IIb, jars with rounded or pointed base) and Type C (Reisner's A-IV, ordinary traditional offering jars). No types of bread platters or bread molds were found in Room 12 (Types D and E, Reisner's types F-XXVI and F-XXV, the latter being the "flower pot" variety, both of which are generally common at the site), indicating that if grain was being processed in this room it was not being subsequently used in the preparation of bread/beer in this location. Nevertheless, the presence of grinding stones in apparent use locations and a wide variety of faunal remains, including a fairly rare *Trionyx*, suggests this room functioned as a specialized processing location for grain and animal foodstuffs. No evidence of cooking of this material was present, which probably accounts for the lack of plant remains from this room.

The adjacent Room 13, directly south of Room 12, is markedly different. It is distinct structurally for having a large circular brick structure (collapsed, not shown in *Fig. 1a*) with a concave bottom. Much burning is evident on the interior. This structure is similar to those described by Samuel (2000) and is interpreted as a baking (presumably bread) facility. The floor associated with this room contained no faunal material and the circular brick structure contained only a single ovicaprid bone. The ceramics associated with this room tend to be large, heavy types, either Type D Bread platters or Type E Bread molds, both of which are associated with bread baking (Arnold 1982; Aston 1996; Nagel 1938). They have been found elsewhere in association with ovens (Peet 1921: 177) and the 12th

Dynasty tomb of Intefiqer shows a platter being used for baking (Davies and Gardiner 1920: 14, plates 8, 9, and 9A). The ceramics and mud brick structure argue for a bakery as the sole function of this room. Its immediate proximity to the grain processing facilities of Room 12, of course, suggests that at least some of the grain processed in Room 12 found its way into the oven next door.

Several sets of deposits from Room 17 (near datum 1156/1004; see *Fig. 1b*) and Unit 1192/1035 showed a tight correlation in artifact types. Three deposits from Room 17 (a floor, a dump, and the contents of a fire-hardened brick pit structure) and a series of dump deposits from unit 1192/1035 had very similar ceramic assemblages, consisting largely of coarse jars (Type C), flower-pot bread molds (Type E), and two kinds of relatively finely made bowls, Meidum bowls (Type H, Reisner's type C-XXXII) and a flat bowl with a contracting mouth (Type L, Reisner's type D-LXXII).

In addition, these same two sets of deposits have similar faunal assemblages: both contain more ovicaprid than pig, which makes them somewhat unique among these deposits. Overall, pigs are the dominant animal present at Kom el-Hisn, contributing approximately 50% of the mammalian assemblages (Cagle 2001: Table 6.9; Redding 1991; Wenke and Redding 1989). This percentage is somewhat higher than that reported at the Neolithic site of Merimde-Benisalame (von den Driesch and Boessneck 1985), generally much higher than other sites of various ages (see Redding 1991, Table 1), and far higher than those reported at Giza by Kokabi (1980) and Redding (1992). Calculating the ratio of ovicaprid to pig at Kom el-Hisn gives a figure of approximately 0.7:1, which is again similar to that at Merimde (0.9:1), and generally much lower than that at other sites, particularly the two samples at Giza by Kokabi (1980) and Redding (1992), which have ovicaprid:pig ratios of 10.5:1 and 51.3:1 (see Redding 1992, Table 2). If one calculates the ratio of ovicaprid to pig for each of the deposits at Kom el-Hisn individually, only seven deposits have a ratio higher than 1.0. Of these, six are from either Room 17 or unit 1192/1035; the other is in Room 22 on the far western edge of the site (not shown in *Fig. 1a* or *Fig. 1b*; see Cagle 2003: 32, Figure 4.1, for location, and Caggle 2003: 253, Area C map, for architectural detail) which has only three identifiable bones. Thus, with this one exception, only Room 17 and unit 1192/1035 contain more ovicaprid than pig.

The structure and contents of the deposits in unit 1992/1035 indicate that it was a topographic depression into which household refuse was dumped along with some slumped wall material from adjacent structures. The tight correlation of these dump deposits with floor and pit deposits in Room 17, along with its close proximity, suggest that some material from Room 17 activities was dumped into the depression in unit 1192/1035. This is not incongruous with activity at Early Dynastic Hierakonpolis noted by Hoffman (1974). At Hierakonpolis, elite areas of occupation had refuse removed from living areas and transported entirely out of the immediate area, while in non-elite areas trash is removed from living areas but may be only transported a short distance to nearby depressions or abandoned structures.

In order to understand the significance of this pattern, it is instructive to compare these two sets of deposits with other deposits that have a similar relationship. Three rooms in this area contained floor deposits with evidence of some *in situ* burning in the form of black ashy debris, and an abundance of burned bone and plant material: Rooms 1, 6, and 8. These surfaces were interpreted as occupation surfaces where some form of cooking fires were in use, but without any sort of containment structure like the brick pit structure in Room 17. Another set of dump deposits in a topographic depression was excavated as unit 1204/1060 (see *Fig. 1a*). These represent a series of directly deposited refuse dumps and slopewash from adjacent structures. Several of the 1204/1060 dump deposits clustered together with the floor deposits from Rooms 1, 6, and 8, suggesting that some of the refuse from these rooms – or at least material very similar to it – was deposited within the depression in unit 1204/1060.

All of these deposits differ in significant ways from those of Room 17 and unit 1192/1035. As indicated earlier, only the Room 17 and 1192/1035 deposits have more ovicaprid than pig remains, while other deposits, including those from unit 1204/1060 and Rooms 1, 6, and 8, have more pig than ovicaprid. Further, an examination of the plant remains also reveals interesting disparities. Plant remains were grouped for this analysis into seven groups: Cereal, Chaff, Weeds, Reeds, Fodder, Other-Unidentified, and Other-Identified. Floral data was available for the floors in Rooms 1, 6, and 8, and for two of the dump deposits in unit 1992/1035. As Moens and Wetterstrom (1988) note, little charcoal has been found at Kom el-Hisn, suggesting that dung or non-woody plant stems were used as fuel for cooking fires rather than wood. Dung fuel is usually prepared by mixing the dung with straw or some other tempering material, forming it into cakes, and allowing it to dry. Such methods are known from modern Egypt and throughout Pharaonic times. The chaff from grain processing could have served as a temper in this context and winnowing debris is also known as a major animal feed (van Zeist and Camparie 1984: 4-5). Very few sheep or goat pellets, which tend to be common when used as fuel (Bottema 1984: 208), have been found. Cattle are thus the most likely source for the generally high densities of berseem clover (*Trifolium alexandriunum*) and winnowing debris found in many samples since cattle are often represented being hand-fed cereal, cereal stalks, straw, and cut fodder (Moens and Wetterstrom 1988: 170) while pigs and ovicaprids were usually allowed to forage on their own.

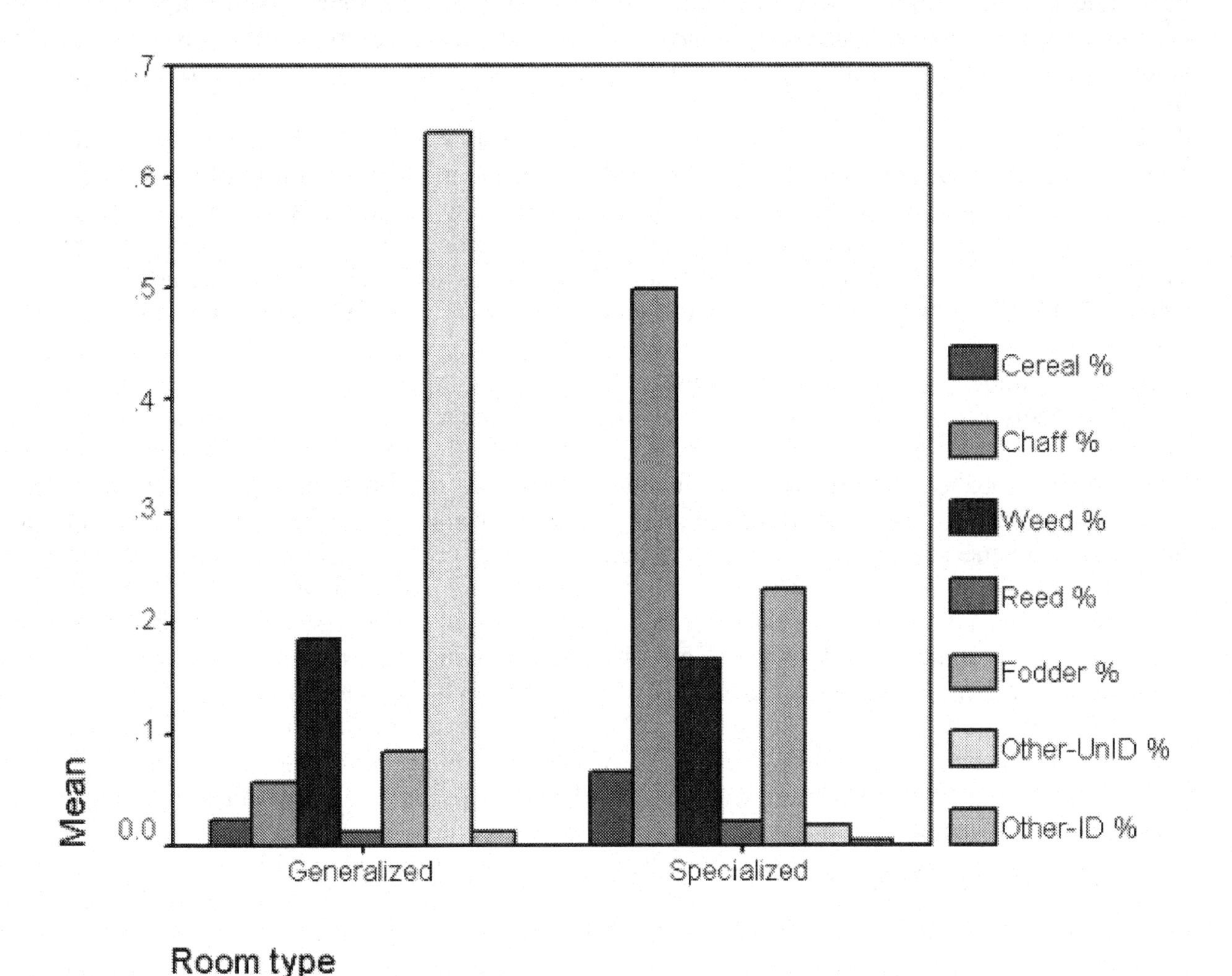

Figure 2: Percentages of plant types from hypothesized Generalized (Rooms 1, 6, and 8) and Specialized (Room 17 via unit 1192/1035 dump deposits) floor deposits.

All three of the floors in Rooms 1, 6, and 8 have very high densities of plant remains, and all are dominated by plants of the Other-Unidentified category (*Fig. 2*). In contrast, the floor and pit from Room 17, which, based on other data outlined earlier, are presumably the source for the 1192/1035 material, are strikingly different from those in Rooms 1, 6, and 8. The main differences in plant remains between these two sets of deposits is the high density of unidentified plants in Rooms 1 and 8, and the higher density of chaff and fodder in the 1192/1035 samples. Clover, a high-quality animal feed, is especially dense in the 1192/1035 samples compared to the others. Since much of this material probably entered the archaeological record as fuel for fires (either directly or incorporated within animal dung), the possibility exists that the differential distributions are a result of different fuel mixtures used in these rooms. The preferred interpretation is that Room 17 was a specialized cooking facility for a particular segment of the Kom el-Hisn community while the other rooms were used for more generalized food preparation. Room 17 is defined by a large number of Type C jars, while Room 1 and Room 8 have more even distributions of different types (particularly bowls). The plant material may indicate some kind of differential access to fuel types, either for functional or social reasons. For example, a variety of plant material may have been used as fuel in Rooms 1 and 8, resulting in a large amount of unidentified remains, while Room 17 had access to a more homogeneous supply of dung fuel.

A floor from an earlier level, Room 18 (Level 4), has a very similar distribution of remains to Room 1 DU-3: very dense cereal grains and unidentified plants, and moderate amounts of other taxa. Clover especially is of similarly low density. The structure of the Room 18 floor is also similar to the floors of Rooms 1 and 8: a series of dark patches (three in the case of room 18) rich in burned organic material. These patches, like those in Room 1 and Room 8, are not as well-defined as the hearth structure in Room 17. Room 18 only contained a single identified pottery type, Meidum bowls, also similar to Room 8.

An examination of artifact diversity is useful in explaining this patterning. Generally, discussion of artifact diversity is couched in the language of "activities", the assumption being that co-occurring sets of artifacts are describing distinct activity sets. Consequently, spatial variation in certain artifact distributions probably reflects different functions or suites of functions carried out in different structures. I further posit that rooms with a greater diversity of artifact types represent more functionally diverse structures. In this case, Room 17, which contains a few abundant ceramic types, would be considered more specialized than others containing a more even distribution of several types.

To test this hypothesis, I constructed a series of simulated random samples using the observed proportions of ceramic types found at Kom el-Hisn, following the procedures outlined in Kintigh (1984) and Schiffer (1989). I then calculated the average richness and standard deviation for each sample size. In this case, richness was computed simply as the number of classes. I also calculated the average and standard deviation for evenness, which was measured using Σp_i^2, where p_i is the percentage of each type in the sample. This provided an expected range of values for richness and evenness for a randomly distributed set of ceramics based on the observed proportions. Consequently, any excavated unit whose richness or evenness index falls outside of two standard deviations can, with 95% confidence, be considered to be non-random. To further explore any observed deviations, I ran the same simulation of random samples for each ceramic type individually. For each sample size I generated an expected proportion (the average, which should and did approximate the percentage of the total assemblage for each type) and standard deviation. The percentage of each type in a deposit is compared to the average for that type and any percentage which falls outside of two standard deviations is considered significant.

The results largely confirmed the results gained from earlier analyses. Taken together, the Room 17 and unit 1192/1035 deposits have higher than expected proportions of Types C and E (crude jars and flower-pot bread molds) and lower than expected proportions of several types of bowls. The deposits from the other units are varied, but most contain lower than expected frequencies of these Type C jars. In addition, no deposits have significantly more types or significantly more even distributions than expected.

This pattern is similar to that observed at Broken K Pueblo in the southwestern U.S. by Schiffer (1989), who found depressed richness values in both floor and "fill" deposits. Schiffer argued that one plausible interpretation of this pattern results from the sherds of reconstructable (whole) vessels being counted as individual sherds rather than as a single vessel. The presence of restorable vessels would tend to increase the number of sherds of that type giving the deposit a relatively lower overall richness value for a given sample size. Reconstructable pots are then taken as evidence that the floor deposits in which these pots occur are "primary" floor assemblages, indicating that the pots are in their locations of use.

I suspect that a similar pattern may be represented at Kom el-Hisn, albeit with different implications for site formation processes. The analyses outlined above show that Room 17 and unit 1192/1035 have similar distributions of ceramics, faunal remains, and plant taxa. Similar, though not as distinctive, patterns also exist between the floor deposits of Rooms 1, 6, and 8 and the dumps making up unit 1204/1060. All of these deposits (except for the floor in Room 6) have depressed richness indices and are more uneven than expected. If the "missed pot" hypothesis were correct, it would imply that some form of functional differentiation is evident within these room structures. Room 17 and its associated disposal area, unit 1192/1035, would be specialized in the use of types C and E, with the result that more whole or nearly whole pots would have been disposed of in these two areas. Similarly, the other rooms would be more specialized in the use of different sets of ceramic types. Rather than being indicators of "primary" or "secondary" refuse as Schiffer suggests, these patterns seem to describe patterns of use and discard in both contexts.

I believe the data presented supports this hypothesis. Room 17 contains a distinct hearth structure not found in other rooms with evidence of *in situ* cooking activities (e.g. the diffuse areas of burning and abundant ash in rooms 1, 6, and 8). Further, the two most abundant types in Room 17 – bread molds and coarse jars – exhibit characteristics suitable for cooking (i.e. resistance to thermal shock). Since both of these vessels were relatively cheap to manufacture it is not surprising that upon breakage they were readily discarded in a nearby dump (unit 1192/1035). These deposits also contained quite distinctive faunal (mostly ovicaprids) and floral remains, the latter indicative of a specialized firing regime using animal (probably cattle) dung as fuel.

While very few reconstructable pots were found, the distribution of vessel elements of Type C jars militates in favor of large numbers of whole or mostly whole vessels in Room 17 and unit 1192/1035. Type C jars are composed of rim sherds originally designated as Type 11, and bases originally assigned to Type 1A. Type 11 rim sherds have a fairly wide distribution in the Level 3 excavation units and rooms, occurring in 12 different rooms or units, of which Room 17 and unit 1192/1035 make up approximately 65% of the number of sherds. Type 1A bases are only found in 4 total rooms or units, of which Room 17 and unit 1192/1035 provide 93% of the total. Clearly, bases are much more concentrated in these two units than in others, suggesting that upon failure the entire vessel was discarded.

In contrast, Rooms 1, 6, and 8 have a very different array of ceramics, concentrating on more finely made bowls which may have served as occasional cooking vessels, but more probably as serving or preparation vessels. The plant taxa from these rooms also suggest a more diverse set of fuels used in the cooking process. The faunal assemblages for these rooms also differ from Room 17, emphasizing pigs, some ovicaprids, and minor components of fish and birds. Due to their depressed richness and uneven distributions, these rooms may also be considered as more specialized around a different set of functions than Room 17. The data suggests that some form of food preparation was carried on in these rooms, but perhaps of a more generalized nature than that found in Room 17. The dominant ceramic types, however, do not allow a more detailed analysis of the whole-pot content of these rooms.

In sum, the overall impression one gains from the evidence provided is of an area of mixed use, much of it related to food storage, processing, and cooking. Different forms of cooking seem to have been practiced, which may be distinguished by the types of fuel used. Room 17 (as derived from the unit 1192/1035 dumps) contained a distinct hearth structure, a specialized ceramic form (Type C jars), and a great deal of chaff and clover, which one might expect if cattle dung were the primary fuel. Other rooms (1, 8, and 18) contained less distinct areas of burning, more generalized ceramic forms, and high densities of unidentified plant types which might result from the use of various plant stems and brush for fuel.

Thus, the preceding analyses show that there is consistent patterning measured across different artifact categories. This suggests that at least some artifacts co-occur discretely in space rather than continuously. The obvious implication of this observation is that, since the artifact classes are assumed to be primarily functional in nature, some degree of functional specialization is being observed.

6.0 Discussion

Based on reports of earlier investigators (Edgar 1915-1919; Griffith 1888; Hamada and el-Amir 1947; Hamada and Farid 1947, 1948, 1950) the structure of the site as a whole probably consisted of a large oval or circular enclosure wall with a principal temple or administrative complex located to the west under the modern village. This administrative district was presumably surrounded by a complex of buildings housing the administrators and their support staff. This is similar to the layout of the villages adjacent to the tomb of Khentkaues (Hassan 1943) and the valley temple of Menkaura (Reisner 1931) at Giza, and the valley temple of Sneferu at Dahshur (Fakhry 1959; Kemp 1989).

The overall structure uncovered thus far at Kom el-Hisn contains two distinct sets of rooms, east and west, separated by a large wall. The east and west sets of rooms may each service a different set of people and the excavations coincidentally straddled two distinct functional areas of each. However, because of the limited extent of the uncovered architecture, these two sets of rooms may be related after all, with access to both obtained by other entrances to the north or south of the excavated areas. If this were the case, and if the columned area of Room 9 is a portion of the habitation area associated with these other rooms, then a case could be made that what we have uncovered are the support areas that service inhabitants of further structures to the east.

This pattern of a columned courtyard surrounded by several service buildings is well established in the Middle Kingdom (Arnold 1989) and has earlier antecedents. The village adjacent to the tomb of Khentkaues at Giza, for example, extended in an L-shaped configuration east of the tomb and consisted of a series of interlocking rectangular rooms. Room blocks usually consisted of a central room with 2-3 ancillary rooms attached to it (Hassan 1943). Two of these rooms contained circular grain pits (one had a single pit, another had four), and the southern or back room in each block served as the kitchen area as indicated by ovens and ash deposits. The village adjacent to the valley temple of Menkaura also had several rooms that contained grain storage structures, most of which were clustered towards the north end of the site, but the functions of other rooms are unknown as is the larger overall functional layout of the site.

A similar dearth of specific functional information is available from Neferirkara's temple complex at Abusir (Borchardt 1909). In this case, the village consisted of an enclosure wall surrounding the temple buildings with the spaces between filled by nine house structures. The houses were presumably the residences and/or administrative offices of the priestly inhabitants who seem to have been resident for only short periods of time, perhaps one month, and had permanent residence elsewhere either in the local area or closer to the central administration. The layout of the rooms indicates that maintenance was not organized (or funded) centrally, but solely through the resources of the temple complex itself.

Overall, Old Kingdom towns seem to have been at least initially planned but then allowed to accrete naturally following the needs of the resident populations. At Edfu, for example (Alliot 1935; Bruyère, et al. 1937), the main settlement area was enclosed by an initial mud brick wall and a second wall was added later, encircling the previous wall with habitation structures filling the space between. Outside of these main occupation areas were a cemetery with numerous mud brick mastabas and an ancillary village to the west. Eventually, a second wall was constructed that enclosed this ancillary village as well, the south side of the wall utilizing the earlier mastabas as part of the wall's structure. The situation is much the same at Elephantine (Bietak 1979b: 108; Kemp 1977, 1983: 99; von Pilgrim 1997) where the main occupation was situated on a granite ridge and enclosed by a mud brick wall which followed the local topography and contained at least one stonelined gateway. There was also an extramural settlement to the west which was enclosed by a later wall.

Similarly, Ayn Asil in the Dahkla Oasis underwent three major construction phases with the first and last each having its own enclosure wall specifically built for it. In Phase I, a large, approximately square, installation defined by a massive mud brick enclosure wall was constructed, the shape of which (Giddy 1987) suggests it is a reflection of direct planning by the central administration. After a period of haphazard settlement (Phase II), the whole area was leveled, former occupation structures being filled in with rubbish, mud brick rubble, sand, and mud brick masonry, and two new enclosure walls were built: a smaller one to the north which followed an irregular path (perhaps enclosing previously existing outlying settlements), and a much larger well-built wall (which Giddy again uses to argue for some kind of central planning) around the main settlement area.

Several Middle Kingdom settlements may also offer some useful comparisons. The valley temple complex of Senusret II (Kahun; ancient name: Hetep-Senusret) near Lahun at the entrance to the Fayum Depression (Petrie 1890, 1891; Petrie, et al. 1923; David 1986) is the largest of the known pyramid towns and large portions have been excavated. The large size combined with textual evidence indicates that Kahun functioned as more than a mortuary complex. Two groups of papyri were found, one group from the temple area having to do with temple functions, and the other within the town proper dealing with the business of the surrounding community. Many of these texts refer to the land holdings and agricultural business of the priests and temple, but it is unknown whether private landowners also did business in the town.

The Kahun settlement itself is roughly square (384 m. x 335 m.) and is surrounded by a large wall with an inner wall separating a strip in the west from the main settlement areas (purpose unknown). The layout is strictly orthogonal or gridlike and is divided into room blocks. Each room block consisted of several parts, basically a central residential core surrounded by groups of chambers. The actual functions of these buildings are determined from the excavations themselves and from house models such as those described by Winlock (1955). The main residential unit was composed of a central entrance court or portico leading to the main residential chambers. One ancillary building was a granary; others functioned as cattle sheds, butcheries, bakery/breweries, and weaving and carpentry work areas. Each room block was probably the residence of a single family together with their supporting workers (soldiers, scribes, household servants) which may be why each room block has its own set of granaries instead of a single centralized granary.

Similar arrangements are found at Abydos (Wegner 1998), the pyramid complex of Amenemhat III at Dahshur (Arnold 1980, 1982; Arnold and Stadelman 1977), Abu Ghalib (Larsen 1935), and Tell el-Daba (Bietak 1979a, 1985, 1995, 1996) in the eastern Delta. All exhibit orthogonal plans oriented along compass points. Abu Ghalib also functioned as a center directing shipments of goods from the Delta to Upper Egypt and as a specialized bead manufacturer.

Other relatively well-known settlements from the Middle Kingdom are the fortified towns and garrisons in Nubia. Buhen (Emery 1963, 1965; Emery, et al. 1979) contains a large heavily fortified inner settlement and an outer, less well fortified area. The inner area, the main fortress, was heavily protected by a large (ca. 9 m. high, 150 m. x 138 m. in area) brick wall and contained a temple, garrison buildings, and several blocks of interlocking rooms and workshops as at Kahun. The outer area (ca. 450 m. x 150 m.) had a much smaller protective wall and, though hardly excavated at all, seems to have been sparsely populated, containing at least one cemetery area to the west. A similar arrangement is found at Shalfak (Reisner, et al. 1967) near the Second Cataract.

The number of rooms per household represented at Kahun is larger than the blocks of 3-4 rooms described for the other Old Kingdom towns around Giza and Abusir. In these cases, the basic habitation unit was relatively small and probably only serviced a few people, perhaps a single family or official, who may have resided there for only part of the year. At Kom el-Hisn, however, there seems to be a larger supported unit. That three separate rooms are devoted to grain storage seems to imply more than a few members of a single family. Food preparation is also divided between the intensive features of Room 17 (sharply defined hearths, cattle dung for fuel, and a higher percentage of ovicaprids) and the more generalized cooking areas represented by Rooms 3 and 8 (diffuse hearth areas, a variety of fuels, and a dominance of pig). Minimally, food preparation was directed at two different groups of people. As Redding (ND) suggests, this could reflect a more elite diet which included more ovicaprids contrasted with a commoner diet predominantly of pig.

Kom el-Hisn seems somewhat analogous to the structures at Middle Kingdom Kahun where large room blocks were devoted to an elite family or set of families and their attendant workers. However, at Kahun, the habitation blocks were more self-sufficient than those of Kom el-Hisn in that they contained numerous areas devoted to a wider range of activities such as weaving, carpentry, etc. While many of these activities may have left no material evidence at Kom el-Hisn, the rooms here tend to be centered around food preparation rather than a complete set of activities. Still, the relatively wide range of activities present in this small area of Kom el-Hisn suggests a more permanent residence and far less dependence on the surrounding communities than that which obtained at the temple towns around Giza.

The evidence thus far presented indicates that Kom el-Hisn was in many ways independent of the central authority while at the same time retaining some dependence on it. Pigs were apparently raised and consumed by some of the resident staff, implying that the residents themselves were not directly involved in intensive agriculture (Harris 1985; Redding 1991), though the presence of numerous sickle blades in various states of use and manufacture argues for some involvement. This militates in favor of their obtaining much of their grain from local fields rather than as assignments from the central government. If the residents were involved in the harvesting of grain, it seems likely that the fields were the property of the foundation. Pious foundations often included tracts of land for support of the cult and this could have included fields for growing food to support the human population as well as grazing land or for growing fodder.

The presence of locally obtained fish and at least one net weight (from Room 23 [not in *Fig. 1a* or *Fig. 1b*], specimen # GS-7) further indicates that some of the residents were involved directly with procuring local resources. However, many other resources, such as marine fish and stone, must have been obtained from a distance. Evidence for ceramic manufacturing is also absent, though manufacturing may have been carried out in other parts of the site. The standardized forms suggest import from central producers for many of the wares. Stone implements, both ground and chipped, were apparently imported into the site in finished, or nearly finished, form, though the inhabitants carried on extensive maintenance of both ground and chipped stone tools (Cagle 2001).

Age and sex ratios of sheep and goats calculated by Redding (1992, ND) indicate that many of these animals were exported rather than consumed locally. Redding has also argued that the ratio of cattle to other mammal remains at Kom el-Hisn and the recently excavated workers' village surrounding the pyramid complex at Giza are complementary. The very low numbers of cattle remains at Kom el-Hisn suggest that if cattle were present in large numbers they may well have been shipped to ceremonial centers such as Giza, where cattle occur in much higher numbers (Redding 1992: 105-106).

Rather than being an isolated island of state control, I would argue that Kom el-Hisn formed something of a middle ground between complete integration with the local economy and total independence of it. This model has textual support from the Middle Kingdom temple complex at

Kahun where some papyri deal with business involving the surrounding community while others involved strictly temple business.

This interpretation may shed light on the question of who the residents actually were. In some of the Old Kingdom pyramid towns, such as Abusir, the priest inhabitants were in residence for short periods of time rather than full-time. The habitation blocks associated with these towns tended to be smaller, with only a few rooms devoted to each household unit. Later in the Middle Kingdom residents inhabited the buildings full-time and the support structures were much larger and more complex. As I have suggested, the room structure presented here seems closer to the Middle Kingdom model, but probably not as elaborate, having more to do with immediate subsistence needs rather than the full range of activities required for a family unit to be entirely self-sufficient. The differential distribution of ovicaprid and pig remains suggests that at least two populations actively inhabited Kom el-Hisn: one group of elite individuals most likely engaged in non-subsistence activities, and another of support staff possibly drawn from the surrounding population, who provided for some of their own subsistence needs through the maintenance of herds of pigs.

This model implies that the residents were not involved in full-time agricultural production and required an ancillary population of support staff to provide subsistence and other domestic needs. Some of the surrounding land was part of the town's property and provided forage for animals and perhaps produce for the local population. This specialization required some dependence on an extra-local economy for some basic goods and services, which I have argued was a combination of direct procurement of local resources (e.g. fishing), use of local labor, and some long-distance trade/exchange nationally. The exact nature of the duties of the elite residents is not a settled issue. The data presented here lends support to the hypothesis that one of the functions of Kom el-Hisn was as a specialized center for cattle production related in some way to the "Estate of the Cattle". The close association with Hathor may also suggest the presence of a cult center at Kom el-Hisn.

This scenario of a major center involved in both the local and national economy lends some support to Trigger's (1993) "territorial state" model and to Kemp's (1983) suggestion that active local economies operated alongside a national system of redistribution of certain goods. While no direct evidence of a strictly appropriative relationship between Kom el-Hisn and the central government has been presented, the facts that it was not wholly self-sufficient in basic goods and services and that it may have been raising and shipping cattle directly to the cult centers at Giza suggest that at least one substantial Old Kingdom community was at least partly left to fend for itself apart from the national state structure.

Dr. Anthony J. Cagle
acagle@drizzle.com

Bibliography

Alliot, M. 1935 *Rapport sur les fouilles de Tell Edfou 1933*, Cairo.

Arnold, D. 1980 Dahschur. Dritter Grabungsbericht. *Mitteilungen des Deutschen Archäologischen Instituts, Abteilung Kairo* 36: 15-17.

Arnold, D. 1982 Keramikbearbeitung in Dahschur 1976-1981. *Mitteilungen des Deutschen Archäologischen Instituts, Abteilung Kairo* 38: 25-65.

Arnold, D. and R. Stadelman 1977 Dahschur. Zweiter Grabungsbericht. *Mitteilungen des Deutschen Archäologischen Instituts, Abteilung Kairo* 33: 15-18.

Arnold, F. 1989 A study of Egyptian domestic buildings. *Varia Aegyptiaca* 5: 75-93.

Aston, D. A. 1996 *Egyptian Pottery of the Late New Kingdom and Third Intermediate Period (Twelfth-Seventh Centuries B.C.). Tentative Footsteps in a Forbidding Terrain.* Studien zur Archäologie und Geschichte Altägyptens 13. Heidelberger Orientverlag, Heidelberg.

Badawy, A. 1967 The civic sense of pharaoh and urban development in ancient Egypt. *Journal of the American Research Center in Egypt* 6: 103-109.

Bietak, M. 1979a Avaris and Piramesse: archaeological exploration in the eastern Nile Delta. In *Proceedings of the British Academy, London*, vol. 65. London, 225-290.

Bietak, M. 1979b Urban archaeology and the "town problem" in ancient Egypt. In *Egyptology and the Social Sciences*, edited by K. R. Weeks. American University in Cairo Press, Cairo, 95-144.

Bietak, M. 1985 Tell El Daba. *Archiv für Orientforschung* 32: 130-135.

Bietak, M. 1995 Connections between Egypt and the Minoan world: new results from Tell El-Dab'a/Avaris. In *Egypt, the Aegean and the Levant: Interconnections in the Second Millennium B.C.*, edited by W. V. Davies and L. Schofield. British Museum Press, London, 19-28.

Bietak, M. 1996 *Avaris, the Capital of the Hyksos. Recent Excavations at Tell El-Dab'a.* British Museum Press, London.

Borchardt, L. 1909 *Das Grabdenkmal des Königs Nefer-ir-keꜣ-Reꜥ*, Leipzig.

Bonnett, H. 1952 *Reallexicon der ägyptischen Religionsgeschichte*, Berlin/New York.

Bottema, S. 1984 The composition of modern charred seed assemblages. In *Plants and Ancient Man*, edited by W. van Zeist and W. A. Camparie, Rotterdam and Boston, 207-212.

Brunton, G. 1927 *Qau and Badari I.* British School of Archaeology in Egypt, London.

Bruyère, B. J., K. Manteuffel, K. Michalowski, and J. Sainte Fare Garnot 1937 *Tell Edfou 1937.* Fouilles franco-polonaises, rapports I, Cairo.

Buck, P. E. 1990 *Structure and Content of Old Kingdom Archaeological Deposits in the Western Nile Delta: a Geoarchaeological Example from Kom el-Hisn.* Dissertation, University of Washington, Seattle.

Buhl, M.-L. 1947 The goddesses of the Egyptian tree cult. *Journal of Near Eastern Studies* 6(2): 80-97.

Cagle, A. J. 1991 *Geoarchaeological Analyses of Room Deposits from Kom el-Hisn, an Old Kingdom Village in the Western Nile Delta, Egypt.* Thesis, University of Washington, Seattle.

Cagle, A. J. 2001 *The Spatial Structure of Kom el-Hisn: An Old Kingdom Town in the Western Nile Delta, Egypt.* Dissertation, University of Washington.

Cagle, A. J. 2003 *The Spatial Structure of Kom el-Hisn: An Old Kingdom Town in the Western Nile Delta, Egypt.* BAR International Series 1099. Archaeopress, Oxford.

Coulson, W. and A. Leonard Jr. 1981 *Cities of the Delta, I. Naukratis.* Undena, Malibu.

Daressy, M. G. 1903 Rapport sur Kom el-Hisn. *Annales du Service des Antiquités de l'Égypte* 4: 282- 283.

David, A. R. 1986 *The Pyramid Builders of Ancient Egypt*, London.

Davies, N. d. G. and A. H. Gardiner 1920 *The Tomb of Antefoker, Vizier of Sesostris I, and his Wife, Senet (No. 60).* The Theban Tomb Series. George Allen and Unwin, London.

Dunnell, R. C. 1978a Archaeological potential of anthropological and scientific models of function. In *Archaeological Essays in Honor of Irving B. Rouse*, edited by R. C. Dunnell and E. S. Hall. Mouton, The Hague, 41-73.

Dunnell, R. C. 1978b Style and Function: A Fundamental Dichotomy. *American Antiquity* 43: 192-202.

Dunnell, R. C. 1983 Aspects of the spatial structure of the Mayo Site (15-JO-14), Johnson County, Kentucky. In *Lulu Linear Punctuated: Essays in Honor of George Irving Quimby*, edited by R. C. Dunnell and D. K. Grayson. Anthropological papers, Museum of Anthropology, University of Michigan #72. Ann Arbor, 109-165.

Edgar, C. C. 1909-1915 Recent discoveries at Kom el-Hisn. In *Le Musée Egyptien III*, edited by M. E. Grebaut. IFAO, Le Caire, 54-63.

Emery, W. B. 1963 Egypt Exploration Society: Preliminary Report on the Excavations at Buhen, 1962. *Kush* 11: 116-120.

Emery, W. B. 1965 *Egypt in Nubia*, London.

Emery, W. B., H. S. Smith, and A. Millard 1979 *The Fortress of Buhen: The Archaeological Report*, London.

Fakhry, A. 1959 *The Monuments of Sneferu at Dahshur I. The Bent Pyramid*, Cairo.

Faulkner, R. O. 1996 *A Concise Dictionary of Middle Egyptian.* Griffith Institute Ashmolean Museum, Oxford.

Giddy, L. L. 1987 *Egyptian Oases: Bahariya, Dakhla, Farafra, and Kharga During Pharaonic Times*, Warminster.

Griffith, F. Ll. 1888 Egyptological notes from Naukratis and the neighbourhood. In *Naukratis Part II. Sixth Memoir of the Egypt Exploration Fund*, edited by E. A. Gardner. Trubner & Co., London, 77-84.

Hamada, A. and M. el-Amir 1947 Excavations at Kom el-Hisn, 1948. *Annales du Service des Antiquités de L'Égypte* 46: 101-111.

Hamada, A. and S. Farid 1947 Excavations at Kom el-Hisn, Season 1945. *Annales du Service des Antiquités de L'Égypte* 47: 195-235.

Hamada, A. and S. Farid 1948 Excavations at Kom el-Hisn, Season 1946. *Annales du Service des Antiquités de L'Égypte* 48: 299-325.

Hamada, A. and S. Farid 1950 Excavations at Kom el-Hisn. Fourth Season 1947. *Annales du Service des Antiquités de L'Égypte* 50: 367-399.

Hannig, R. 1995 *Die Sprache der Pharaonen. Großes Handwörterbuch Ägyptisch-Deutsch (2800-950 v. Chr.)*, Mainz.

Harris, M. 1985 *Good to Eat: Riddles of Food and Culture.* Simon and Schuster, New York.

Hassan, S. 1943 *Excavations at Giza IV (1932-1933)*, Cairo.

Helck, W. 1974 *Die altägyptische Gaue*, Wiesbaden.

Helck, W. and E. Otto 1980 *Lexikon der Ägyptologie* 3. Otto Harrassowitz, Wiesbaden.

Hoffman, M. A. 1974 The social context of trash disposal in an Early Dynastic Egyptian town. *American Antiquity* 39: 34-50.

Hoffman, M. A., H. A. Hamroush and R. O. Allen 1986 A model of urban development for the Hierakonpolis region from Predynastic through Old Kingdom times. *Journal of the American Research Center in Egypt* 23: 175-187.

Kanawati, N. 1977 *Egyptian Administration in the Old Kingdom*, Warminster.

Kanawati, N. 1980 *Governmental Reforms in Old Kingdom Egypt*, Warminster.

Kemp, B. J. 1977 The early development of towns in Egypt. *Antiquity* 51: 185-200.

Kemp, B. J. 1983 Old Kingdom, Middle Kingdom, and Second Intermediate Period c. 2686-1552 B.C. In *Ancient Egypt: A Social History*, edited by B. G. Trigger, B. J. Kemp, D.O'Connor and A. B. Lloyd. Cambridge University Press, Cambridge, 658-769.

Kemp, B. J. 1989 *Ancient Egypt: Anatomy of a Civilization*. Routledge, London.

Kintigh, K. 1984 Measuring archaeological diversity by comparison with simulated assemblages. *American Antiquity* 49: 44-54.

Kirby, C. J. 1998 Preliminary report on the survey of Kom el-Hisn, 1996. *The Journal of Egyptian Archaeology* 84: 23-43.

Kokabi, V. M. 1980 Tierknochenfunde aus Giseh / Ägypten. *Annalen des Naturhistorischen Museums in Wien* 83: 519-537.

Larsen, H. 1935 Vorbericht über die schwedischen Grabungen in Abu Ghâlib. *Mitteilungen des Deutschen Archäologischen Instituts, Abteilung Kairo* 6: 41-87.

Lichtheim, M. 1976 *Ancient Egyptian Literature. A Book of Readings. Volume II: The New Kingdom*, Berkeley/Los Angeles/London

Moens, M.-F. and W. Wetterstrom 1988 The agricultural economy of an Old Kingdom town in Egypt's west Delta: insights from plant remains. *Journal of Near Eastern Studies* 3: 159-173.

Mond, R. and O. H. Meyers 1937 *Cemeteries of Armant*, London.

Montet, P. 1946 Tombeaux des Ière et IVme Dynasties à Abou-Roach. *Kemi* 8: 157-227.

Montet, P. 1957 *Géographie de l'Égypte Ancienne I*, Paris.

Nagel, G. 1938 *La céramique du Nouvel Empire à Deir el-Médineh Tome 1*. Documents de fouilles publié par les membres de l'Institut Français d'Archéologie Orientale du Caire 10. IFAO, Cairo.

Peet, T. E. 1921 Excavations at Tell el-Amarna. *Journal of Egyptian Archaeology* 7: 169-185.

Petrie, W. M. F. 1886 *Naukratis I*, London.

Petrie, W. M. F. 1890 *Kahun, Gurob, and Hawara*, London.

Petrie, W. M. F. 1891 *Illahun, Kahun, and Gurob*, London.

Petrie, W. M. F. 1900 *The Royal Tombs of the First Dynasty 1*, London.

Petrie, W. M. F., G. Brunton and M. A. Murray 1923 *Lahun II*, London.

Porter, B. and R. L. B. Moss 1934 *Topographical Bibliography of Ancient Egyptian Hieroglyphic Texts, Reliefs, and Paintings IV, Lower and Middle Egypt*, Oxford.

Redding, R. W. 1991 The role of the pig in the subsistence system of ancient Egypt: A parable on the potential of faunal data. In *Animal Use and Culture Change*, edited by P. J. Crabtree and K. Ryan. MASCA Research Papers in Science and Archaeology, vol. 8, Supplement, 20-30.

Redding, R. W. 1992 Egyptian Old Kingdom patterns of animal use and the value of faunal data in modeling socioeconomic systems. *Paleorient* 18(2): 99-107.

Redding, R. W. ND The vertebrate fauna from the excavations at Kom el-Hisn. In *Unpublished report*.

Reisner, G. A. 1931 *Mycerinus*, Cambridge, MA.

Reisner, G. A. 1932 *A provincial cemetery of the Pyramid Age: Naga-Ed-Der Part III*. Egyptian Archaeology VI. University of California Press, Berkeley.

Reisner, G. A. 1942 *A History of the Giza Necropolis I*. Harvard University Press, Cambridge.

Reisner, G. A. 1955 *A History of the Giza Necropolis II*. Harvard University Press, Cambridge.

Reisner, G. A., N. F. Wheeler and D. Dunham 1967 *Oronarti Shalfak Mirgissa, Second Cataract Forts II*, Boston.

Samuel, D. 2000 Brewing and Baking. In *Ancient Egyptian Materials and Technology*, edited by P. T. Nicholson and I. Shaw. 537-576. Cambridge University Press, Cambridge.

Schiffer, M. B. 1989 Formation processes of Broken K Pueblo: some hypotheses. In *Quantifying Diversity in Archaeology*, edited by R. D. Leonard and G. T. Jones. Cambridge University Press, Cambridge, 37-58.

Silverman, D. P. 1988 *The Tomb Chamber of Ḫsw the Elder: The Inscribed Material at Kom el-Hisn, Part I: Illustrations*. Eisenbrauns, Winona Lake.

Sjoberg, G. 1960 *The Preindustrial City: Past and Present*. The Free Press, Glencoe, IL.

Sterling, S. and R. J. Wenke 1997 Attribute scale seriation of Old Kingdom Egyptian ceramics. Paper presented at the Annual Meeting of the Society for American Archaeology, Nashville, TN.

Strudwick, N. 1985 *The Administration of Egypt in the Old Kingdom*. KPI, London.

Toussoun, O. 1922 *Mémoire sur les anciennes branches du Nil: Epoque Ancienne*. Mémoires à l'Institute d'Égypte 4 (Part 1).

Trigger, B. G. 1993 *Early civilizations: Ancient Egypt in Context*. The American University in Cairo Press, Cairo.

van den Brink, E. C. M. 1987 A geo-archaeological survey in the north-eastern Nile Delta, Egypt; the first two seasons, a preliminary report. *Mitteilungen des Deutschen Archaologischen Institut, Abteilung Kairo* 43: 7-31.

van den Brink, E. C. M. 1988 The Amsterdam University Survey Expedition to the northeastern Nile delta (1984-1986). In *The Archaeology of the Nile Delta: Problems and Priorities*, edited by E. C. M. van den Brink. Netherlands Foundation for Archaeological Research in Egypt, Amsterdam, 65-114.

van den Brink, E. C. M. 1992a *The Nile Delta in Transition: 4th-3rd Millennium B.C.* Israel Exploration Society, Jerusalem.

van den Brink, E. C. M. 1992b Preliminary report on the excavations at Tell Ibrahim Awad, seasons 1988-1990. In *The Nile Delta in transition: 4th-3rd Millennium B.C.*, edited by E. C. M. van den Brink. Israel Exploration Society, Jerusalem, 43-68

van Zeist, W. and W. A. Camparie 1984 *Plants and Ancient Man*, Rotterdam and Boston.

von den Driesch, A. and J. Boessneck 1985 *Die Tierknochenfunde aus der neolithischen Siedlung von Merimde-Benisalame am westlichen Nildelta.* Insitut für Paläoanatomie, Domestikationshforschung und Geschichte der Tiermedizin der Universität München und Deutsches Archäologisches Institut, Abteilung Kairo, München.

von Pilgrim, C. 1997 The town site on the island of Elephantine. *Egyptian Archaeology* (#10): 16-18.

Wegner, J. 1998 Excavations at the town of Enduring-are-the-places-of-Maa-Kheru-in-Abydos. A preliminary report on the 1994 and 1997 seasons. *Journal of the American Research Center in Egypt* 35: 1-44.

Wenke, R. J. 1997 City-states, nation-states, and territorial states: the problem of Egypt. In *The Archaeology of City-states: Cross-cultural Approaches*, edited by D. L. Nichols and T. H. Charlton. Smithsonian Institution Press, Washington and London, 27-49.

Wenke, R. J. and D. J. Brewer 1992 The Archaic-Old Kingdom Delta: the evidence from Mendes and Kom el-Hisn. In *Haus und Palast im alten Ägypten*, edited by M. Bietak. Sonderdruck, Wien, 265-285.

Wenke, R. J., P. E. Buck, H. A. Hamroush, M. Kobusiewicz, K. Kroeper and R. W. Redding 1988 Kom el-Hisn: Excavation of an Old Kingdom Settlement in the Egyptian Nile Delta. *Journal of the American Research Center in Egypt* XXV: 5-34.

Wenke, R. J. and R. W. Redding 1985 Excavations at Kom el-Hisn, 1984. *Newsletter of the American Research Center in Egypt* 129: 1-11.

Wenke, R. J. and R. W. Redding 1986 Excavations at Kom el-Hisn, 1986. *Newsletter of the American Research Center in Egypt* 135: 11-17.

Wenke, R. J. and R. W. Redding 1987 Provincial Socio-economic Structure of Early Pharaonic Egypt. In *Proposal Submitted to the National Science Foundation.*

Wenke, R. J. and R. W. Redding 1989 Early Pharaonic Cultural Integration of the Nile Valley and Delta. In *Proposal Submitted to the National Science Foundation.*

Wilson, J. A. 1960 Egypt through the New Kingdom: Civilization without cities. In *City Invincible: A Symposium on Urbanization and Cultural Development in the Ancient Near East*, edited by C. H. Kraeling and R. M. Adams. University of Chicago Press, Chicago, 124-164.

Winlock, H. E. 1955 *Models of Daily Life in Ancient Egypt*, Cambridge, MA.

Zibelius, K. 1978 *Ägyptische Siedlungen nach Texten des Alten Reiches*, Weisbaden.

The Egypt Archaeological Database and the Birth of Computerized Inter-site Analysis[1]

Matthew Joel Adams

I. Introduction

The enormous volume of data collected by archaeologists world-wide is recorded in a variety of formats unique to each individual site excavator. This paper-and-pencil, non-uniform method of recording is not conducive to central storage or to comparisons across different excavation seasons or multiple sites. The amount of material excavated at one site in a single season makes it extremely tedious and difficult to sift through all the written records in order to perform a particular analysis. Doing this on a regional scale is virtually impossible, yet inter-site analysis is essential for sound interpretation of cultural material. There has been a gradual shift in recent years from recording data into notebooks by hand to also recording the data directly into a Relational Database Management System (RDBMS). Many excavation teams have made this change because storage in an RDBMS permits excavators to record data on a daily basis in a digital format. This allows data to be retrieved at any time and to be archived efficiently. The ability to retrieve and analyze data in mid-excavation gives the team better control of the excavation and provides it with daily checks on its progress. Implementation of RDBMSs has also made it possible for archaeologists to include regional geographic data such as digital maps, satellite imagery, and survey data into their analyses.

This is a move in the right direction, but it does not exploit the full potential of an RDBMS and does not fulfill the immediate and long-term needs of the field. Most archaeologists would agree that the amount of excavation data that actually survives into the final report is less than 5%. The material included in the final publication is determined solely by what the excavator has chosen to include. For example, only six plates of pottery were presented in the publication of four seasons of excavation at Deir el-Ballas, Egypt.[2] Only 89 vessels and sherds were published from the 1979-1980 excavations at Mendes.[3] There are, of course, restrictions on what can be provided in a paper publication, and no one is to blame for this constraint. Archaeologists certainly recognize the *theory* that omitting any data is scientifically and professionally unacceptable, but practicalities such as time and money inhibit such comprehensiveness. Unfortunately, final reports are necessarily limited publications which leave researchers with a severely restricted ability to analyze individual excavations and to compare data across sites. Sample size is naturally limited by preservation, and this is then cut smaller by excavation units and reduced *again* in the publication of final reports. Ultimately, secondary interpreters derive their interpretations from a sample of a sample of a sample. Archaeological theorists have dealt overmuch with the problem of representative samples in archaeology, and yet the fact of the matter remains: preserved material is our only sample of the ancient world. The less we cut into that sample the higher our accuracy will be. As a result, research conducted using site reports is scientifically flawed at a basic statistical level.

---

[1] I owe a great debt to Professor Donald B. Redford of the Pennsylvania State University for allowing this project to be affiliated with his site, Mendes, for access to site data, and for his support of the ideas proposed in this paper. I must also express much gratitude to Jonathan David, without whose constant proofreading and suggestions this paper would not have achieved coherency suitable for publication. I would also like to thank Professor Donna Peuquet (PSU) for her advice on RDBMS design. The opinions herein do not necessarily represent the views of these individuals.

[2] P. Lacovara, *Deir el-Ballas: Preliminary Report on the Deir el-Ballas Expedition, 1980-1986*. ARCE Reports, vol. 12. Winona Lake: Eisenbrauns, 1990.

[3] K. L. Wilson, *Cities of the Delta, II: Mendes*. ARCE Reports vol. 5. Malibu: Undena Publications, 1982.

There is one way that this issue can be resolved: *everything* must be published. All excavated material needs to be made available to all researchers at all times in a convenient, fast, and easy to use manner, but not in isolated single-site databases. I therefore propose the creation of the Egypt Archaeological Database, an inter-site database of all excavations. This paper presents the preliminary results of a developed prototype RDBMS project with great potential to be developed into the Egypt Archaeological Database, a new tool for Egyptology. The paper is also an entreaty for scholarly support, feedback, and collaboration in its continued implementation and development.

II. The Benefits of the Egypt Archaeological Database

The benefits of a database that stores all data from excavations in Egypt, past and present, up-to-date from season to season, are immediately apparent. All excavation notebooks, finds, sample results, plans, drawings, pottery details, etc. would be accessible to all Egyptologists. Data would be universally available immediately from its input on site. Currently, archaeologists work in relative isolation, no archaeologist knowing exactly what is being done at other sites. With this database, the nature of archaeological dialog, currently conducted from isolated hillocks via site reports, would change. Archaeology would be conducted in a much more communal environment with more standardized methodology and terminology.

The Egypt Archaeological Database would facilitate regional analysis and open the archaeological record to the whole academic community through the Internet. Currently, if a researcher wants to study the temporal and regional distribution of a certain pottery type, he must leaf through all the site reports available, look for the type, catalog the localities and strata, and then perform the analysis. Data collection alone may require months and months of research. This will never be necessary with the Egypt Archaeological Database. An interested researcher could perform an unlimited number of queries. For example, all Bes figurines from Lower Egyptian sites could be displayed with a few key-strokes. The response would be the hundreds of drawings and attribute information of all the figurines found to date in the Delta, all of which could be scrutinized together by performing additional queries to isolate traits, styles, or other information in which the particular researcher is interested. A researcher could query the database for pottery types, and the database would retrieve them and display whatever information is pertinent, whether that be sites, strata, loci in which the type was found, ware detail, measurements, drawings, or publication locations. In such a scenario, collection of data can be completed in a single day, and the researcher is able to devote much more time to his analysis. Reducing catalog time for research projects will increase the productivity of archaeologists, as well as ensuring accuracy and comprehensiveness for each individual study.

The database will also facilitate the paper publication process.[4] Currently, site analysis can literally take decades. The archaeological database is a better window into the data for the excavation leader than a pile of notebooks. A slew of statistical tools that will be built into the database will help archaeologists, for example, to quantify pottery seriation at their sites, define strata, manage samples, and conduct statistical analyses. Additionally, publication tools will assist the excavator in compiling lists of artifacts and pottery plates.

At the most basic level, the database provides essential central storage for archaeological data. Illegibly written field notebooks and other data records rotting in storage basements of universities and museums will no longer be the norm. All data will be in one location available for all, preserved electronically in an efficient and complete format that will be available indefinitely. Currently, it can be difficult for a researcher to offer a reinterpretation of a site that he has not excavated personally.

[4] This essay does not advocate the elimination of paper reports, it merely proposes the use of a database as the primary medium for publication of raw data and suggests that "site reports" be media for interpretive analysis by the primary excavator.

Without access to data other than the notebooks and final reports, secondary interpretations are often skewed. Scholarship progresses most effectively through iterative analysis. This database permits a dramatic increase in the ability of outside researchers to reanalyze past excavations.

III. Wookie[5]: The Mendes Excavation Relational Data Management System

Mendes lies in the Eastern Nile Delta and has been excavated intermittently throughout the last century by several different expeditions. The current expedition began ten years ago, spearheaded by Donald B. Redford of the Pennsylvania State University. Like other archaeologists before him, Redford uses a personal recording system that is theoretically similar to others, but uses different terminology and emphasizes certain data types over others. It became apparent through my work at Mendes that an RDBMS would not only provide a means for the electronic entry and manipulation of archaeological data, but would also standardize the entry format and dramatically increase the efficiency of analyses. Additionally, it would facilitate the painstaking process of site publication by providing artifact lists and query results. Further, the entire database could be provided on a ready-made CD-ROM along with the traditional paper publication, allowing researchers outside the excavation team to develop equally viable interpretive analyses informed by all data available.

The current functional prototype of Wookie provides the capability to do all of these things, but has also been designed to be open-ended. This means it can integrate other recording systems into the data model, facilitate future upgrades, and accommodate data from previous expeditions to the site of Mendes to be stored in the database and queried alongside the recent excavation's data. The ability to incorporate material from earlier excavations into current research is a valuable, important, and essential part of archaeological publication. Wookie has been designed with this in mind, and the facility will be implemented when it is taken into the field for the summer 2003 season, where post-developmental testing will be completed. Additionally, the database is currently populated with approximately 500 records from the 2001 season of excavation. The data model was developed during the course of my work at Mendes in the last two excavation seasons, and constructed by a team assembled in early fall of 2002.[6] Although designed as a functional archaeological recording database for use at Mendes, its open-ended design allows the evolution of Wookie into the Egypt Archaeological Database.

The core features of Wookie are the following:

- An operational database with a schema conducive to the recording methods applied at Mendes, storing all data produced from the site.
- GIS[7]/temporal underpinnings within the database, with attributes such as altitude above sea level, latitude, longitude, and time periods of specific strata, to maintain both spatial integrity and a spatial-temporal link.
- Several stock queries allowing users to view oft-accessed data in the system.
- A user-friendly, WWW-based graphical user interface (GUI) accessible via any standard Internet browser.

[5] Wookie, the official name for the Mendes database, originated as a term of endearment by the programming team. The team consisted of Justin Plock, Robin Smail, William Them (all Penn State students), and myself.

[6] As Penn State IST students, the team I assembled to help construct the RDBMS did so in exchange for using the project as a requirement for a course approved by Professor Donna Pequet of the Geography Dept. at The Pennsylvania State University. The students had great experience with programming and software debugging. J. Plock, who had the most experience with SQL did much of the coding. W. Them and R. Smail provided a great service in helping to refine my model, debug the early versions of the system, and input data.

[7] GIS = Geographic Information Systems

- Security roles within the GUI and database offering differing levels of access and ability. The roles consist of administrators, site supervisors, data enterers, and researchers.
- Ability to attach any external file type to any database entity, including photos, illustrations, documents (word processor reports or .PDFs of publications[8]) etc., with accessibility within the GUI.

A technical discussion of Wookie follows, directed towards database programmers working closely with archaeologists in order to outline the design and construction of our data model. It is intended as a guide for excavation teams wishing to develop a database of their own and to describe the structure of the proposed Egypt Archaeological Database. Those interested in results rather than technical details should feel free to skip ahead to section IV, where the discussion of the Egypt Archaeological Database will resume.

The database was created and runs on the RDBMS MySQL (http://www.mysql.com/), an open-sourced, highly reliable and robust system operating on SQL standards. The web-based GUI was coded entirely in PHP (http://www.php.net/), an open-sourced web scripting language allowing for database connectivity.

Entity	Definition
Bone	Miscellaneous human and animal remains.
Burial	Full bodied burials.
Features	Non-movable finds.
Finds	Movable artifacts.
Lab	Items that will be sent to a lab for further analysis.
Locus	Discrete soil layers.
Locus Neighbor	Spatial relationship of loci.
Photos	Miscellaneous photos.
Pottery	Sherds or vessels.
Seasons	Seasons at each site.
Sites	Specific location of the dig (e.g., Mendes, East Karnak, etc.).
Units	Squares and trenches.
User Roles	Associated roles for the users of the database.
Users	List of valid users of the Wookie Archaeological Database.
Walls	Discrete architectural units.

Table 1. Initial Entities for Wookie Archaeological Relational Database Management System.

[8] As archaeologists are, after a fashion, historians, it may behoove an excavator to keep all of his notes in the database, providing an historical account of the excavation for future researchers. It would be possible to keep that material hidden from anyone but that excavator until such time as he deems it appropriate to reveal such material.

Bone
- Bone ID
- Locus ID
- Type
- Burial
- Quantity
- Notes

Burial
- Burial ID
- Locus ID
- Grave
- Bone ID
- Notes

Features
- Features ID
- Unit ID
- Associated Loci
- Evacuated As

Finds
- Find ID
- Locus ID
- North Baulk
- West Baulk
- Type
- Description
- Material
- Length
- Width
- Height
- Thickness
- Diameter
- Notes
- Inscription

Lab
- Lab ID
- Locus ID
- Type
- Quantity
- Notes

Locus Neighbor
- Locus Neighbor ID
- Locus ID
- Position

Locus
- Locus ID
- Unit ID
- Locus Name
- Description
- Notes
- Trench
- Level
- Period
- Opened Date
- Closed Date

Photos
- Photo ID
- Wall ID
- Find ID
- Site ID
- Feature ID
- Locus ID
- Roll Number
- Photo Number
- Photo
- Date Taken

Pottery
- Pottery ID
- Locus ID
- Description
- Material
- Period
- Rims
- Bases
- Body
- Plate
- Notes

Season Supervisor
- Season Spvr ID
- Season ID
- User ID

Seasons
- Season ID
- Site ID
- Start Date
- End Date
- Year
- Lead Director

Sites
- Site ID
- Site Name
- History
- GPS

Units
- Unit ID
- Unit Name
- Season ID
- Archive
- Field Report
- Latitude
- Longitude
- Description

User Roles
- User Role ID
- Role Name
- User ID

Users
- User ID
- User Name
- Password
- Season ID
- Unit ID
- First Name
- Last Name
- CV
- Photo
- Position
- DOB
- Notes
- Phone Number
- Address 1
- Address 2
- City
- State
- Zip
- Country ID
- Emergency Name
- Emergency Phone

Walls
- Wall ID
- Unit ID
- Contiguous Walls
- Material
- Preserved Courses
- Foundation Trench
- Plaster
- Width
- Rows
- Notes
- Description

Table 2. Wookie Database Conceptual Model, showing entities refined with attributes.

The data model for the Wookie database was created following relatively standard processes for creating a relational database.[9] The first step was to create a preliminary list of entities (*Table 1*) in order to define and conceptualize the tables to be used within the Wookie database. Most of the entities were derived directly from the Mendes excavation, either as entities that defined the site itself, or are typically found in a site, or are in some way related to the recording criteria. Others were entities necessary to adapt archaeological recording to the digital medium.

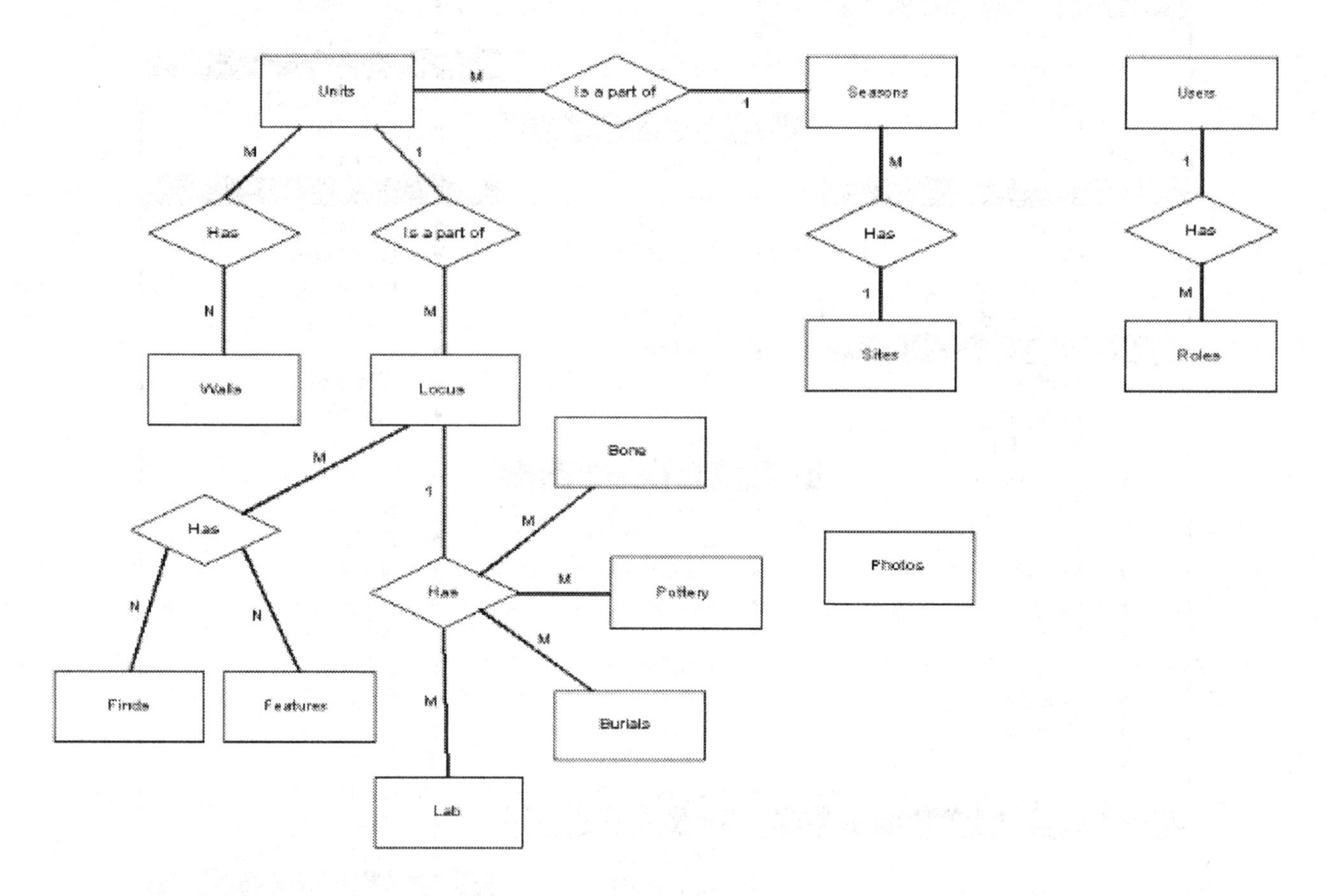

Figure 1. The Entity-Relationship (E-R) Diagram for the Wookie Database.

The next step was to identify the attributes that defined these initial entities, as seen in the Conceptual Model (*Table 2*).[10] This Conceptual Model became the baseline and was integral in enabling the modeling of the entity relationships within the database. The relations illustrated in Entity Relationship (ER) diagram (*Fig. 1*) are the foundation for the proposed Egypt Archaeological Database.[11] In terms of hierarchical structure, the site is the highest entity and all others can extend from it in "one-to-many" relationships. A site, such as Mendes, can have many seasons, and many units are a part of each season. By taking normalization to the third form, we were able to structure the tables in a logical way so that data interdependency is reduced, although this did require creating additional tables in several instances. In fact, all primary keys now exist as auto-generated numerical

[9] See M. F. Worboys, *GIS: A Computing Perspective*. Philadelphia: Taylor & Francis, Inc., 1995; M. N. DeMers, *Fundamentals of Geographic Information Systems*. 2nd ed. New York: John Wiley & Sons, Inc., 2000; M. Zeiler, *Modeling Our World: The ESRI Guide to Geodatabase Design*. Redlands, CA: ESRI Press, 1999.

[10] The attributes of these entities have grown considerably since this phase. This table is an example of the process only.

[11] For the sake of simplicity, the relationships between the Photos entity and all others have not been diagrammed, as it is related to most tables.

keys in order to lessen the dependency on multiple field primary keys. Some additional tables were necessary just to be able to join data within separate tables. The resulting Table Model (*Fig. 2*) demonstrates most accurately the final database structure upon which the programming is written. Comparing the original Conceptual Model (*Table 2*) to the final Table Model (*Fig. 2*) illustrates the expansion that this project has undergone, and the increasing complexity of the programming involved.

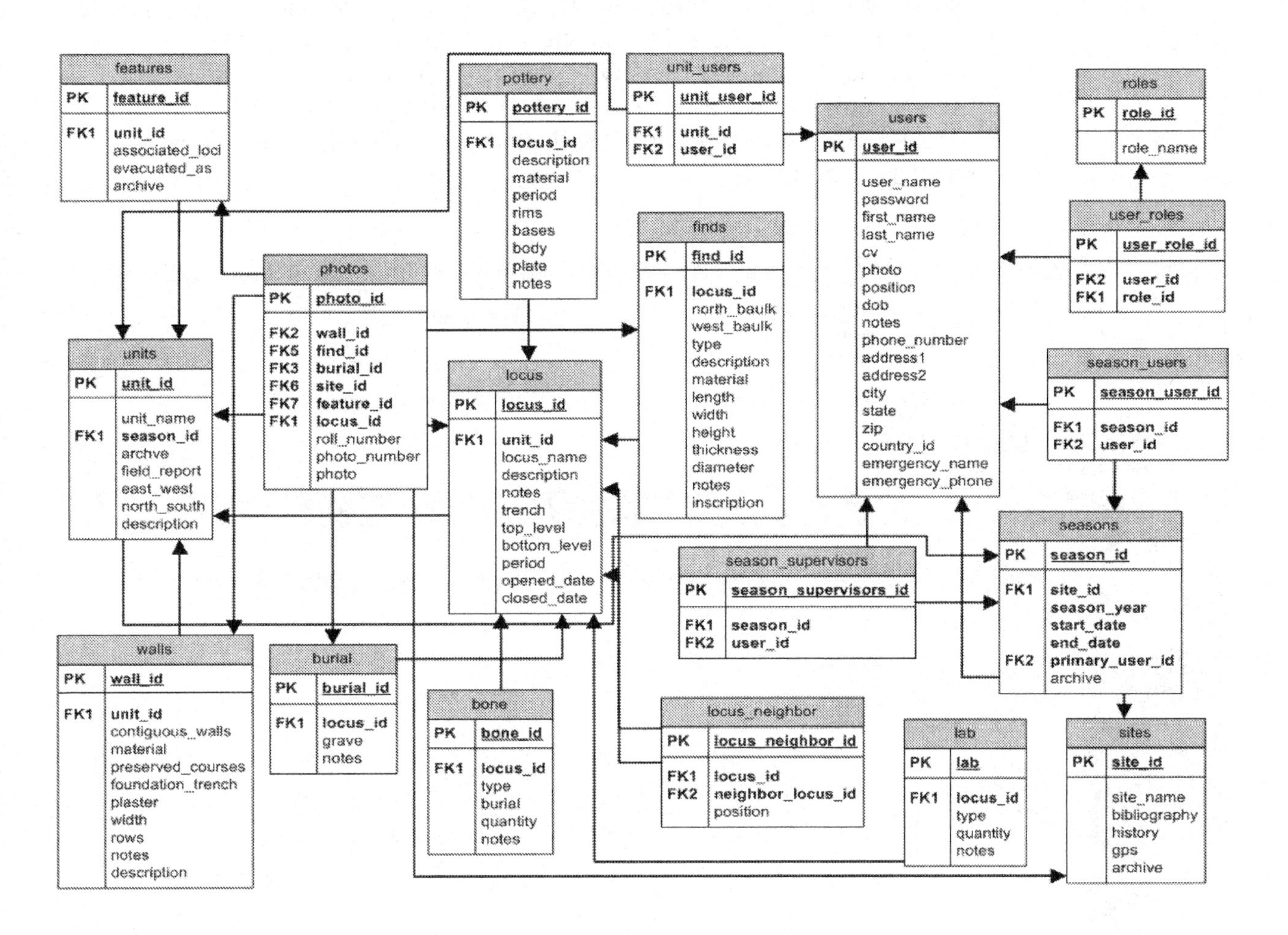

Figure 2. The Table Model of the Wookie Archaeological Database. Primary keys (PK) and Foreign keys (FK) are indicated where applicable.*

The GUI contains several stock queries accessible from pull down menus (*Fig. 3*). The query page, useful yet simple, offers options for text searching, for searching pottery typologies, and for listing entities from any season and/or unit. The stock queries are built into the GUI, but unique and more complex SQL queries can be written via MySQL. Future updates to the database will feature additional abilities for more complex querying, significantly expanding the basic utility of the database as a whole.

Following is an example of a query showing the SQL underlying the GUI display. The example shows the usability of the interface and the querying system, while displaying the pottery typology feature. By assigning numbers to pottery types, the database can facilitate ceramic analysis.

Query Criteria: List all pottery photos of type: 0 Show Photos: YES. (*Fig. 4*)

SELECT p.photo_id, p.file_name, pb.pottery_id, po.locus_id, su.unit_id, su.season_id, p.pottery_type, p.diameter, s.season_year, u.unit_name, l.locus_name

FROM photos p LEFT JOIN photos_bind pb ON p.photo_id=pb.photo_id LEFT JOIN pottery po ON pb.pottery_id=po.pottery_id LEFT JOIN locus l ON po.locus_id=l.locus_id LEFT JOIN seasons_units su ON l.season_unit_id=su.season_unit_id LEFT JOIN seasons s ON su.season_id=s.season_id LEFT JOIN units u ON su.unit_id=u.unit_id

ORDER BY p.pottery_type asc

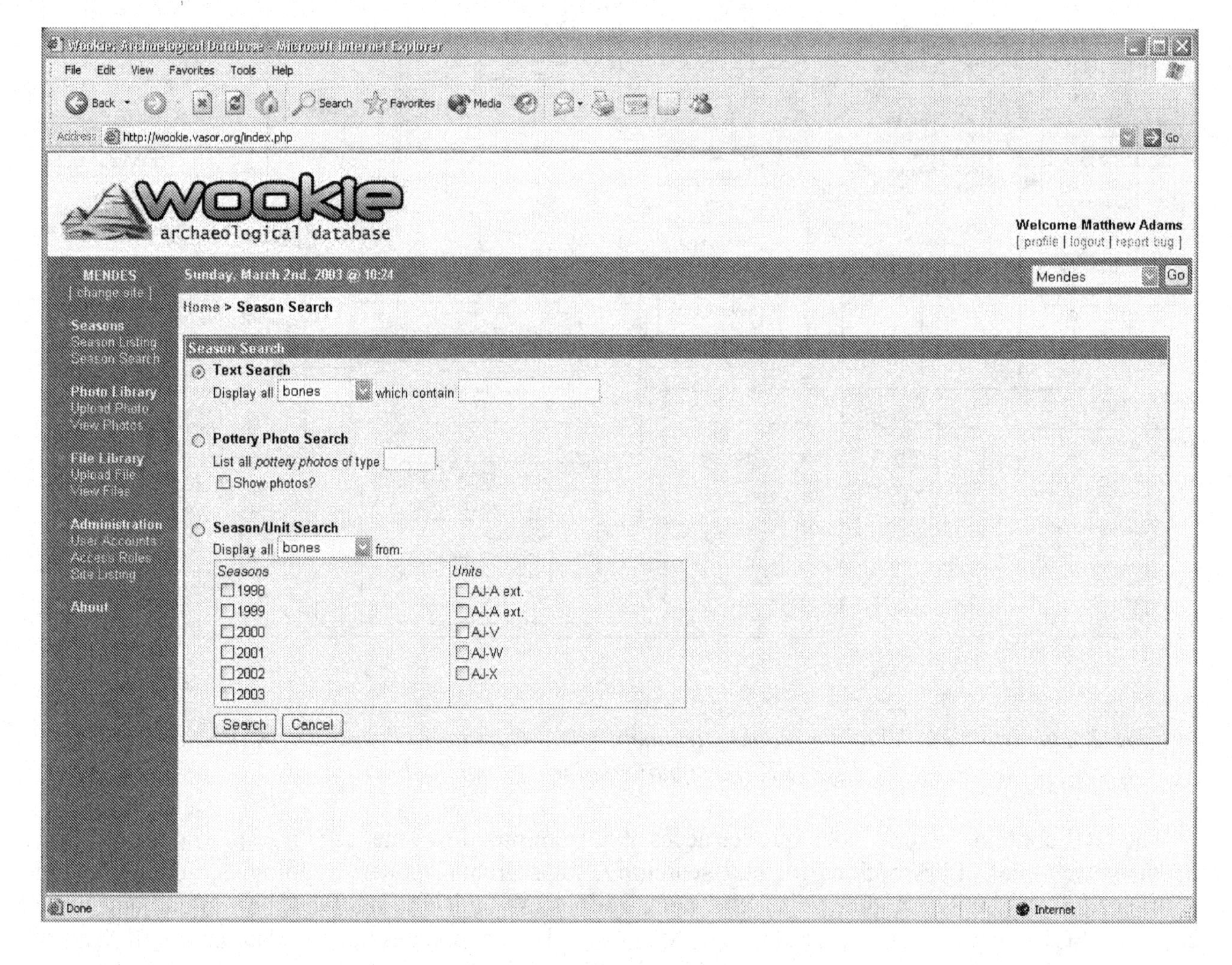

Figure 3. The query page.

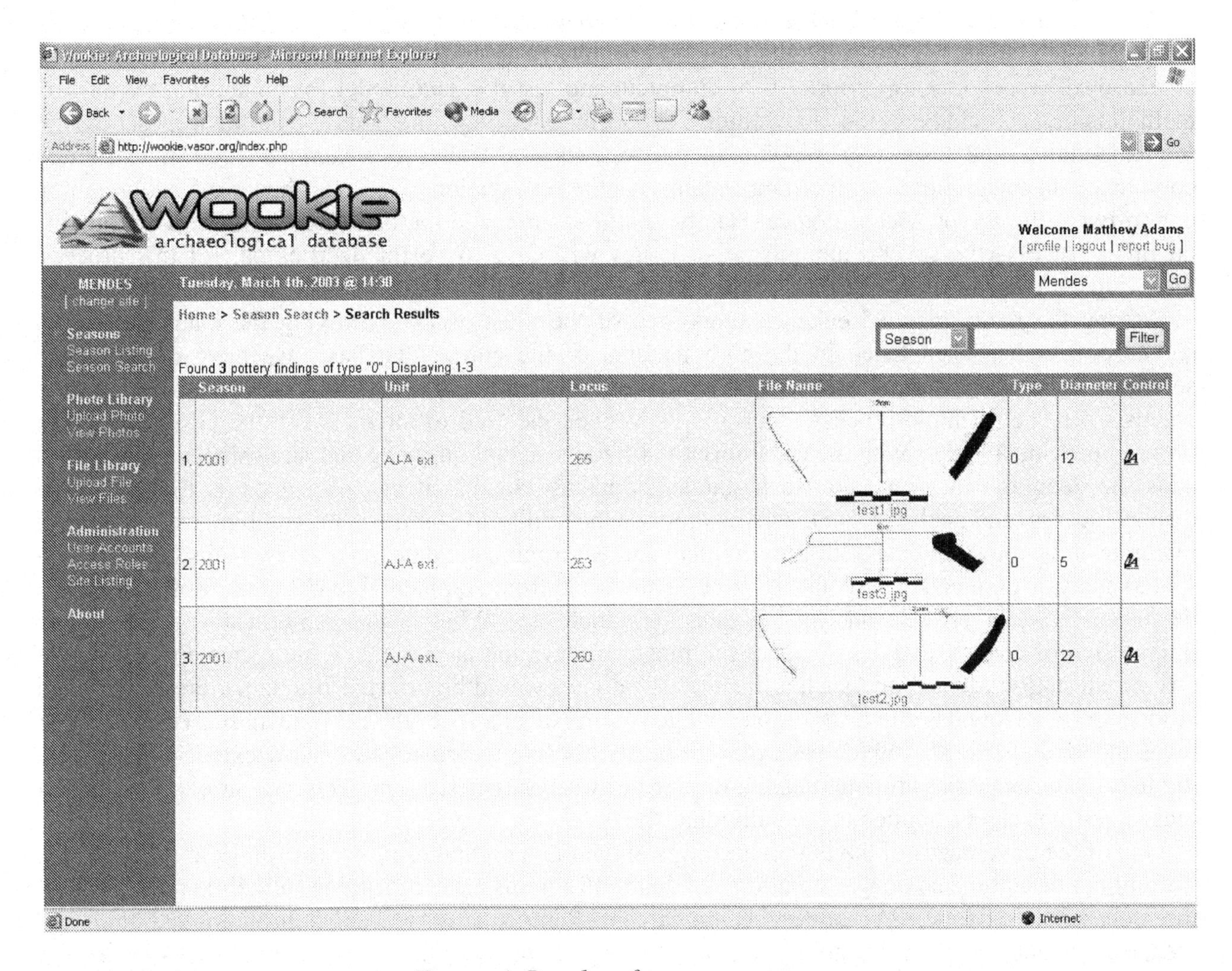

Figure 4. Results of query on pottery.

While variations in methods of classification prohibit the construction of a database system that is standard and usable by every excavation,[12] it is possible to link databases from each site so that they can be searched together. The Egypt Archaeological Database will be employed as a central database for hierarchically managing other excavations' databases by way of relational linkages. It will provide the analytical tools that researchers may need, while allowing the excavators to retain most of their current field methodology.

IV. The Evolution of Wookie into the Egypt Archaeological Database

The relatively narrow project of digitally storing Mendes data has suggested a much broader application. Successful development of a model that takes into account the variation of on-site recording methodologies between different excavations at one site (Mendes) implies that the same model would be able to handle multiple sites in multiple seasons. In short, a model could be developed on this basis to manage all of the archaeological data in Egypt, from expeditions both past and present.

[12] I. Benenson & I. Finkelstein, "The Megiddo Excavation Data Management System", in I. Finkelstein, D. Ussishkin, & B. Halpern, *Megiddo III: The 1992-1996 Seasons*. Jerusalem: Graphit Press, 2000, pp. 14-24.

As demonstrated above, Wookie is currently a viable RDBMS that can be used effectively for practical applications. However, our team is currently implementing several enhancements of the database software. Our top priority is to complete and test the data model for combining excavation methodologies, including the integration of surface survey methods. Following successful implementation and testing of Wookie in excavations at Mendes this summer (2003), work will commence on integrating the recording system of other excavations. We will begin officially calling our software the Egypt Archaeological Database at that time, and the database will be made available on line. Necessarily, several administrative issues will have to be finalized as part of this process, including database security and access, user management, and user interest.

Among the many planned enhancements, one of the most urgent is linking the database to GIS software. As has been listed in the core features of the current database, we have designed the software to accept temporal and spatial information. Therefore, integrating GIS software into the database will be a simple process. Our goal is to code the GUI to interact with the GIS software in order to provide all users with the capability to utilize its spatial querying and analysis features. A GIS adds the geographical dimension to a standard database so that it can use maps as tools for data display, access, and analysis.[13] This gives the user the ability to compare and combine existing maps, integrate archaeological data with geographical and environmental data, create new maps, and perform spatial analysis on them. These maps can be of any scale from continental to site size. With the proper database, GIS can provide answers to questions such as: What is the geological makeup of the intersection of river X and Y? What is the range of elevation at site X? Which sites are related via size or attestation of such-and-such artifacts? What sites would be affected if a dam were constructed at location X? How has site distribution changed from Naqada III to the Early Dynastic Period? Who owns the land at site X? Which sites were excavated before a certain date? How extensively? Can I dig here? These queries are intentionally directed towards larger-scale projects, but similar versions of these questions can be applied at the individual site level.

One of the advantages of having archaeological data in a computer rather than in field notebooks is the amount of automated analysis that can be done. Programs can be written that will handle much of the traditional legwork. As pottery is the archaeologist's most valuable chronological tool, the database currently includes the ability to store drawings of pottery within the site framework noting the stratigraphic details of their discovery. The present author plans to develop ways for users to compile publishable plates of selected pottery from these drawings and to perform statistical analyses on them with ease and simplicity. The stratigraphy of an archaeological site is of comparable importance to the archaeologist. A Harris Matrix is somewhat complicated to create, but quite useful in reconstructing the stratigraphic history of a site. There are very specific rules to the design of the matrix that a computer could automatically generate based on the data stored in the database. Methods of encoding this into the database functionality are currently being developed. Along the same lines, many charts, graphs, and tables can easily be generated based on the data already stored. We plan to set up functionality to provide for these needs within the final integrated system. These features will aid significantly in analyzing of the archaeological data as well as in the publication process, which is historically a long, tedious, and fallible procedure.

With the amount of valuable data that will be stored in the database, the possibilities for statistical analyses are tremendous. Features will be written into the software to allow users to compile data and perform various mathematical operations on them. These functions can be as simple as creating charts

[13] For GIS in archaeology, see: K. M. S. Allen, S. W. Green, and Z. B. W. Zubrow, *Interpreting Space: GIS and Archaeology*. Philadelphia: Taylor & Francis, Inc., 1990; M. Gillings, D. Mattingly, and J. van Dalen, *Geographical Information Systems and Landscape Archaeology*. AML series #3. Oxford: Oxbow Books, 1999; M. Gillings and A. Wise. *GIS Guide to Good Practice*. Oxford: Oxbow Books, 1990. For GIS in general, see Worboys (1995) & DeMers (2000), mentioned in note 9, and respective bibliographies.

and graphs or calculating percentages, or as complex as chi-squared tests and regression analyses.[14] Additionally, the quantification of uses and access of space in buildings can be a valuable tool for assigning cultural characteristics to a settlement. Hillier and Hanson have successfully shown that structures have a "genotype" determined by use of space and access.[15] Spatial syntax diagrams expressing these "genotypes" constructed based on Hillier and Hanson's established rules can be valuable for analyzing architecture. The automation of spatial syntax diagram construction for complete excavated structures is one of the features that will be pursued in the evolution of Wookie into the Egypt Archaeological Database.

A particularly useful feature for the Egypt Archaeological Database would be adapting it to exploit wireless connectivity with Personal Digital Assistants (PDA) to allow excavators to enter data immediately into the RDBMS from any remote location. A square supervisor could fill out digitized recording sheets from his unit, a field supervisor could save stratigraphic notes, or a survey team could upload data from the survey area, updating the database immediately. This will save time and preserve accuracy, as it eliminates data having to be recorded multiple times, which could result in losing subtle findings during the secondary recordings. As a future possibility for the database, the team will be exploring the viability of such a component.

Additionally, some other modifications to the database will include connection with an advanced image processor that would enable particular strata in digital representations of section drawings and site sketches to be linked with corresponding data in the database. We will also be investigating efficient ways to bring together lab results from analyses such as ^{14}C, principal component, and petrography with their associated loci and stratigraphy within the database. These are only a few of the additions currently being incorporated. Additional enhancements will certainly be implemented as needed or as suggested by collaborators.

V. Theory and Application

An RDBMS of all of the archaeological sites in Egypt would be the most valuable archaeological tool since the theories of stratigraphy and pottery typology were introduced in the late 19th and early 20th centuries. Many factors contribute to the current difficulty in comparing data across sites. First, archaeological reports tend to be published relatively slowly, if they get published at all. Second, the written reports contain only a tiny fraction of the actual data excavated from a site. In most cases only the major finds and relative stratigraphy of the site are discussed. The real deficiency is the lack of exact detail in the records presented. In essence, any type of inter-site analysis conducted via paper reports lacks scientific accuracy and precision because sample size (i.e. material available) is skewed and biased. This RDBMS would allow a researcher to search all excavations quickly for the occurrence and context of a particular artifact. The database could also provide schematic and cross-section drawings of artifacts for a complementary visual analysis. Moreover, if the database is kept up-to-date on site, current excavation material would be available at all times, whereas paper reports take three to ten years for processing and publication.[16] Thus, anyone researching from the database will have the most current data available.

[14] For examples of statistical analysis that can be automated in the database, see S. Shennan, *Quantifying Archaeology*. Iowa City: University of Iowa Press, 1988 (1997: 2nd Edition).

[15] See B. Hillier & J. Hanson, *The Social Logic of Space*. Cambridge: Cambridge University Press, 1984, pp. 143-175. E. B. Banning (University of Toronto) has recently used this technique to demonstrate affinities between concepts of domestic space of Western Asiatic structures and so-called "Hyksos" structures at Tell ed-Dab'a and Tell el-Maskhuta in his paper, "The Spatial Organization of Middle Bronze Age 'Asiatics' in Northeastern Egypt," given at the 2002 ASOR meeting in Toronto.

[16] In some recent cases, reports have been published almost 100 years after the actual excavation by scholars who did not even excavate the site! J. Garstang's excavations at Hierakonpolis around 1900, for example, were only published in 1990

The concept of the Egypt Archaeological Database permits the realization of several shifts in archaeological theory. In the preface of *The Archaeological Process*, Ian Hodder describes archaeologists as digging "'sheltered' from outside criticism and untroubled by reflexive analysis."[17] Hodder calls for reflexivity and engagement between archaeologists in order to promote a more productive and evolving dialogue regarding archaeological field method, data collection techniques, and the conceptualization of archaeology.[18] The most direct way to stimulate such a discourse is to place all data on the table, so that collection methods and interpretations can be scrutinized by others in the field. The Egypt Archaeological Database would be that table, a natural checks-and-balances system for archaeology.[19]

The database would offer an important opportunity for mid-excavation interpretation. As has been observed by Hodder and others,[20] archaeological interpretation occurs in all parts of excavation and cannot be separated from the data collection. The field archaeologist must interpret the type of deposit that he is dealing with before he can make a decision about how to excavate it. The Egypt Archaeological Database would offer excavators an opportunity to query and interpret data on a daily basis. This would provide the excavator with potential for reassessment of ongoing procedures, as well as increased adaptability to the inherent uniqueness and unpredictability of archaeological localities.

The Egypt Archaeological Database will fully open the archaeological record to all researchers, advancing the field in a very innovative direction. The archaeological community will still depend upon the excavational competence of individual archaeologists, but it will no longer have to rely solely upon the subjective interpretation of each archaeological team. The raw data collection will be open for all to examine and study. This transparency will also be a corrective check on the methodology of archaeologists, in that their work will now be on display and much more subject to critique, elevating the standard for excavation quality and integrity. In addition, the Egypt Archaeological Database will also promote the development and evolution of archaeological theory and methods. The continued viability of any science requires that its theories and methods are adaptable, and archaeology is no exception. The data considered vital to analysis today would have been thrown away fifty years ago. Before the 1950s, archaeo-botanical samples were not taken. With the advent of Carbon-14 dating, these remains are now considered highly valuable. The Egypt Archaeological Database inherently takes into account this type of continuing change, allowing all types of data to be stored whether deemed useful by the excavator or not. Fifty years from now, archaeologists will find much data pertinent in ways that we cannot imagine today, and they will be able to go back to old excavations through this database to acquire the data needed to reevaluate past excavation material.

Currently, complete data sharing in Egyptian archaeology is non-existent. Archaeologists are constantly excavating and collecting masses of data, all of which needs to be sorted and limited by the lead excavator to determine what will be included in a written report. In theory, this is professionally unacceptable, and violates the principles upon which archaeology is founded, but, as of yet, no archaeologist can avoid the restrictions placed on him by traditional publication and funding. The Egypt Archaeological Database would offer a solution to this problem: a medium through which *all*

---

by Professor Barbara Adams at the University College, London, who was forced to rely on private journals and informal notebooks to provide archaeological data. She has accomplished an admirable and much-needed feat, but this is far from the ideal situation, from a scientific perspective.

[17] I. Hodder, *The Archaeological Process: An Introduction.* Oxford: Blackwell, 1999, x.

[18] I. Hodder, op. cit., xi.

[19] The database does shift the duties of the lead excavator slightly. His interpretations will no longer be the final word on the site, and his primary duty will be to ensure that the site is excavated properly, and he will have the incentive that his data is being watched. The published "final report" will become more of a first hand interpretive analysis, rather than the definitive answer.

[20] I. Hodder, op. cit., pp. 80-104, for references.

archaeological data collected can be easily stored, searched, and analyzed. Furthermore, this project offers data access not just to the excavator, but to all interested professionals. This concept may be difficult for many of my colleagues to accept, but it will certainly be the path for the future of archaeology as a viable science.

Matthew Joel Adams
mja198@psu.edu
The Pennsylvania State University

Digital Egypt for Universities: Koptos in the Second Intermediate Period

Wolfram Grajetzki

The online-learning project Digital Egypt for Universities (hereafter DEU) is a collaboration between CASA (Centre for Advanced Spatial Analysis) and the Petrie Museum of Egyptian Archaeology, both part of University College London. The Petrie Museum, with about 80,000 objects, is one of the most important collections of Egyptian archaeology outside Egypt. The range of objects covers the Palaeolithic, Neolithic, Early Dynastic Period, Pharaonic Period, and the Greek, Roman and Islamic Periods. A project to photograph the whole collection for delivery on the web was completed in spring 2002. Using these copyrighted images, DEU is assembling resources for university courses and providing contextual information for many of the objects in the Petrie Museum. Beside the digitised pictures, the museum holds the copyright on Petrie's notebooks and tomb cards, on many photos taken by Petrie on his excavations, and on the publications of the British School of Archaeology in Egypt. DEU is a three year project funded by JISC (Joint Information Systems Committee), which started in August 2000 and will finish in July 2003. To date we have covered the Early Dynastic and Old, Middle and New Kingdom Egypt and we are currently (summer 2002) working on material of the Late Period. The DEU site is not described in further detail in this article, but everyone is invited to investigate the site for themselves.[1] The focus of this article is a group of relief fragments for which reconstructions may be found on DEU. The full reassembly and colour photographs can be found on the web.[2]

One important part of DEU is the series of 3D reconstructions for different sites or monuments. For several sites DEU will present more than one version. Egypt is rich in well-preserved monuments familiar to any tourist, from the pyramids of Giza to the temples of southern Upper Egypt and Nubia. In the Near East, in general, buildings are only preserved to the foundation. Therefore reconstructions in 3D on paper have always played a very important part in studying architecture.[3] In Egyptology, such models have never been as prominent in research on architecture. There are important exceptions. The isometric drawings by Emery of the original forms of the Early Dynastic tombs at Saqqara[4] should be mentioned. Two alternate versions for the valley temple of the sun temple of Userkaf[5] already anticipate the principle of multiple reconstructions we propose in DEU.

Many sites excavated by Petrie are so heavily destroyed that it is very difficult to gain even a vague idea of the original appearance of these places in ancient times. The Labyrinth at Hawara is a perfect example. More survived from other sites, but Petrie never tried to reconstruct buildings to the same extent as other people.[6] A case in point is the elite cemetery at Qaw el-Kebir. Petrie excavated here and

---

[1] http://www.petrie.ucl.ac.uk/digital_egypt/Welcome.html

[2] http://www.petrie.ucl.ac.uk/digital_egypt/koptos/reliefs/2inter.html

[3] See for example the reconstructions in E. Heinrich. *Die Tempel und Heiligtümer im Alten Mesopotamien*, Denkmäler Antiker Architektur 14, Berlin 1982.

[4] W. B. Emery. *Great Tombs of the First Dynasty II*, London 1954, pl. XXXIX.

[5] H. Ricke. *Das Sonnenheiligtum des Königs Userkaf. I. Der Bau.* Beiträge zur Ägyptischen Bauforschung und Altertumskunde, 7. Kairo 1965, plan 5.

[6] A most impressive series of reconstructions is offered by A. Badawy. *A History of Egyptian Architecture. I. From the Earliest Times to the End of the Old Kingdom.* Giza 1954; *II. The First Intermediate Period, the Middle Kingdom, and the Second Intermediate Period,* Berkeley and Los Angeles 1966; *III. The Empire (the New Kingdom).* Berkeley and Los Angeles 1968.

published the plans of the structures he found,[7] while Steckeweh, working at the site, published both plans and a reconstruction.[8] The Steckeweh version is often reproduced in other publications,[9] giving the impression that we have exact knowledge of the original form of the buildings. However, a closer look at the reconstructions and the surviving remains reveals many questions. Did the causeway really have a roof? Was there a royal-style valley temple, or just some kind of small landing stage? What kind of structure is envisaged with the pylon-like building at the end of the last courtyard?

Another example of a reconstruction of an ancient Egyptian building is the mortuary temple of Mentuhotep II in Deir el-Bahari. The first excavator, Naville, reconstructed a pyramid on top of the building.[10] Arnold proposed a square building,[11] while Stadelmann reconstructed a small mound.[12] This is one of the few cases where Egyptologists are fully aware of the uncertainty of our knowledge.[13]

DEU will offer several reconstructions of sites excavated by Petrie. Various possible versions of a structure will be published on the web, drawing attention to what we do not know as well as to what we know. The first 3D models, which were made by Narushige Shiode for DEU using VRML (Virtual Reality Modeling Language), are models of palace facade tombs at Tarkhan. While the plan of these tombs is quite well preserved, it might be debated whether the tombs were once painted in bright colours or left white or even in mud colour.[14] Therefore polychrome and plain versions of these tombs are published on the web.[15]

The reconstructions discussed in this article concern relief blocks from Koptos dated to king Nubkheperra Intef (17th Dynasty), and are among a series belonging probably, though not certainly, to the main Min temple. They were uncovered by Petrie in 1893-1894 and brought to University College London. The Min temple in Koptos, including the Roman temple, was already badly destroyed when Petrie excavated it. He was not even able to draw or reconstruct a more detailed plan of the latest building. The problem was even greater for the earlier temple buildings. Only loose blocks were found, with no clear relation to any original architectural setting. Several fragments, including parts of three colossi, demonstrate the importance of the temple in Pre- and Early Dynastic times.[16] Few fragments survived from Old Kingdom royal activity.[17] There is rather more evidence for royal activity in the Middle Kingdom. A number of blocks of the 17th Dynasty were found reused in the later buildings. The blocks, which all belong to one or more structures of King Nubkheperra Intef, were mostly found reused in the pavement of the 18th Dynasty temple, which was built or rebuilt by Thutmosis III. Eighteen inscribed slabs were already published by Petrie in 1896.[18] Some other blocks not mentioned by Petrie are now in the Petrie Museum and in Berlin; scholars are reminded that the 1896 publication was selective. King Nubkheperra Intef is one of the better attested kings of the 17th Dynasty. He also

[7] W. M. Flinders Petrie. *Antaeopolis, The Tombs of Qau*, BSAE 51, London 1930.

[8] H. Steckeweh. *Die Fürstengräber von Qaw*. Leipzig 1936, frontispiece.

[9] W. S. Smith. *The Art and Architecture of Ancient Egypt. Revised with additions by W. K. Simpson*. New York 1984, 190, fig. 184.

[10] E. Naville. *The XIth Dynasty temple at Deir el-Bahari II*. London 1910, pl. XXIV.

[11] D. Arnold. *Der Tempel des Königs Mentuhotep von Deir el-Bahari. I. Architektur und Deutung*. Mainz 1974, frontispiece.

[12] R. Stadelmann. *Die ägyptischen Pyramiden, Vom Ziegelbau zum Weltwunder*. Mainz 1991 (second edition), 232, fig. 74.

[13] Smith, op. cit. 159, figs. 149-150; B. J. Kemp. *Ancient Egypt*. London 1989, 104, fig. 38.

[14] The colouring on such tombs was found still well preserved on some tombs at Saqqara. W. B. Emery. *Great Tombs of the First Dynasty III*. London 1958, pls. 6-8.

[15] http://www.petrie.ucl.ac.uk/digital_egypt/tarkhan/tarkhangreattombs/index.html

[16] For reconstructions of the early temple: Bruce Williams. "Narmer and the Coptos Colossi." *JARCE* XXV (1988), 35-59; Barry J. Kemp, with assistance of Andrew Boyce and a geological report on the stone by Jones Harrel. "The Colossi from Early Shrine of Coptos in Egypt." *Cambridge Archaeological Journal* 10:2 (2000), 211-242.

[17] W. M. Flinders Petrie. *Koptos. With a chapter (the classical inscriptions) by D. G. Hogarth*. London 1896, passim; Dorothea Arnold (ed.). *Egyptian Art in the Age of the Pyramids*. New York 1999, 136, no. 101; 444, no. 167.

[18] W. M. Flinders Petrie. *Koptos*, pls. VI-VII.

built in Abydos; remains of a temple built by him have been found near Thebes, and his Theban tomb has recently been relocated.[19]

Relief-decorated blocks from the Second Intermediate Period are also known from other sites in Upper Egypt. The importance of the blocks from Koptos lies in the number which survived and the potential for piecing together at least some of the scenes as they originally appeared on individual walls. Furthermore, it may be possible to gather an idea of the structure to which these blocks once belonged, by comparison with other Middle Kingdom temples.

Two groups of blocks can be distinguished: 1. blocks in raised relief; 2. blocks in sunken relief. Most of the latter are now in the Petrie Museum (University College London), while the other blocks are distributed between Berlin, Manchester, Oxford and Philadelphia. The blocks for which the present location is not known might have been items retained in Egypt at the division of finds, or left on site:[20]

Raised relief:
Pl. VI

1. Present location unknown
2. Manchester 1762
3. UC 14780
4. Philadelphia E941
5. Berlin 12489
6. Oxford Ashmolean Museum 1894.106

Sunken relief:
Pl. VI

7. Present location unknown
8. UC 14781
9. Berlin 12488
10. UC 14787
11. Present location unknown
12. UC 14781

Pl. VII

13a. Philadelphia E942
13b. Philadelphia E942
14. Berlin 12486[21]
15. Manchester 1773
16a. UC 14784
16b. UC 14784
17. Philadelphia 938-40 (three fragments)
18. Oxford Ashmolean Museum 1894.106

Fragments from Koptos additional to those published by Petrie:

19. Berlin 12487[22]
20. UC 14790 (two joining blocks)[23]

---

[19] List of attestations in K. S. B. Ryholt. *The Political Situation in Egypt during the Second Intermediate Period c. 1800-1550 B.C.* Carsten Niebuhr Institute Publications 20. Copenhagen 1997, 394-395, file 17/4. For the tomb, see D. Polz, in Archaeologie-online 5 (2001) – http://www.archaeologie-online.de/magazin/thema/2001/05/d_1.php

[20] Plates refer to Petrie, op. cit.

[21] G. Roeder. *Aegyptische Inschriften aus den Königlichen Museen zu Berlin I. Inschriften von der ältesten Zeit bis zum Ende der Hyksoszeit*. Leipzig 1913, 137.

[22] Roeder, op. cit. 137.

21. UC 14788[24]
22. Relief built into a bridge near Koptos[25]

It is possible to reconstruct with some certainty three walls of a small chapel or of a room in a temple from the sunken reliefs. Petrie and Stewart arranged the reliefs in a certain order in their publications, implying that they already had some specific scenes in mind.[26]

Scene 1:
The reliefs no. 16a and 16b might have formed part of the (exterior?) rear wall of the chapel, showing the standing king in the middle between two gods, crowning the king. On one side is the falcon headed "Hor-shenu, foremost of Hut-nisut"[27] (left) and on the other side stands Horus Behdeti (Horus of Edfu).[28]

Scene 2:
Blocks no. 8, 10 and 12 might belong to a left wall, according to the position of the principal deity.[29] On the right side stands Min and in front of him the king followed by Isis (identified by the inscription) (*Fig.1*).

Scene 3:
Nos. 11, 21 and 22 belong to a scene which can be attributed to a right wall from the orientation of the figures. On the left stands Min; before him is the king, and behind the king stands a goddess, maybe again Isis.

All three of these walls are surrounded by a compartmented border and have a *kheker* frieze at the top. Only fragment no. 7, which shows hieroglyphic inscriptions, cannot be placed with any degree of certainty in any of the scenes, though it may belong to the right wall slightly above fragment no. 11.

It is more difficult to arrange the blocks with the raised relief. Nos. 2 and 5 belong together and are part of a scene showing Min, the king and the *meret*-boxes.[30] The scene might belong to a left wall (assuming the blocks come from one or two chapels with one scene occupying one whole wall). For the other blocks we can only be sure at this time that these blocks must also have belonged originally to two different walls. Nos. 3 and 6 come from the left end of two different scenes showing Min (no. 3 without *kheker* frieze, may be from a totally different scene/wall, or from the rear, which was slightly differently arranged). No. 6 might belong to a right wall of a chapel. Block no. 1 is the upper edge of a

---

[23] H. M. Stewart. *Egyptian Stelae, Reliefs and Paintings from the Petrie Collection. Part Two: Archaic to Second Intermediate Period.* Warminster, 1979, no. 83.
[24] Stewart. *Egyptian Stelae*, 19, no. 84 (without illustration).
[25] A. H. Gardiner. "A Monument of Antef V from Coptos." *PSBA* 24 (1902). 204-205.
[26] Stewart. *Egyptian Stelae*, pls. 16-17.
[27] "Hut-nisut" – "chapel of the king". Is this the name of the building? It is unlikely that it refers to the city Hutnisut in the eighteenth nome of Upper Egypt, for which see J. Brinks, J. Dittmar, F. Gomaà, P. Jürgens, W. Schenkel, *GM* 79 (1984), 76.
[28] Similar scenes in Middle Kingdom temples showing a male and a female god crowning the king: Mont and Junit (Junit is not named; el-Tod, F. Bisson de la Roque, *Tôd (1934 à 1936).* Cairo 1937, FIFAO XVII, 74, fig. 26, pl. XVIII, D. Wildung. *Sesostris und Amenemhet*, München 1984, 54, fig. 48); Tjenet and Month (el-Tod, Bisson de la Roque, *Tôd (1934 à 1936),* 79, fig. 32, pl. XXII, Wildung, *Sesostris und Amenemhet*, 55, fig. 49); in the New Kingdom: E. Naville. *The Temple of Deir el Bahari III.* London 1898, pl. LXIV (Seth and Horus).
[29] The god faces out. H. G. Fischer. *The Orientation of Hieroglyphs. Part I. Reversals.* Egyptian Studies 2. New York 1977, 41.
[30] A. Egberts. *In Quest of Meaning. A Study of Ancient Egyptian Rites of Consecrating the Meret-Chests and Driving Calves.* Leiden 1995, 10-11, pl. 11a (it is so far the earliest attestation of the ritual).

left end. It might belong with no. 2. Finally, no. 4 shows Mont with the sun disc on his head and with a feather crown. The god wears a double uraeus at his head. It is likely that the god stands in front of the king.

Block 13a and 13b, 14, 17 and 18 must belong to different doorways. The three fragments under no. 17 clearly belong to the two jambs of one doorway with one vertical hieroglyphic inscription on each. No. 18 might belong to another doorway with two vertical lines on each jamb.

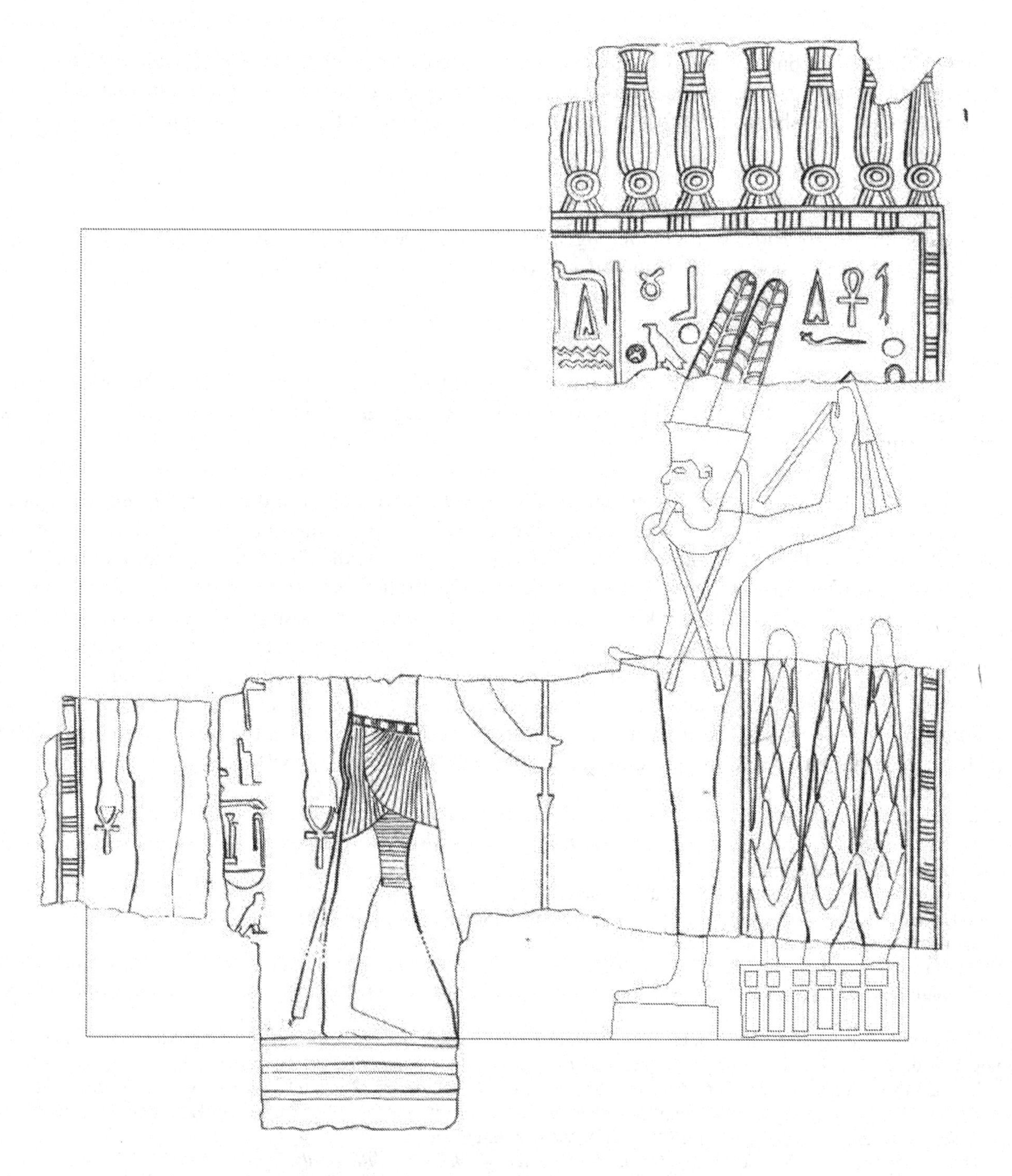

Fig. 1 – Wall scene (no. 2) with the god Min

Fig. 2 - Reconstruction of one doorway: Fragments 14 and 17

No. 15 is a door lintel; it is not clear which lintel belongs to which jamb. Since the lintel no. 14 is shorter than no. 13 it is possible that no. 14 is the lintel of the smaller door jamb (no. 17 with only one vertical line). Gardiner, who published a further fragment from Koptos with names and titles of King Nubkheperra Intef, suggested that his fragment might belong with no. 14.[31] Sadly he only reproduced the hieroglyphs of the fragment and therefore it is not possible to check his idea.

In addition there are three other unplaced fragments of sunken relief: nos. 9, 12 and 15. Of special interest is no. 9 which is an edge fragment. On one side the framing which was described above (*kheker* frieze and colour band) is still visible. On the other side there are two lines of hieroglyphic inscriptions. This fragment of a doorway might therefore belong to one of the wall scenes described above. Hence it seems possible that these walls belonged to small chapels with three decorated walls and each with a decorated doorway as entrance.

Two different options for reconstructing these chapels are proposed here and can be seen on DEU.[32]

1. Many reliefs found at Koptos belong to building activity from the reign of Senusret I, who therefore must have added to the temple, and perhaps even rebuilt the whole temple of Min.[33] In the first reconstruction it is assumed that the main temple of Senusret I had been ruined by the time of the Second Intermediate Period, and was rebuilt under king Nubkheperra Intef. In favour of this scenario one might cite the stela of king Rahotep from Koptos, who said that he found something in ruins, which often is taken as reference to the temple, and it seems likely that he also added to the Koptos temple.[34] The problem is that it is not certain how literally that statement should be taken and whether it refers to the entire temple. As an additional complication, it is not really known whether Rahotep reigned before or after Nubkheperra Intef.[35] Assuming that the Middle Kingdom temple was really destroyed in the Second Intermediate Period, the reliefs of king Nubkheperra Intef would then have been part of the decoration of the principal sanctuary, which must accordingly have been entirely rebuilt under that king. Three chapels standing next to each other might be assumed: a central sanctuary with raised reliefs, and two other sanctuaries, one of them decorated with sunken relief. Three sanctuaries in a temple are very common in the Middle Kingdom. Well-documented examples include a temple in Ezbet Rushi[36] or the Renenutet temple at Medinet Maadi.[37]

2. For a second reconstruction, in contrast to the preceding, it is assumed that the main building of Senusret I was still standing in the Second Intermediate Period. The reliefs of king Nubkheperra Intef would then have come from two or three small chapels constructed as additions to an already existing temple building. A similar arrangement was revealed by excavations at Elephantine and dates to king Wahankh Intef.[38] Here the king had built a subsidiary chapel in the courtyard of the Satet temple.

---

[31] Gardiner. *PSBA* 24 (1902), 204-205.

[32] After this article was written and submitted (2002), a publication with a reconstruction of the Koptos temple appeared that proposes a totally different solution: C. Eder, *Die Barkenkapelle des Königs Sobekhotep III. in Elkab. Beiträge zur Bautätigkeit der 13. und 17. Dynastie an den Göttertempeln Ägyptens*. Elkab 7, Turnhout 2002.

[33] The literature includes notably: Adolophe Reinach. *Catalogue des Antiquités Égyptiennes*. Chalon-sur-Saone. 1913; Marc Gabolde. "Blocs de la porte monumentale de Sésostris Ier à Coptos. Règne de Sésostris Ier (circa 1990 av. J.C.)," in *Bulletin des Musées et Monuments Lyonnais* 1990 n. 1-2; see also now the exhibition catalogue: *Coptos. L'Égypte antique aux portes du désert*. Lyon, Museés des beaux-arts, 3 février - 7 mai 2000.

[34] UC 14327: Stewart. *Egyptian Stelae*, pl. 15.1; Ryholt. *Political Situation*, 392-393, file 17/1 (with further literature).

[35] C. Bennett. "The Date of Nubkheperre Inyotef." GM 147 (1995), 19-27; Ryholt. *Political Situation*, 167-171.

[36] M. Bietak, Josef Dorner. "Der Tempel und die Siedlung des Mittleren Reiches bei Ezbet Rusdi." *Ägypten und Levante* VIII (1998). 15, fig. 4.

[37] A. Vogliano. *Secondo rapporto degli scavi condotti dalla Missione Archeologica d'Egitto della R. Università di Milano nella zona di Madinet Madi*. Pubblicazioni della Regia Università di Milano, Milano 1937.

[38] W. Kaiser, M. Bommas, H. Jaritz, A. Krekeler, C. v. Pilgrim, M. Schultz, T. Schmitz-Schultz, M. Ziermann. "Stadt und Tempel von Elephantine. 19./20. Grabungsbericht." *MDAIK* 49 (1993). 148-149, abb. 6, 8; W. Kaiser. *Elephantine. Die antike Stadt*. Cairo 1998, 21.

Finally, a set of blocks is known from a small building of king Khaankhra Sobekhotep at Abydos, which must either come from a chapel on its own or might have been part of a larger structure, which is now lost. Nevertheless the scale of the blocks naming this king suggests a relatively small scale building activity.[39]

All reconstructions can be seen at http://www.petrie.ucl.ac.uk/digital_egypt/koptos/index.html

Wolfram Grajetzki
University College London

Fig. 3 - chapels, 3D models: Naru Shiode

[39] E. Bresciani. "Un edificio di Kha-anekh-Ra Sobek-hotep ad Abido (Mss Acerbi, Biblioteca Comunale di Mantova)." *Egitto e Vicino Oriente* 2 (1979), 1-20.

Otto Friedrich von Richters Expedition in Unternubien im Jahre 1815

Sergei Stadnikow

Otto Friedrich von Richter wurde am 6. August 1791 in Neu-Kusthof (es lag in der Provinz Livland des damaligen russischen Zarenreiches), dem heutigen Vastse Kuuste, in der Nähe von Dorpat (jetziges Tartu) in der Familie des livländischen Landrats Otto Magnus von Richter geboren[1]. Sein Hauslehrer war seit 1803 G. Ewers, der später, in den Jahren 1818-1830, Rektor der Dorpater Universität war. Dank Ewers erwachte in dem jungen Otto das Interesse für die Antike. Es dauerte nicht lange, bis er bereits recht gut Latein und Griechisch beherrschte. Nachdem er sich einige Kenntnisse auch vom Neugriechischen erworben hatte, fuhr er im Herbst 1809 nach Heidelberg, um dort unter der Leitung von Prof. F. Wilcken die Grundlagen des Persischen und Arabischen zu studieren. 1812 ging Richter über die Schweiz und Italien nach Wien. Die dortige Bibliothek bot ihm günstige Möglichkeiten zur Vervollkommnung seiner Studien. In diesem Abschnitt seines Lebens wurde er stark von Friedrich Schlegel, dem berühmten deutschen Schriftsteller, Sprach- und Literaturwissenschaftler und führenden Theoretiker der deutschen Vorromantik, beeinflusst, der damals ebenfalls in Wien lebte.

Im Sommer 1813 kehrte der junge Mann über Böhmen, Schlesien und Polen in die Heimat zurück, weilte hier aber nur ein Jahr lang. Bereits am 31. Oktober 1814 segelte O. F. von Richter von Odessa nach Istanbul (Konstantinopel), um die östlichen Kulturen unmittelbar zu erleben. Zugleich hatte er die Hoffnung, im Orient zu einer neuen Welterkenntnis zu gelangen, dies um so mehr, als das zeitgenössische Europa mit seinen Ansichten ihm langweilig und einseitig erschien[2]. Das war die romantische Welthaltung, die häufig ihre Ideale in Vergangenheit sah.

---

[1] O. F. v. Richter, *Wallfahrten im Morgenlande. Aus seinen Tagebüchern und Briefen dargestellt von J. P. G. Ewers und einem Beitrag von K. Morgenstern.* Berlin, 1822, S. V; vgl.: *Genealogisches Handbuch der livländischen Ritterschaft*, Görlitz, 1932, S. 173; J. von Recke, K. Napiersky, *Allgemeines Schriftsteller- und Gelehrtenlexikon der Provinzen Livland*, Esthland und Kurland, Bd. 3, Mitau, 1831, S. 544; *Deutsch-baltisches biographisches Lexikon 1710-1960*, Köln-Wien, 1970, S. 631; *Who was Who in Egyptology* (Hrg. M. L. Bierbrier), London, 1995, S. 357; S. Stadnikow, „Otto Friedrich von Richter und Ägypten", in *Altorientalische Forschungen*, Bd. 18, 1991, S. 195-203. In meisten Untersuchungen und Nachschlagewerken ist Otto Friedrich von Richters Geburtsjahr unkorrekt angegeben (1792); nach neuesten Archivforschungen wurde er jedoch 1791 geboren.

[2] O. F. v. Richter, *Wallfahrten...*, S. VI-VII. Das war die erste Seereise des O. F. von Richters (Das Estnische Historische Archiv in Tartu, die Bibliothek, KS-17, S. 6-7). Unter der Bestandnummer KS-17 befindet sich in der Archivbibliothek das Tagebuch von Richters. Auf der ersten Seite des noch nicht publizierten Tagebuchs beschreibt der junge Orientalist kurz vor der Abfahrt, bereits an Bord des Schiffes, seine romantische Stimmung („...in der Fremde will ich alle Tage etwas Neues sehen") und seinen krankhaften physischen Zustand wegen einer Erkältung (op. cit., S. 1). Im Laufe der ganzen Studien- und Forschungsreise, seit dem Abschied von Odessa, führte er die Tagebücher. In bezug auf den ägyptischen Teil der Levante-Expedition gibt es im Estnischen Historischen Archiv in Tartu zwei Tagebücher. Eines von ihnen, das kleinere, ist im Archivbestand als F. 1388, Nr. 1 geführt. In diesem Reisetagebuch befinden sich etwa ein halbes Hundert Skizzen, Pläne, Nilkarten, Erläuterungen, Zeichnungen von Tempeln, Grabanlagen usw. sowie aus dem täglichen Leben der Einheimischen und die Kommentare zu ihnen. Unter anderem hatte O. F. von Richter den Namen des makedonischen Königs und Herrschers von Ägypten Philippus Arrhidäus von einer hieroglyphischen Inschrift im erhaltenen Pronaos des Thot-Tempels zu Hermopolis kopiert (Das Tagebuch von O. F. v. Richter, F. 1388, Nr. 1, S. 13). Das zweite Reisetagebuch wurde von mir im Februar 2000 in der Archivbibliothek unter der Nummer KS-17 gefunden. In dem Tagebuch gibt es keine Zeichnungen, aber kann man von einer handschriftlichen Expeditionsbeschreibung (wenigstens zum Thema Ägypten und Unternubien) sprechen. Den umfangreichen Text von 357 Seiten ergänzen einige Tempelanlagenpläne. Das Reisetagebuch selbst beginnt mit Eintragungen über die Seefahrt von Odessa nach Istanbul im Spätherbst 1814 und endet am 4. August 1815 in Kairo, kurz vor der endgültigen Trennung von Ägypten. Es scheint, dass dieser zweite Tagebuchband eine durchdachte ausführliche und besinnliche Darlegung der Reiseeindrücke sowie auch Forschungsergebnisse enthält, die

Abb. 1. Otto Friedrich von Richter.

In der Hauptstadt der Osmanen begann Richter bei einem Mullah seine Kenntnisse im Persischen und Arabischen zu vervollkommnen und Türkisch zu lernen. Im gastfreundlichen Haus des schwedischen Gesandten Nils Gustaf Palin freundete der Balte sich mit dem Prediger der Gesandtschaft, Sven Fredrik Lidman (1784-1845), an und beschloss, mit ihm gemeinsam eine längere wissenschaftliche Reise nach Ägypten zu unternehmen. Lidman war 1811 zum Dozenten des Arabischen an der Universität zu Uppsala ernannt worden, fuhr aber dennoch im selben Jahr nach Istanbul und wurde Prediger an der Gesandtschaft; G. Ewers bezeichnet ihn allerdings in seinem unten angeführten Werk als Gesandtschaftssekretär.

So traten Richter und Lidman am 30. März 1815 in Begleitung eines weitgereisten Dieners, des Armeniers Kirkor (Gregor), an Bord eines griechischen Schiffes die Segelfahrt nach Alexandria an. Nach Besichtigung der Inseln Lesbos und Rhodos trafen sie am 12. April in Alexandria ein. Die Wissenschaftler konnten dem neuen Herrscher Ägyptens, Mohammad Ali, der von 1805-1849 regierte,

---

hauptsächlich abends in gotischer Schrift niedergeschrieben wurden. Als Romantiker schenkte der Orientalist den Naturschilderungen ziemlich viel Aufmerksamkeit. Dieses Manuskript soll als eine akademische Edition im Jahre 2003 veröffentlicht werden. Als 1822 Prof. G. Ewers die Briefe und Tagebücher Richters herausgab, wurde der ägyptisch-nubische Teil ausgelassen. Aus welchem Grunde? Das kann man nicht mehr eindeutig klären. In sprachlich-kompositioneller Hinsicht ist das zweite Tagebuch heute mit gewissen Schwierigkeiten lesbar. Aus indirekten Quellen geht hervor, dass dem Redakteur des Werkes G. Ewers beide Tagebücher bekannt waren. Vielleicht kann man das Weglassen von Ägypten und Nubien mit dem vorwiegend wissenschaftlichen Charakter der Darlegung der Expedition nach dem Nilland erklären. Zu Beginn des 19. Jahrhunderts waren Reisebriefe von Hellas oder Levante in Europa populär geworden, aber sie mussten für gebildete Leser auch interessant sein. Ein reiner Fachtext wäre zu sachlich-trocken gewesen. Es ist nicht unmöglich, dass Prof. G. Ewers diese Tatsache bei der Redaktionsarbeit im Auge hatte. Den Bibliotheks- und Archivmaterialien zufolge hat es in Privatbibliotheken der Deutschbalten zahlreiche landeskundliche Bücher gegeben (zum Thema siehe Cl.-E. Savary, *Zustand des alten und neuen Egyptens in Ansehung seiner Einwohner, der Handlung, des Ackerbaues, der politischen Verfassung etc.*, Bd. 1-2, Berlin, 1786-1789; S. Witte, *Ueber den Ursprung der Pyramiden in Egypten und der Ruinen von Persepolis*, Leipzig, 1789; F. Ch. Gau, *Neu entdeckte Denkmäler von Nubien, an den Ufern des Nils, von der ersten bis zur zweiten Katarakte. Gezeichnet und vermessen im Jahre 1819, und als Fortsetzung des grossen französischen Werkes über Egypten*, Lfg. 1-13, Gotta, 1821-1828; V. Denon, *Vivant Denon's Reise in Nieder- und Ober-Aegypten, während der Feldzüge des Generals Bonaparte*, Berlin, 1803).

zahlreiche Empfehlungsschreiben vorlegen. Dank den Empfehlungen erteilte der Pascha ihnen eine offizielle Genehmigung für Reisen durch das Land, so dass die Lokalbehörden ihnen keinerlei Hindernisse in den Weg legten. Im Gegensatz zu vielen anderen Reisenden jener Zeit trugen sie während der ganzen Expedition europäische Kleidung. Hussein, der Gebieter Assuans, begleitete sie höchstpersönlich durch die Wüste bis nach El-Heifa (Philae) beim ersten Nilkatarakt[3]. Ihr Reiseweg hatte sie zuvor an den Denkmälern von Luxor, Kom Ombo und Assuan vorbeigeführt, die durch Publikationen der Expedition Napoleons mit Abbildungen und Schilderungen, u. a. von V. Denon, in Europa bekannt geworden waren[4]. Am 1. Juni 1815 waren die Reisenden bereit, nilaufwärts nach Nubien zu ziehen. Im Gepäck hatten sie unter anderem die Veröffentlichungen von F. L. Norden und W. R. Hamilton[5]. Nubien galt zu jener Zeit als ein sehr wildes Land, in das sich nur wenige Europäer wagten[6]. Es war größtenteils eine Terra incognita; die dort befindlichen Denkmäler aus dem alten Ägypten waren nahezu wenig erforscht. So kam es, dass wir O. F. von Richter und S. F. Lidman zu den ersten Altertumsforschern an den Baudenkmälern Nubiens zählen können. Ihr Hauptziel war es, die Lage des ehemaligen Königreiches von Meroe weit im Süden ausfindig zu machen[7]. Leider ging dieser Wunsch aus Gründen, die nicht von ihnen abhingen, nicht in Erfüllung. Die Forscher führten Tagebücher, Richter auf Deutsch und Lidman in schwedischer Sprache.

Schon am zweiten Tag fuhren die Orientalisten nilaufwärts an Debod, an Dehmit, Kertassi, Taffeh und Kalabsche vorbei. Am darauffolgenden Tag wurde der auf dem Westufer liegende Tempel von Dendur und nach dem Wadi Abiad das Gebiet von Merieh passiert. Am vierten Tag erreichten die Reisenden über Sabagura mit seiner Festung und gegenüberliegenden Gerf Hussein den Tempel von Dakka und die Festung Ghaban (modernes Kubban), Gurtah (heutiges Qurta)[8] und den Tempel zu Garb es-Sebua[9]. Am gleichen Abend passierte das Segelboot mit Richter und Lidman den Tempel von

[3] G. Ewers, „Richter's Reise durch Aegypten und Nubie“, in *Dörptische Beyträge für Freunde der Philosophie, Literatur und Kunst*, Bd. 2, 1819, S. 451. Auf Arabisch heißt Husseins Amtstitel „Kaschef“. Gründlicher zum Thema siehe Das Estnische Historische Archiv in Tartu, die Bibliothek, die Bestandnummer KS-17, S. 219-221 (das Tagebuch von O. F. v. Richter).

[4] *Description de l'Égypte ou Recueil des observations et des recherches qui ont été faites en Égypte pendant l'expédition de l'armée française, publié par les ordres de Sa Majesté l'empereur, Napoléon le Grand*, Paris, 1809-1828. Vom militärisch-politischen Standpunkt aus betrachtet konnte die Expedition nicht alle erhofften Ziele erreichen, war aber für die Gestaltung der Ägyptologie als Wissenschaft von fundamentaler Bedeutung. Man verschaffte sich Sammlungen und verfasste Beschreibungen. Außerdem wurde der Stein von Rosette entdeckt. Die Entzifferung der altägyptischen Schrift von J.-F. Champollion im Jahre 1822 hatte den eigentlichen Grundstock für die Ägyptologie gelegt.

[5] F. W. Hinkel, „Otto Friedrich von Richters Reise in Unternubien im Jahre 1815“, in *Altorientalische Forschungen*, Bd. 19, 1992, S. 232. In seinem Tagebuch erwähnt Richter besonders oft W. R. Hamilton (Das Estnische Historische Archiv in Tartu, die Bibliothek, KS-17, S. 322). Die Werke von Herodot, Diodor von Sizilien und Strabo waren den Reisenden auch gut bekannt. Zum Thema siehe: F. L. Norden, *Voyage d'Égypte et Nubie*, Bd. 1-2, Kopenhagen, 1750-1755; W. R. Hamilton, *Remarks on Several Parts of Turkey*, Bd. 1 Aegyptiaca, London, 1809.

[6] Einer von ihnen war Johann Ludwig Burckhardt, der 1813 und 1814 in Nubien weilte. Aus seiner Feder stammt das umfangreiche Buch *Reisen in Nubien, von der Londoner Gesellschaft zur Beförderung der Entdeckung des Inneren von Afrika herausgegeben*, Weimar, 1820 (Übersetzung des 1819 erschienenen englischen Originals). Zur Geschichte der früheren Nubien-Reisen der Europäer siehe die Übersicht von F. W. Hinkel, *Otto Friedrich von Richters...*, S. 230-235).

[7] B. Peterson, „Über Altertümer in Nubien. Sven Fredrik Lidmans Reise im Jahre 1815“, in *Orientalia Suecana*, Bd. 23, 1974, S. 45.

[8] Vgl.: F. W. Hinkel, „Otto Friedrich von Richters...“, S. 233; Das Estnische Historische Archiv in Tartu, die Bibliothek, KS-17, S. 229.

[9] In der Eintragung vom 4. Juni zählte Richter noch auf: Scheir-Gab Kustombäch, Wadi Adascheb, Schark, Gab Handschari, Bardeh, Gebel Hayut, Allaghi, Gisireh Sirat, Ufeddin, Wadi Alenghera, Charab, Scherich Scheraf, Moharraqa, Siala, Ichmterraschid, Bardeh, Abu Schager, Ghemerat, Wadi Abdel Ghit, Wadi Hamd Hassein, Schack Sabua. Diese Orte hatten die Reisenden persönlich besucht oder angeschaut (siehe Das Estnische Historische Archiv in Tartu, die Bibliothek, KS-17, S. 229-231). Den Tempel zu Garb Sabua (modern Wadi el-Sabua) beschreibt O. F. von Richter auf folgende Weise: „Vor dem Tempel stehen bis auf die verstümmelten Gesichter wohl erhalten zwei vorschreitende Colosse von Sandsteinen. Sie haben Büste, Hieroglyphen am Gürtel und auf der hohen Rückenlehne. Vom Dromos [in griechischer Sprache bedeutet dieses Wort „der Weg“ - S. S.] sieht man noch etwa sechs grosse bärtige Androsphinxe vom groben grauen Sandstein. Vor

Ufeddin, dem modernen Maharraqa, die Festung Ikhmindi auf dem Westufer und noch eine Reihe kleinerer Siedlungen und Dörfer. Einen Tag später wurde Kurusko (Korosko) und am nächsten Spätabend (7. Juni), nach dem Sonnenuntergang, Ed-Derr erreicht[10]. Von dort wollten die Männer trotz der Feindseligkeiten der lokalen Herrscher doch weiter nach Süden segeln: „Man rechnet von Eshnah bis Derr 4 Tagereisen zu Lande auf Dromedaren, von Derr bis Ibrim 2 Stunden, bis zu den grossen Cataracten 4 Tagereisen, also dass Derr der halbe Weg ist. Von Derr bis Dongolah 20 Tagereisen"[11].

Im Laufe des dreitägigen Aufenthaltes in Derr und seiner Umgebung machte Richter eine ziemlich gründliche Beschreibung des Felsentempels Ramses II.[12] zu Derr, aus der manche Auszüge hier angeführt seien:

> „Man tritt durch die Thüre (7) oder durch eine der zwei Nebenthüren in den offenen Hof oder Halle E. Die Seitenwände sowohl als die zwischen den eben genannten drei Thüren sind sehr zerstört, ihr Zustand ist aus meiner Zeichnung zu ersehen. Man trifft darin zuerst die Reste von 8 viereckigen Pfeilern (6) die nicht aus dem Felsen gehauen, sondern hingestellt und stückweise aufgebaut zu seyn scheinen, wie aus meiner Zeichnung zu ersehen. Sie scheinen ein künstliches Dach, wenn jemals eins, getragen zu haben, sie sind ohne Hieroglyphen. Dann kommt man an 4 andere Pfeiler welche ein Gesimse tragen. Dieses war durch eine natürliche Felsendecke mit dem übrigen Tempel verbunden, und bildete eine kleine bedeckte Vorhalle. Der Felsen ist zum Theil eingestürzt und liegt vor der Thür, wie in meiner Zeichnung zu sehen. Jetzt fällt das Licht zwischen den Säulen und der Wand hinein. An diesen Pfeilern lehnten sich sitzende Colosse, sie ruhten auf einem viereckigen Piedestal und sind von unten auf bis an die Knie leidlich erhalten: der Körper

---

dem Eingange des Propylon zwei umgestürzte Colossen, den ersten gleich doch weit grösser und besser gearbeitet [Ramses II. - S. S.]. Das Propylon besteht aus grossen zusammengesetzten Steinen, auf denen einige grosse Gestalten und kleine Hieroglyphen nur skizziert und nie vollendet scheinen. Die Mauer des darauffliegenden Hofs ist zum Theil umgestürzt, zum Theil, wie der Sekos [in griechischer Sprache bedeutet dieses Wort „der Innenhof" - S. S.] ganz mit Sande angefüllt, dass kaum der obere Rand der Mauer hervorragt. Inwendig bildeten auf den Seiten 12 viereckige Pfeiler zwei Gallerien. An diese Pfeiler lehnen sich Hermen mit Hirtenstab und Dreschflegel in den gekreuzten Armen. Die Hieroglyphen in dem Propylon, die Gallerie, der Nil und die wilden arabischen Gebirge im Hintergrunde" (Das Estnische Historische Archiv in Tartu, die Bibliothek, KS-17, S. 231-232). Zur Frage der Beschreibung dieses Tempels, vgl. die Beschreibung von S. F. Lidman (siehe B. Peterson, „Über Altertümer in Nubien...", S. 52-53). Zum Thema siehe: D. Arnold, *Lexikon der ägyptischen Baukunst*, München-Zürich, 1997, S. 277-278; F. Ch. Gau, *Antiquités de la Nubie, ou Monuments inédits des bords du Nil, entre la première et la deuxième cataracte*, Paris, 1821-1827, Taf. 42- 47; Prisse d'Avennes, *Histoire de l'art égyptien*, Bd. I, Paris, 1878, Taf. 48; F. Daumas, *Ce que l'on peut entrevoir de l'histoire de Ouadi es Sebua en Nubie*, Le Caire, 1966, S. 23ff.; R. Gundlach, „Sebua", in *Lexikon der Ägyptologie* (Hrg. W. Helck, W. Westendorf), Bd. V, Wiesbaden, 1984, Sp. 768-769; I. Hein, *Die ramessidische Bautätigkeit in Nubien*, Wiesbaden, 1991, S. 17-20; S. Farid, *Excavations of the Antiquities Department at El-Sebua*, Cairo, 1967, S. 61ff.; B. Porter, R. Moss, *Topographical Bibliography of Ancient Egyptian Hieroglyphic Texts, Reliefs and Paintings*, Bd. VII, London, 1951, S. 53ff.; H. Gauthier, *Le temple de Ouadi es-Seboua*, Le Caire, 1912; F. W. Hinkel, *Auszug aus Nubien*, Berlin, 1978, S. 19; S. Stadnikov, „Die Wanderungen des deutsch-baltischen Orientreisenden Alexander von Üxküll 1822-1823", in *Göttinger Miszellen*, Heft 146, 1995, S. 82; J. L. Burckhardt's *Reisen in Nubien...*, S. 147-149.

[10] Das Estnische Historische Archiv in Tartu, die Bibliothek, KS-17, S. 235. Vgl. auch F. W. Hinkel, „O. F. von Richters Reise... ", S. 235. In der Tageseintragung sind noch folgende Ortschaften erwähnt: Wadi Aba Hadjadi, Hushaja, Charaba, Wadi Hamadun, Drivan, u.a.

[11] Das Estnische Historische Archiv in Tartu, die Bibliothek, KS-17, S. 237. Vgl. F. W. Hinkel, „O. F. von Richters Reise...", S. 235.

[12] Zum Thema siehe: D. Arnold, *Lexikon...*, S. 66; Porter, Moss, *Topographical...*, Bd. VII, S. 91; F. Ch. Gau, *Antiquités...*, Taf. 50-52; I. Hein, *Die ramessidische...*, S. 23-25; A. M. Blackman, *The Temple of Derr*, Cairo, 1913; T. Säve-Söderbergh, „Derr", in *Lexikon der Ägyptologie* (Hrg. W. Helck, E. Otto) Bd. I, Wiesbaden, 1973, Sp. 1069-1070; U. Monneret de Villard, *La Nubia Mediovale*, Bd. I, Cairo, 1935, S. 101; L. Habachi, *Features of the Deification of Ramses II*, Abhandlungen des Deutschen Archäologischen Instituts in Kairo, Bd. 5, Kairo, 1969; S. Curto, *Nubia*, Novara, 1965, S. 255-261; H. Bonnet, *Reallexikon der ägyptischen Religionsgeschichte*, New York-Berlin, 1971, S. 157.

ist zerstört; sie scheinen hohe Mitren[13] getragen zu haben, das ist in der Zeichnung dargestellt. In der Wand hat man später eine Nebenthür gebrochen, und in der westlichen Wand eine Nische.

An der westlichen Wand ist eine Schlacht abgebildet, wie die zu Karnak und Medinet Aba[14], wo der König in Riesengestalt auf einem Wagen mit grossen Pferden über Haufen Todter und Verwundeter von kleiner Gestalt, einherfährt; seine Pfeile erfüllen die Luft[15]. Vor ihm erscheinen drei vielleicht weibliche Gestalten, die linke Hand aufhebend. In der Rechten halten sie etwas das einem Stocke oder Schwerdte gleicht und führen Gebundene. Darüber erscheinen eine Kuh und ein laufender Hund unter einem Palmbaum. Darüber dieselbe Gestalt, als ob [auf] gedachte Führer von Gefangenen kniend, die linke Hand gegen den Kopf aufhebend, die rechte halb ausgestreckt, unter ihr ein laufender Hund. Vor ihr ein Knabe, den eine grössere Gestalt ihr zuzuführen scheint, der ihr mit offenen Armen entgegen eilt. Ich habe dieses gezeichnet[16]. Ganz oben scheint eine grosse Gestalt, wie an der gegenüber liegenden Wand, der Gottheit ein Opfer zu bringen, doch sind die Figuren, wie vieles Andere dieser Wand, sehr verstümmelt. Zwischen und über den Gemälden, sieht man gewöhnlichen Hieroglyphen, die an vielen Stellen noch mit Kalk bedeckt sind, dessen Malerei man sieht.

Die Wand gegenüber enthält ein ähnliches Schlachtgemälde. Hier aber erscheint der feindliche Anführer fliehend zu Fuss und zwar fast eben so gross als der Sieger. Im zweiten Felde bringt der siegreiche König dem Gotte mit dem Sperberkopfe[17] eine Menge Gefangene dar, deren Hände und Arme in allerhand marternde Stellungen gebunden sind. Darüber bringt er dem Gotte mit der Mitra ein Opfer dar. Dazwischen eine Menge Hieroglyphen wie die vorigen.

An der Wand sind die Darstellungen auf beiden Seiten der Thüre ungefähr dieselben. Die Hauptfigur ist der König, im Begriffe vier knienden Menschen, deren Haare er in der linken Hand hält, mit der rechten die Köpfe abzuschneiden. Sie strecken die Hände in verschiedenen Richtungen aus. Hinter ihm ist eine Art Standarte aus vier Armen mit verschiedenen Attributen. Vor ihm steht der Gott mit dem Sperberkopfe, worauf eine Kugel und reicht ihm mit der Rechten eine Sichel dar. Im zweiten Felde näher der Thüre bringt der König dem Gotte mit dem Widderkopfe[18] und der Mitra, der in der linken den Nilschlüssel[19] in der Hand hält. Hier trägt der König auf dem Kopfe eine Kugel mit zwei Schlangen, an deren Halse der Nilschlüssel hängt. Im correspondierenden Felde auf der anderen Seite der Thüre, bringt der König den Nilschlüssel dem Gott dar, der Sonne und Mond auf dem Kopfe trägt. Unter diesen grossen Bildern sind auf jeder Seite neun kleine Figuren, links tragen sie Federn, rechts Sistrums. Ganz oben an der Wand sind verschiedene Opfer und dazwischen eine Menge Hieroglyphen.

---

[13] Die Doppelkrone von Ober- und Unterägypten.

[14] Karnak und Medinet Habu.

[15] Die Kriegsmotive sind in ramessidischen Tempeln Ägyptens und Nubiens häufig dargestellt. O. F. von Richters Vergleich der Tempel zu Derr mit den Tempelanlagen von Karnak und Medinet Habu ist kunsthistorisch sehr korrekt, was auf die gute wissenschaftliche Intuition hinweist. Man darf nicht vergessen, dass im Jahre 1815 die Ägyptologie als Wissenschaft noch nicht entstanden war (siehe Anmerkung 4).

[16] Vgl. Das Estnische Historische Archiv in Tartu, Bestand F. 1388, Nr. 1, S. 23 A.

[17] Der Sonnengott Re-Harachte.

[18] Amun-Re.

[19] In der Tat das Lebenszeichen „anch".

An den Pfeilern sieht man verschiedene Opfer an besondere Gottheiten, unter anderen an einen Priap mit Sonne und Mond auf dem Kopfe, das Gebälke ist mit gewöhnlichen Hieroglyphen von unten und an den Seiten geziert.

Dann tritt man in die grosse Halle D, sie ruht auf sechs viereckigen Pfeilern. An der Wand des Eingangs, links von der Thür, sieht man Isis und Osiris sitzen; sie stützt die Rechte, ihn umarmend, auf seiner rechten Schulter. Ein Priester führt ihnen, Hand in Hand, zwei Menschen zu die Nilschlüssel halten. Der Priester trägt eine Mitra, die letzte Figur einen Sperberkopf. An derselben Wand, rechts an der Thür, stehen zwei Menschen auf Nilschlüsseln und giessen aus Gefässen, Nilschlüssel über eine in der Mitte stehende Figur aus. Näher der Thür der König mit Isis.

An der Wand sieht man oben die geflügelte Kugel auf einem Tempel ruhend, der Tempel auf einem Boote. Den Tempel umgeben viele kleine Figuren, unter anderen aufrecht stehender Lotus. Eine Menge grosser Menschengestalten mit kahlen Köpfen und langen Kleidern, tragen das Boot[20] dem Könige entgegen, der ihnen Lotus entgegenreicht mit beiden Händen. Hinter den Trägern geht ein Mensch mit einer Art runder fächerförmigen Standarte. Unter den Trägern tragen zwei die kleine runde Mütze, mit der Schlange an der Stirne und Bärenhaut um die Schultern. Der erste trägt unter der Bärenhaut ein kurzes und ein langes durchsichtiges Gewand, der zweite eine vielfach gefaltete Tunica. Im zweiten Felde wird dem Priap[21] und der Isis ein Opfer von Früchten gebracht. Im dritten Felde steht der König mit dem Flagellum[22] unter einem Baume, hinter ihm eine Gestalt mit Ibiskopf, Scepter und Schreibstift[23]. Vor ihm Priap mit langem Barte und andere Göttergestalten. Das Gesimse hier und ausserhalb besteht aus Siegeln mit zwei Schlangen und anderen Hieroglyphen, Kugeln darüber, doch in der Mitte über den Pfeilern ist eine ohne Symmetrie fortlaufende Hieroglyphenreihe. Die Decke in der Mitte besteht aus Kugeln mit himmelblauen Flügeln.

Auf der hinteren Wand bringt der König, den Nilschlüssel haltend, einen anbetenden Aquacephalus[24] dem Osiris und der Isis dar, die auf blau und roth quer gestreiften Thronen sitzen. Auf der anderen Seite der Thüre zum Sekos[25] dasselbe. Auf der letzten Wand sieht man dasselbe Boot wie im 1., aber in umgekehrter Richtung, der Thür zugehend. Der König hält ein Weihrauchfass entgegen. Im zweiten Felde erscheint die Gottheit auf Nilschlüssel thronend, hinter ihr ein Priester. Davor ein bärtiger Orus[26], mit dem Flagellum auf dem Lotus in anbetender Stellung kniend. Hinter ihm drei Figuren mit Vögelköpfen, in den Händen Scepter mit Wiedehopfköpfen umgekehrt haltend.

An den Pfeilern empfängt der König Nilschlüssel von vielen Göttern, mit anderen steht er Arm in Arm etc.

Durch die Thüre tritt man in das Sekos A. In der Thüre sieht man dieselben Vorstellungen als auf oben erwähnten Pfeilern. Die ganze Breite der hinteren Wand nimmt eine Stufe ein, darauf ruhen verstümmelte sitzende Gestalten, haut Relief. In der Wand hat man zwei, in der Wand (4) eine Nische gesprengt. An der Wand hält der König in einer Hand das Rauchfass, mit der anderen begiesst er den Lotus. Vor ihm ein grosser Tempel, auf demselben ein Boot, und auf diesem wieder ein Tempel, dahinter eine Fahne, Lotus, eine

---

[20] Die Priester tragen die Gottesbarke.
[21] Der Fruchtbarkeitsgott Min.
[22] Die Peitsche.
[23] Der Weisheits- und Schriftgott Thot.
[24] Eine Fischart.
[25] Altgriechisch *sèkos,* „eingezäunter Raum", „Innenhof (eines Tempels)".
[26] Wahrscheinlich hatte von Richter hier den Gott Horus im Auge.

Figur mit einem Scepter, worauf ein Wolfskopf. Dahinter bringt man drei Vasen einer sitzenden bärtigen Gestalt ohne Arme. An der Wand ist derselbe Tempel, aber davor ein Priester in einem langen Gewande, den Nilschlüssel haltend. Hinter dem Tempel ein grosser Palmenbaum und eine Gestalt die nach dem sitzenden Osiris den Arm ausstreckt.

Aus der Halle D führen zwei kleine Seitenthüren in die Zimmer C und B. In C ist die hintere Wand sehr verstümmelt. Man sieht Opfer an verschiedene Gottheiten, unter anderem steht zwischen dem Könige und der sitzenden Gottheit ein gebeugter Lotus.

Das Zimmer B enthält dasselbe, aber an der Wand sitzen drei Gottheiten beisammen.“[27].

Am 9. und 10. Juni machten die Orientalisten, trotz der unsicheren Lage in der Nähe von Derr, mit dem einheimischen beritten Geleit Ausflüge in die Gegend von Qasr Ibrim (die Festung Ibrim, die Ansiedlungen und Dörfer Genenah, Ahabah, Kafri, Mhegeh, etc.)[28].

Ibrachim, der Sohn von Muhammed Ali, stellte den Reisenden nubische Ruderer zur Verfügung[29]. Dadurch wurde die Fahrt auf dem Nil recht angenehm, und sie konnten verhältnismäßig weit nach Süden vordringen, bis nach Qasr Ibrim nördlich von Abu Simbel[30]. Dies blieb der äußerste Punkt ihrer Expedition. Ein weiteres Vordringen wäre höchst gefährlich gewesen, weil, wie G. Ewers auf Grund eines Briefes von Richter feststellte, in diesem Gebiet gerade die Feindseligkeiten zwischen drei dort im Namen des Paschas regierenden und durch ihre Gewalttätigkeiten berüchtigt gewordenen Brüdern ausgebrochen waren[31].

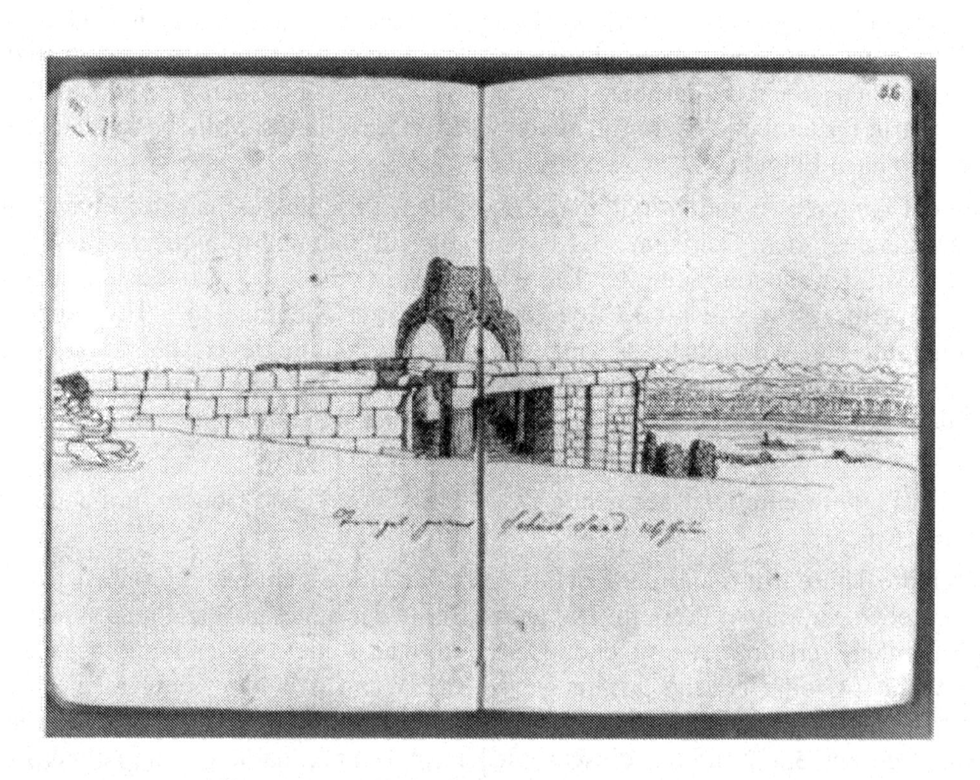

Abb. 2. Tempel zu Scheich Saad, 11. Juni, 1815.

[27] Das Estnische Historische Archiv in Tartu, die Bibliothek, KS-17, S. 240-244.
[28] Das Estnische Historische Archiv in Tartu, die Bibliothek, KS-17, S. 245-252.
[29] G. Ewers, „Richter's Reise... “, S. 451; S. Stadnikow, „Otto Friedrich von Richter... “, S. 197.
[30] Vgl. B. Peterson, „Über Altertümer... “, S. 45-46.
[31] G. Ewers, „Richter's Reise... “, S. 197.

Nach der Meinung des Dr. F. W. Hinkels, waren O. F. von Richter und S. F. Lidman die siebenten und achten Europäer, denen es gelang, diesen nur von J. L. Burckhardt 1813 überschrittenen südlichsten Punkt zu erreichen[32]. In Ibrim zeichnete Richter die dortige mächtige Festung auf den Felsen.

Am 11. Juni 1815 begann mit dem Besuch des Tempels von „Scheich Saad", dem modernen Amada, auf dem Westufer, der Rückweg. Im Skizzen- und Tagebuch hielt Richter den Grundriss des Tempels in einer Skizze fest, nummerierte in ihr die einzelnen Wandteile und zeichnete eine Ansicht der Ruinen, die den Rest des auf ihr errichteten Adobengewölbes aus der auf die christliche Zeit zurückgehenden Wiederverwendung des Pharaonischen Tempels als Kirche zeigt (*Abb. 2*). Zusätzlich wurden die Winkel zu den wichtigsten Geländepunkten gemessen und im Tagebuch festgehalten[33]. Im Verlauf der Rückreise machte O. F. von Richter am 11. Juni noch drei Zeichnungen: eine von Kurusko (mod. Korosko), von Ghiruw aus gesehen, eine von der Felsinsel zwischen Ward el-Abadia und Schemt Amar, und eine vom Gebirge Sabadorat (Jebel Sabadorat) bei Semjari. Am letztgenannten Orte machten die Wanderer eine Pause, und die Bewohner erzählten, ein Krokodil habe den Tag zuvor zwei Menschen gefressen, die zum Wasserschöpfen an den Nil gegangen waren. Dennoch schliefen die Reisenden am sandigen Ufer von Angora[34].

Am Montag, den 12. Juni, um 10.30 Uhr morgens landeten die Wissenschaftler in Garb Sabua (modern Wadi es-Sebua) und besuchten zweimal den örtlichen Tempel, dessen gründliche Beschreibung mit Zeichnungen sowie Anlageplan Richter auch anfertigte[35].

Hier sind einige Textstellen:

> „Die Gestalt der Colosse ist aus der Zeichnung zu ersehen. Sie lehnen sich an eine hohe Rückenlehne, auf welcher zwei verticale Streifen von Hieroglyphen zu sehen sind. Diese sind einander völlig gleich, und durch[aus] auf beiden Colossen dieselben. Sie tragen hohe Mitren, wie die zu Luxor, das Gesicht ist bärtig und verstümmelt. Das Haar am Nacken in einen Zopf gebunden und hängt über die Schulter gegen die Brust herab - oder sind diese quergestreiften Lappen ein Kopfputz. Die Ohren sind frei, gross, und eckig. Unter den Lappen geht ein breites, gestreiftes Halsband durch, woran Bartlocken hängen. Die Arme hängen dicht am Körper geschlossen herab, die Beine sind vorschreitend, beide mit dem linken. Um die Hüften geht ein Stück Zeug, oder eine Schürze; es scheint ein längliches Stück Zeug, das man von hinten herum genommen hat, vor dem Bauche in Falten zusammen genommen, mit einem Gürtel befestigt und das Ende lang herabhängen lassen, dieses Ende ist wie an Scheren und Shawls geziert. In der Mitte läuft ein Streif von Hieroglyphen, an deren Seiten zwei kleine verticale Streifen, durch horizontale abgetheilt. Am Ende hängen sechs gute Schlangen, von vorne gesehen, mit breiter aufgerichteter Brust. Der Gürtel, aus mehreren horizontalen Streifen, mit verticalen Durchschnitten, hat als Schloss ein Siegel mit Hieroglyphen; von dem ein Isiskopf mit Katzenohren herabhängt. Zwischen den Beinen der Figuren sieht man noch die Falten des Tuchs.
>
> Die Sphinx hab ich aus mehreren Teilen zusammengesetzt, da keiner ganz vollständig erhalten war. Der Kopf ist glatt und ein Loch darin, wahrscheinlich um den Kopf zu befestigen. Das Haar ist eben wie beim Colossen. Vor der Stirne trägt er eine gerundete Schlange[36]. Der Bart ist quer geringelt. Das Halsband ist dasselbe. Um die Schultern läuft ein gefalteter Kragen, der vielleicht die Löwenmähne darstellen soll. Der Körper ist

---

[32] F. W. Hinkel, „Otto Friedrich von Richters Reise... " , S. 235.

[33] F. W. Hinkel, „Otto Friedrich von Richters Reise... ", S. 235-236.

[34] Das Estnische Historische Archiv in Tartu, die Bibliothek, KS-17, S. 257.

[35] Das Estnische Historische Archiv in Tartu, die Bibliothek, KS-17, S. 257-262, die Tageseintragung vom 12. Juni 1815.

[36] Die Uräusschlange.

grossartig und kühn in ganzen Massen entworfen, ohne weitere Details. Die grossen Colosse sind umgestürzt. Ihr Körper gleicht denen oder der Kopfputz ist verschieden, er scheint gelocktes Haar vorzustellen, von einer Binde umgeben, und wohl erhalten, so viel man vermuten kann, da er fast ganz verschüttet ist.

Die Wand enthält eine colossale männliche Figur, Intaglio und Relief, auf dem Kopfe eine Federmütze, Schlangen an der Stirne. Er hebt die rechte Hand auf, als wolle er schlagen, in der Linken hält er zwei Menschen bei den Haaren. Ihm gegenüber steht der Gott mit der Federmütze und dem Nilschlüssel in der Linken, er reicht ihm mit der Rechten eine Sichel dar. Darunter, an den Seiten, dazwischen und darüber Hieroglyphen. Unter dem ausgeschweiften Karnies, das mit länglichen Blättern geziert zu seyn scheint, läuft eine Friesverzierung von Cynecephalen[37] und Harpokraten[38] in anbetender Stellung.

Die Wand (11) enthält dieselbe Darstellung. Der Gott hat einen Sperberkopf, darüber drei Vasen auf den Kugeln ruhen. Die hauende Figur ist umgekehrt, sie hält das Opfer mit der Rechten“[39].

Am Abend, um 6 Uhr, kamen die Altertumsforscher in die Ortschaft Medigh an, die am östlichen Nilufer lag. O. F. von Richter beschreibt das Ankommen auf folgende Weise:

„Ich weiss nicht ob aus Scherz, oder aus welcher Ursache, der Janitschar [der die Reisegesellschaft begleitete. – S.S.] warf einem Araber ein Stück Erde an den Kopf und dieser ergriff die Flucht, ob er gleich mit einer Lanze bewaffnet war. Indessen kam er nach einiger Zeit wieder, und brachte die sauere Milch, weshalb der Streit entstanden war. Er hatte einen breiten vorstehenden Mund mit grossen weissen Zähnen, einen ungeheuer breiten starken Haarwuchs, der ihm auf beiden Seiten in langen geringelten Locken, wie eine Perücke, die doch nicht bis zu den Schultern herabging. Seine Farbe war schwarzbraun. Es waren noch mehrere von demselben Volke zu Fuss und auf Dromedaren da, und darunter ein sehr schöner schwarzer Mann. Sie waren vom Stamme Allaghi und wahrscheinlich vom Dorfe gleiches Namens. Ich dachte an die Goldmine Allaki und fragte nach dem Namen des Berges, sie wiesen mich aber nach ihrem Dorfe und kannten nicht einmal das Rothe Meer. Sie erzählten von ihren Crocodilljagden; zwei bis drei Mann mit Lanzen und Hacken sind nötig und kaum hinlänglich um ein Crocodill ans Land zu ziehen. Endlich wollte einer uns Alterthümer zeigen, es war aber nichts als Fundamente von Häusern aus rohen Steinen. Wir fuhren im Mondschein weiter und schliefen bey Abu Schager am westlichen Ufer. Unser Djellab war in Sabua zurückgeblieben, und hatte uns seinen Bruder gelassen“[40].

Um 5 Uhr abends, am 13. Juni, erreichten die Orientalisten den Tempel zu Ufeddin (das alte Hierasykaminos und heutige El-Maharraqa)[41]. Dieser unvollendete römische Isis- und Serapis-Tempel lag an der Südgrenze des römischen Weltreichs[42]. Auch hier skizzierte Richter den Tempelgrundriss

---

[37] Die Wesen, die einem Hund ähneln.

[38] O. F. von Richter benutzte oft die griechischen Namen für ägyptische Gottheiten.

[39] Das Estnische Historische Archiv in Tartu, die Bibliothek, KS-17, S. 261.

[40] Das Estnische Historische Archiv in Tartu, KS-17, S. 262.

[41] Das Estnische Historische Archiv in Tartu, KS-17, S.262-263.

[42] Zum Thema siehe: D. Arnold, *Lexikon der...*, S. 145; W. Helck, „Hierasykaminos“, in *Lexikon der Ägyptologie* (Hrg. W. Helck, W. Westendorf), Bd. II, Wiesbaden, 1977, Sp. 1186-1187; F. Ch. Gau, *Antiquités...*, Taf. 40-42; Porter, Moss, *Topographical...*, Bd. VII, S. 51ff.; H. Bonnet, *Reallexikon...*, S. 300; R. Lepsius, *Denkmäler aus Ägypten und Äthiopien*, 5. Textband, Leipzig, 1913, S. 78ff.

und die Tempelansicht. Dem Grundriss wurden die Erläuterungen und Maßangaben beigegeben[43]. Zusätzlich kopierte der energische Wissenschaftler vier unbekannte griechische Inschriften[44].

Seine Untersuchungen fasste er folgendermaßen zusammen:

„Der ganze Tag war ausserordentlich heiss, wir rückten nur langsam fort, und erreichten den Tempel von Uffeddin erst um 5 Uhr Abends. Er wird von dem gegenüber liegenden Berge auch Brihheh el Moharraka genannt. Man trifft zuerst eine formlose Mauer die einem unvollendeten Propylon zu gehören scheint. An einer Wand desselben hat man, scheint es, angefangen, Figuren und Hieroglyphen auszuarbeiten und es unvollendet gelassen, denn sie haben keinen Zusammenhang. Oben, in der Mitte, sieht man eine weibliche sitzende Figur mit langem über die Schulter herabfallenden Haar, das Gesicht en face dem Beschauer zugewandt, dahinter ein Baum der sie beschattet. Diese Gruppe hat sehr das Ansehen eines römischen Bas Reliefs. Ein Kind reicht ihr mit beiden Händen eine Vase dar. Über beide Hieroglyphen. Links weiter unten eine Isis[45] mit Halbmond und Kugel auf dem Kopf und Scepter in der Hand auf Hieroglyphen zuschreitend. Gegenüber, dem Tempel zu, 9 Reihen verticaler Hieroglyphen, darüber eine sitzende Isis mit dem Nilschlüssel. Nach unten zu die Wand ganz unvollendet und in der Mitte sieht man mehrere unvollendete Figuren. Ich überzeugte mich bald, dass dieser Tempel aus einer Zeit herrührt, wo Römer den aegyptischen Styl kopierten, z. B. unter Hadrian. Das erhellt sowohl aus der Bauart, als aus den Inschriften[46]. Von den formlosen Trümmerhaufen kommt man an eine lange zerstörte Mauer die zur Wand des Tempels führt, diese ist ganz platt umgefallen, so dass fast alle Steine ihren Zusammenhang behalten haben. Der Haupteingang war bey. Diesen hat man in späterer Zeit, als das Gebäude wahrscheinlich eine christliche Kirche war, vermauert, und eine Nische daraus gemacht. Jetzt sind sie offen, aber auswendig sehr hoch von der Erde, denn der ganze Tempel ist nicht nur verschüttet sondern fast das ganze Fundament ist entblösst. Das Ganze bildete ein längliches Viereck mit einer bedeckten Halle umher, in der Mitte offen. Die Säulenreihe war eingemauert und hatte eine Thüre bey. Bey war eine runde Wendeltreppe um einen runden Pfeiler, die auf das Dach führte. Kornische und Säulen aegyptisch, letzte mit unvollendeten Capitälen. Bey 4 über den zwei Säulen war mit rother Farbe die lange Inschrift 2. an den Fuss geschrieben. Das Zeichen am Ende der letzten langen Zeile stand hinter jeder in einiger Entfernung senkrecht unter einander. Über den Friese, am Kranze, auf einer viereckigen Tafel, die glaube ich, mit der Zeit eine Flügelkugel werden sollte, waren die Worte 1. Die Inschrift 3. fand sich nebst einem Löwenkopf en relief, den ein Arm hielt und andere Fragmente sind an der umgestürzten Mauer. War dieses etwa das

---

[43] Das Estnische Historische Archiv in Tartu, die Bibliothek, KS-17, S. 263-267; Das Estnische Historische Archiv in Tartu, Bestand F. 1388, Nr. 1, S. 29-30. Vgl. B. Peterson, „Über Altertümer in Nubien... “, S. 53-54.

[44] Das Estnische Historische Archiv in Tartu, die Bibliothek, KS-17, S. 264-265; Das Estnische Historische Archiv in Tartu, der Bestand F. 1388, Nr. 1, S. 30 B. Vgl. B. Peterson, „Über Altertümer in Nubien... “, S. 54. F. W. Hinkel verweist auf die Meinung von U. Monneret de Villard, der „den Umstand (beklagt), dass von den frühen Reisenden (griechische) Inschriften vom Tempel, die den alten Namen des Ortes mit ‚Maurage‘ angeben sollen, erwähnt aber nicht kopiert wurden und heutzutage verloren seien“ (F. W. Hinkel, „Otto Friedrich von Richters Reise.. “, S. 236, Anm. 20).

[45] Die Verbundenheit zwischen der Göttin Isis und dem Tempel zu Ufeddin ist von O. F. von Richter ganz richtig bemerkt worden.

[46] Eine im Prinzip richtige kunsthistorische Einschätzung (zusätzlich siehe Anmerkung 41). Im Jahre 130 n. Chr. besuchte der römische Kaiser Hadrian (117-138) Ägypten; im Verlauf seiner Regierung entstanden dort neue Tempel, so das nach ihm benannte Tor auf der Insel Philae.

Heiligtum? Die Inschrift 4. war nebst einigen anderen coptischen und griechischen an einer Säule mit rother Farbe“[47].

In Qurta verbrachten die Reisenden die Nacht auf dem Lande, und am 14. Juni, nach 9 Uhr, gelangten sie nach Dakke mit seinem Tempel[48]. Damals, am Anfang des 19. Jahrhunderts, war er bewohnt: Männer, Weiber, Kinder, Kälber, usw. Wie gewohnt, machte Richter auch hier eine Grundrissskizze des Heiligtums, zeichnete und schilderte diesen Tempel[49].

Noch am gleichen Abend erreichten die Altertumsforscher den Felsentempel zu Garb Girscheh oder Garb Hussein (heute Gerf Hussein): „Sein grober Porticus liess nicht viel erwarten, um so mehr überraschte die Grösse der sechs Colosse im Inneren dieses Felsentempels. Wir kehrten im Mondschein bei Fackellicht zur Barke zurück und schliefen die Nacht da“[50].

O. F. von Richter skizzierte den Tempeleingang[51] und war der erste, der den Grundriss des unter Ramses II. erbauten Tempels zeichnete und für seine Beschreibung benutzte[52].

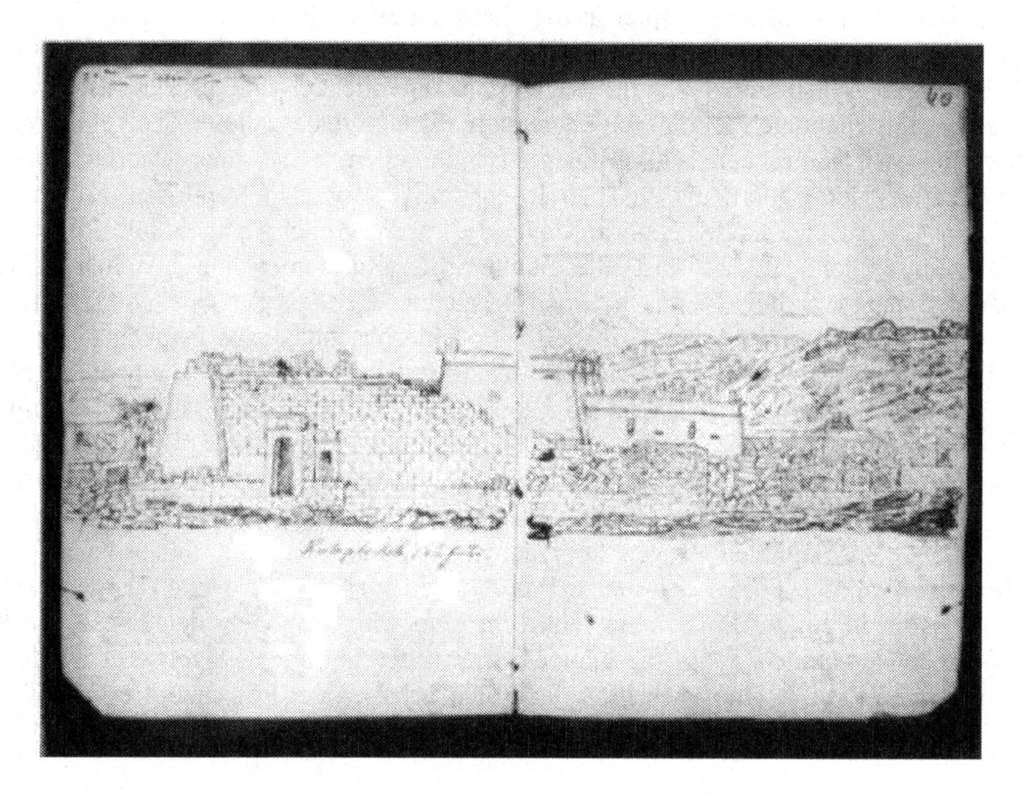

Abb. 3. Die Ruinen des Tempels von Kalaptscheh, 16. Juni, 1815.

[47] Das Estnische Historische Archiv in Tartu, die Bibliothek, KS-17, S. 262-264.

[48] Zum Thema siehe: D. Arnold, *Lexikon der ägyptischen...*, S. 58-59; F. Ch. Gau, *Antiquités...*, Taf. 33-36; G. Roeder, *Der Tempel von Dakke*, Kairo, 1930; E. Bresciani, „Dakke“, in *Lexikon der Ägyptologie* (Hrg. W. Helck, E. Otto), Bd. I, Wiesbaden, 1973, Sp. 988; I. Hein, *Die ramessidische...*, S. 11-12; H. Bonnet, *Reallexikon...*, S. 145ff.

[49] Das Estnische Historische Archiv in Tartu, die Bibliothek, KS-17, S. 267-281; vgl. B. Peterson, „Über Altertümer in Nubien... “, S. 54-57.

[50] Das Estnische Historische Archiv in Tartu, die Bibliothek, KS-17, S. 286; vgl. F. W. Hinkel, „Otto Friedrich von Richters Reise...“, S. 236.

[51] Das Estnische Historische Archiv in Tartu, der Bestand F. 1388, Nr. 1, S. 34 B, vgl. F. W. Hinkel, „Otto Friedrich von Richters Reise...“, S. 236, 238.

[52] F. W. Hinkel, „Otto Friedrich von Richters Reise... “, S. 234-235. Siehe auch B. Peterson, „Über die Altertümer in Nubien...“, S. 56-57.

Am 15. Juni, um 10 Uhr vormittags, kamen die Wissenschaftler mit Gefolge zum Tempel von Garb Dendur (modernes Dendur), „...der uns bis vier Uhr beschäftigte“[53]. Auch hier zeichnete Richter die Ansicht und den Grundriss des Tempels sowie beschrieb diese Stätte[54].

In der Nacht waren die Orientalisten in Abahar, um am Morgen, am 16. Juni, die von F. Norden erwähnten Altertümer zu suchen. Um 2 Uhr nachmittags fuhren O. F. von Richter und S. F. Lidman mit einem Boot bis zu den Ruinen des Tempels von Kalaptscheh (heute Kalabsche) und arbeiteten dort den ganzen Tag (*Abb. 3*). Natürlich machte Richter eine detaillierte Tempelbeschreibung und den Tempelplan.

Am 17. Juni vollendeten die Forscher die Untersuchung der Tempelruinen und danach besichtigten sie den bei Kalabsche belegenen Felsentempel Beit el-Wali.

Am 18. Juni, bei Sonnenaufgang, stiegen die Reisenden zu den Ruinen von Hindaw hinauf. Richter beschrieb ganz kurz zwei dortige Tempel (in Taffeh) und bestimmte richtig ihre Bauzeit. Und natürlich zeichnete er die Ansicht der Heiligtümer.

Einige Stunden später, um 10 Uhr, gelangten die Wissenschaftler nach Kardas (modern Kertassi), wo Richter die Tempelsäulen und Mauerreste skizzierte (*Abb. 4*). Zusätzlich schrieb er eine kurze Notiz zum Tempel.

Abb. 4. Kardas, 18. Juni, 1815.

[53] Das Estnische Historische Archiv in Tartu, die Bibliothek, KS-17, S. 288.

[54] Das Estnische Historische Archiv in Tartu, die Bibliothek, KS-17, S. 288-293; Das Estnische Historische Archiv in Tartu, der Bestand F. 1388, Nr. 1, S. 34-36. Vgl.: B. Peterson, „Über Altertümer in Nubien...“, S. 57; S. Stadnikov, „Die Wanderungen...“, S. 82; J. L. Burckhardt's *Reisen...*, S.147-149. Zum Thema siehe D. Arnold, *Lexikon der ägyptischen ...*, S. 66.

Am 19. Juni, nach einer kurzen Bootsfahrt, erreichte man den Tempel zu Debad (modern Debod). Hier bereitete O. F. von Richter eine umfangreiche Beschreibung des ganzen Tempels mit Grundriss vor. Am nächsten Tag landeten die Altertumsforscher auf der Insel Philae. Die Nubien-Expedition hatte insgesamt fast drei Wochen gedauert.

O. F. von Richter hatte die Absicht, die nubischen Tempel mit den bedeutendsten Werken der persischen und indischen Baukunst aus kunsthistorischer Sicht zu vergleichen. S. F. Lidman dagegen stellte sich die Aufgabe, eine Geschichtsbeschreibung Nubiens zu erarbeiten. Sein Tagebuch, zusammen mit Richters Zeichnungen, sollte dabei als Grundlage dieser Arbeit dienen. Leider ist nur ein Exzerpt aus diesem Manuskript bis heute publiziert. Würde es vollständig veröffentlicht worden sein, wäre es zu einem bahnbrechenden Werk in der Erforschung der nubischen Baudenkmäler geworden. Aber auch in der Gegenwart wäre eine Publikation des gesamten Materials von großer Bedeutung für die Ägyptologie und Nubienkunde, da seit 1815 viele nubische Altertümer entweder stark beschädigt oder völlig zerstört worden sind und diese wie auch Zeichnungen anderer Reisender einen nicht mehr vorhandenen Bauzustand bezeugen können.

Lidman und Richter hatten gehofft, die Ergebnisse ihrer Forschungsreise in den ägyptischen Teil Nubiens später in Rom einem gebildeten Publikum vorstellen zu können. Diese Hoffnung, wie auch Richters Wunsch, seine Orientalistikstudien in Paris fortsetzen zu können, gingen leider nicht in Erfüllung. Aber Richters Wanderungen in Levante und Griechenland setzten sich doch fort bis zu seinem frühen Tod 1816 in Izmir (Smyrna).

Sergei Stadnikow, M. A. (Ägyptologie)
sergei.stadnikov@mail.ee
Estonia

La construction « *jw sḏm=f* prospectif » dans Kagemni

Henri Doranlo

Travaillant sur la « Nature et fonction des auxiliaires énonciatifs en moyen égyptien, analyse linguistique de l'auxiliaire dans son rapport à l'énonciation », dans le cadre d'un Diplôme EPHE, j'ai eu la surprise de rencontrer dans « L'Enseignement pour Kagemni » le cas exceptionnel d'une construction « *jw sḏm=f* prospectif »: *jw gr=k m r(ꜣ)=k* (R° 2, ligne 1).

Texte[1]:

Translittération:

jmj pr rn=k

jw gr=k m r(ꜣ)=k

njs.t(w)=k

m ꜥꜣ jb=k ḥr ḫpš

m ḥr(y)-jb wḏꜣ m ẖrd.w=k

sꜣw jtn=k

n rḫ.n=tw ḫpr.t

jrr.t nṯr ḫft ḫsf=f

« The Instruction Addressed to Kagemni » traduit par Miriam Lichtheim[2]:

> « *Let your name go forth*
> (II, 1) *While your mouth is silent.*
> *When you are summoned, don't boast of strength*
> *Among those your age, lest you be opposed.*
> *One knows not what may happen,*
> *What god does when he punishes.* »

« Enseignement pour Kagemni » traduit par Pascal Vernus[3]:

> « *Fais que ton nom soit évoqué.*
> (R° 2) *Mais tu resteras discret dans ton propos quand [de sorte qu'] on te convoque;*

---

[1] Alan H. Gardiner, « The Instruction addressed to Kagmeni and his Brethren », *JEA* 32, 1946, 71-74; Kurt Sethe, *Lesestücke*, 42-43 (version abrégée, disponible sur le site de S. Rosmorduc: http://webperso.iut.univ-paris8.fr/~rosmord/hieroglyphes/hieroglyphes.html)

[2] Miriam Lichtheim, *Ancient Egyptian Literature*, vol. 1, University of California Press, 1975, 60.

[3] Pascal Vernus, *Sagesses de l'Egypte pharaonique*, Imprimerie Nationale, 2001, 57-58.

Ne sois pas outrecuidant pour des raisons de supériorité physique
Au milieu de ceux de ta génération.
Garde-toi de faire de l'opposition; on ne peut savoir ce qui peut arriver,
Ce que fait le dieu quand il punit. »

Concernant la forme *jw gr=k m r(ꜣ)=k*, P. Vernus précise[4]: « Probablement un emploi exceptionnel de *jw* « cotextuel » devant le prospectif, à distinguer de ses emplois originels référant à la situation d'énonciation; voir P. Vernus, *Les parties du discours en moyen égyptien, autopsie d'une théorie* (Cahiers de la Société d'Égyptologie, 5 Genève, 1997, p. 28). La construction du passage est discutée aussi dans D. Römheld, *Wege der Weisheit*, 1989, p. 167, n. 83. »

P. Vernus oppose ici un *jw* « contextuel » et un *jw* « cotextuel »: le point de référence par rapport auquel est validée la phrase est « contextuel » quand défini par la situation de l'énonciateur, mais « cotextuel » quand défini par ce qui précède dans l'énoncé.

En dépit de certaines divergences terminologiques, les linguistes s'accordent à définir ainsi l'énonciation[5]:

- la référence est absolue quand elle est datée (ex. « J.-Ph. Lauer s'est éteint le 15 mai 2001 à Paris à l'âge de 99 ans »);
- la référence est relative:
  - par rapport à la situation quand le repère coïncide avec le moment d'énonciation (R = ME);
  - par rapport au cotexte quand le repère ne coïncide pas avec le moment d'énonciation (R ≠ ME).

Remarque: Ch. Bally[6] nomme « absolue » la référence déictique que la plupart des auteurs qualifient de « relative »; et il est d'usage, dans les grammaires, d'opposer « temps absolu » (R = ME) et « temps relatif » (R ≠ ME) dans le système d'aspect/temps/mode: « *jw* has the effect of relating the statement to the sphere of interest and to the time of the speaker thus to relate the action to what is called absolute time. » (G. Englund)[7].

Par ailleurs, la notion de « co(n)texte » est employée au sens de « cotexte » *vs* « contexte », dans une visée élargie de l'énoncé (cotextuel) qui précède immédiatement une phrase incidente lorsque le contexte général participe au repérage temporel.

Illustration des modèles énonciatifs (Sinuhe R 5-12)[8]:

- a) exemple de référence absolue:

rnp.t-sp 30 ꜣbd 3 ꜣḫ.t sw 7 ꜥr nṯr r ꜣḫ.t=f

[4] Pascal Vernus, *Sagesses...* , 61, n. 23.
[5] Catherine Kerbrat-Orecchioni, *L'énonciation*, Armand Colin, Paris, 1999, 45.
[6] Charles Bally, « Les notions grammaticales d'absolu et de relatif », in *Essais sur le langage*, éd. de Minuit, Paris, 1969, 191.
[7] Gertie Englund, *Middle Egyptian, an Introduction*, Uppsala, 1995, 36, d'après la définition, souvent citée, de Hans J. Polotsky, *Egyptian Tenses*, The Israel Academy of Sciences and Humanities, Proceedings, vol. 2, no. 5, Jerusalem, 1965, §35. Discussion: James P. Allen, « Tense in Classical Egyptian », in *Essays on Egyptian Grammar,* YES 1, New Haven, 1986, 4 et n. 18.
[8] Aylward M. Blackman, *Middle-Egyptian Stories, The Story of Sinuhe*, Bibliotheca Aegyptiaca II, Bruxelles, 1972.

nsw-bity sḥtp-jb-rꜥ sḥr=f r p.t (R 5-7)

Année de règne 30, troisième mois de l'inondation, jour 7: ascension du dieu vers son horizon. Le roi du sud et du nord, Séhetepibrê, il s'éleva au ciel (...)

- b) exemple de référence relative par rapport à la situation (R = ME : *jw* = T_0 (simultanéité) par rapport à la datation absolue)[9]:

jw ẖnw m sgr jb.w m gmw (R 8-9)
La Résidence était dans le silence, les esprits dans l'affliction (...)

- c) exemple de référence relative par rapport au cotexte (R ≠ ME : *jst* = T_{-1} (antériorité) en arrière-plan par rapport au co(n)texte)[10]:

jst rf sbj.n ḥm=f mšꜥ r tꜣ tjmḥj.w (R 11-12)
Or, sa majesté avait expédié une armée en terre de Libye (...)

Commentaire: H. J. Polotsky[11] désigne ce *jw* appartenant à l'énonciation (référence coïncidant avec le moment de l'énonciation), comme un « nynégocentrisme » (position de l'énonciateur (ego) dans l'espace et le temps formant le triptyque « moi-ici-maintenant »). Le terme est emprunté à J. Damourette et E. Pichon[12] qui ont opposé référence « nynégocentrique » *vs* « allocentrique », c'est à dire référence « déictique » *vs* « cotextuelle ». Cette opposition rend compte de la distinction entre le *jw* « standard » de l'ancien égyptien et le *jw* « évolutif » apparu en moyen égyptien (et qui aboutira au *jw* circonstanciel suivi du pronom suffixe en néo-égyptien)[13]:

- *jw* est un repère déictique/contextuel quand il coïncide avec le moment d'énonciation (R = ME).
- *jw* est un repère non-déictique/cotextuel quand il est distinct du moment d'énonciation (R ≠ ME).

Remarque: au sens strict, le terme « déictique » ne s'applique qu'aux deux premières personnes JE-TU (TU devient JE quand il parle), la troisième personne IL étant délocutée, c'est à dire ne participant pas directement au discours. Mais c'est toujours JE qui dit: « IL ».

Exemple de référence relative à la situation d'énonciation (*jw* déictique):

[9] Antonio Loprieno, *Ancient Egyptian, A linguistic introduction*, Cambridge University Press, 1995, 165, donne cet exemple (90), pour illustrer la notion grammaticalisée de "*parataxis, i.e. the linkage between main clauses*".

[10] Antonio Loprieno, *Ancient Egyptian...* , 165, donne cet exemple (85), pour illustrer la notion grammaticalisée de "*hypotaxis, i.e. a semantic, rather than syntactic dependency of a sentence on the discourse nucleus*".

[11] Hans J. Polotsky, « Les transpositions du verbe en égyptien classique », in *Israel Oriental Studies* 6, 1976, 36.

[12] J. Damourette et É. Pichon, *Des mots à la pensée. Essai de grammaire de la langue française*, Tomes 2, Paris, 1930.

[13] Sur l'évolution historique du morphème *jw*, cf. James P. Allen, *Tense...* , 13-15; Antonio Loprieno, *Ancient Egyptian...* , 174.

(*le directeur des artisans, le dessinateur, Irtysen, qui dit:*)

jw(=j) rḫ.kw(j) sšt3 n mdw-nṯr (Louvre, C. 14, 6-7)[14]
« *Je connais les secrets des hiéroglyphes* » (litt.: *de la parole divine*)

L'omission du pronom suffixe (*=j*) est cohérente avec le caractère « anaphorique » du pronom qui renvoie au locuteur propriétaire de la stèle, Irtysen, précédemment nommé: *jr.ty=sn ḏd jw(=j) rḫ.kw(j)...*

> « C'est évidemment à partir des cas où il était suivi d'un pronom suffixe que *jw* a évolué du contextuel au cotextuel, puisqu'un pronom, parce qu'il est anaphorique, tend à faire prévaloir le lien avec ce qui précède. » (P. Vernus).[15]

Approfondissant l'analyse de Polotsky, J. P. Allen[16] clarifie le rôle de *jw* dans la relation énonciation-énoncé (Speech-Utterance):

> « ...the value of *jw* was originally to signal that the verb-form or construction following has specific relevance beyond that of the Utterance itself. In doing so, *jw* makes an additional statement about the sentence with which it is used. »

Exemple des deux emplois de *jw*:

jw wp.n=f r(ꜣ)=f r=j

jw=j ḥr ẖ.t=j m-bꜣḥ=f[17]

Il a ouvert sa bouche vers moi,
alors que j'étais sur mon ventre devant lui.

a) R = ME:
jw [wp.n=f r(ꜣ)=f r=j]
Il a ouvert sa bouche vers moi: l'antériorité dans le récit (= T_{-1}) est repérée par *jw* (moment de l'énonciation = T_0).

b) R ≠ ME:
jw[=j ḥr ẖ.t=j m-bꜣḥ=f]
alors que j'étais sur mon ventre devant lui: est relatif au cotexte, le repère *jw* ne coïncidant pas avec le moment d'énonciation.

---

[14] Winfried Barta, *Das Selbstzeugnis eines altägyptischen Künstlers, Louvre C 14*, Berlin, 1970 (disponible sur le site de S. Rosmorduc indiqué en note 1).
[15] Pascal Vernus, *Cahiers de la Société d'Égyptologie* 5, 28, n. 91.
[16] James P. Allen, *Tense...* , 15.
[17] Shipwrecked Sailor 81-82: Aylward M. Blackman, *Middle-Egyptian Stories, The Shipwrecked Sailor*, Bibliotheca Aegyptiaca II, Bruxelles, 1972 (disponible sur le site de AEL [Ancient Egyptian (Language) Listserver]: http://www.rostau.org.uk/AEgyptian-L/).

> « Here *jw wp.n=f* has (absolute) past tense. The situation *jw=j ḥr ẖ.t=j*, however, is hardly concomitant with the (narrator's) moment of speaking; this is true regardless whether the clause is analysed as independent or circumstantial syntactically. » (J. P. Allen) [18]

Si donc le moment de l'énonciation définit le présent linguistique, toutes les indications temporelles ne sont pas repérées par rapport au moment de l'énonciation (contexte), mais peuvent être aussi repérées dans l'énoncé (cotexte). C'est le moment où l'énonciateur parle qui ordonne la chronologie de l'énoncé, d'où, « *à ce moment-là* », un nouveau repère temporel peut lui être antérieur ou postérieur.[19]

Il est reconnu que *jw* a la particularité de présenter ce qui suit comme un fait objectivement validé: « The speaker assigns the truth value « true » to the proposition » (Th. Ritter)[20]. M. Malaise et J. Winand[21] conviennent que cet auxiliaire qui sert à former le véritable indicatif « ne peut s'accommoder du prospectif ou du subjonctif, puisque le premier est essentiellement une forme dépendante, et le second une forme modale, bien différente donc de l'indicatif ».

D'autre part, l'énonciation étant un acte personnel de production de l'énoncé par un énonciateur, un énoncé porte subjectivement la marque linguistique de son émetteur: « *jw* profiles a subjectively construed grounding relation of *relevance* between ground and scene. Relevance is, of course, the most basic and abstract form of subjective tracking and, in this sense, 'presents' the scene to the ground (...), but clearly has much wider application than a purely temporal or truth-based interpretation ».[22] M. Collier montre ainsi combien le champ d'application du morphème *jw* portant sur la phrase (Utterance), décrite par Allen, « is not just the ground participants and their temporal setting, but also the physical context of the memorial and in particular the theme of self-presentation under 'discussion'. »

La problématique de *jw* étant posée, comment analyser la construction *jw sḏm=f* prospectif dans Kagemni?

Après quelques propos généraux en forme de maxime, la sagesse de Kagemni s'ouvre sur ce précepte fondamental « *Ne parle pas* » (au moment T_0), suivi par des appels à la modération comportementale en société. Au terme de quoi « *Fais que ton nom soit évoqué* » (au moment T_{+1}). Autrement dit, moins le prétendant à une bonne conduite se montrera loquace, plus les autres seront amenés à parler de lui en bien.

jmj pr rn=k « *Let your name go forth* » ou « *Fais que ton nom soit évoqué* » (à T_{+1}), tient lieu de cotexte immédiat à *jw gr=k m r(ꜣ)=k.*

Etant exclu qu'il s'agisse d'une référence relative par rapport à la situation d'énonciation (R = ME), la phrase se présente comme une référence relative par rapport au cotexte (R ≠ ME) donnant à *jw* le sens de « *à ce moment-là* », « *alors* » (par rapport à T_{+1}).

Le discours s'enchaîne ainsi: *jmj pr rn=k jw gr=k m r(ꜣ)=k njs.t(w)=k*

Litt.: « *Fais que sorte ton nom. A ce moment-là, tu feras silence dans ta bouche quand on te convoquera* (autre traduction possible: *de sorte qu'on te convoque*). »

Soit: « *Fais que ton nom soit évoqué. Alors, tu garderas le silence quand on te convoquera* ».

- Première visée chronologique: « *Ne parle pas* » (au moment T_0) aboutit à « *Fais que ton nom soit évoqué* » (au moment T_{+1}).

---

[18] James P. Allen, *Tense...* , 13.

[19] Cf. Dominique Maingueneau, *L'Enonciation en linguistique française*, Hachette, 1999, 33-42.

[20] Thomas Ritter, « On Particles in Middle Egyptian », *Lingua Aegyptia* 2, 1992, 133.

[21] Michel Malaise et Jean Winand, *Grammaire raisonnée de l'égyptien classique*, Aegyptiaca Leodiensia 6, Liège, 1999, §408, 246.

[22] Mark Collier, « Grounding, cognition and metaphor in the grammar of Middle Egyptian », *Lingua Aegyptia* 4, 1994, 81-87.

- Deuxième visée chronologique incidente: prenant appui (à T_{+1}) sur la réalisation de « *Fais que ton nom soit évoqué* », « *à ce moment-là* »[23], (simultanément au moment T_{+1}), « *tu garderas le silence* » (sera réalisé à T_{+2}) « *quand* » (à ce moment T_{+2}) « *on te convoqua* ».

Sur cette deuxième visée chronologique, incidente, l'auteur (ou le copiste) de l'Enseignement pour Kagemni a sans doute fait une utilisation inattendue de *jw* associé à un prospectif pour exprimer cette simultanéité conjointe dans le futur.

Compréhension du texte: le bénéfice d'une bonne réputation (acquise sur un passé vertueux) se traduit par une convocation en haut lieu (promesse d'un bel avenir) à la condition de garder le silence (au sens de « rester discret dans son propos » comme le propose P. Vernus).

On notera que le moyen égyptien dispose de trois auxiliaires séquentiels pour le prospectif[24]:

- *ḫr*, sans valeur temporelle, exprime une contrainte externe, une conséquence inévitable;
- *kꜣ*, sans valeur temporelle, est purement séquentiel, de sens neutre;
- *jḫ*, avec valeur de futur, exprime une contrainte interne, la volonté du locuteur.

Aucun de ces auxiliaires ne convenait au discours tenu. Au mieux, le scribe aurait pu écrire **jḫ gr=k m r(ꜣ)=k* (*ainsi tu garderas le silence*), mais le sens eût été différent.

Le recours au morphème *jw* était donc nécessaire pour faire de cette phrase un prospectif incident, comme quoi « l'énonciation (...) construit sa référence à travers ses opérations propres. »[25]

Addenda:

Après avoir pris connaissance de cet article, Pascal Vernus m'a indiqué un autre exemple exceptionnel dans « L'Enseignement de Ptahhotep », ligne 59, version scolaire L 2[26] (XVIIIe dynastie). Le cas a été remarqué par A. Gardiner: « The geminating *3ae inf. gmm.tw.s* is an isolated exception. »[27]

La version de Prisse[28] (fin XIIe dynastie), qui fait référence, donne:

(*une belle parole...*)

jw gm.t(w)=s m-ꜥ ḥm.wt ḥr bnw.t
on la trouve chez des servantes (affairées) aux meules.

Quant à la variante L 2:

(...)

jw gmm.tw=s (...)

[23] « Il s'agit d'un rapport psychologique, puisque l'événement n'est situé au présent que par les incidentes qui lui sont attribuées. » (David Cohen, *L'aspect verbal*, PUF, Paris, 1989, 93-94).
[24] Pascal Vernus, *Future at Issue. Tense, Mood and Aspect in Middle Egyptian...*, YES 4, 1990, 114.
[25] Dominique Maingueneau, *L'Enonciation...* , 44.
[26] D'après le papyrus BM 10509.
[27] Alan H. Gardiner, *Egyptian Grammar*, Griffith Institute, Oxford, 1988, 385, § 462 et n. 6 et 7.
[28] Gustave Jéquier, *Le papyrus Prisse et ses variantes*, Paris, 1911 (version abrégée, disponible sur le site de Matt Whealton: http://members.aol.com/mwhealton/pthgly.htm).

P. Vernus propose de traduire ce passage: « *de fait, c'est même chez des servantes (affairées) aux meules qu'on la trouve.* »[29], analysant *jw gmm.tw=s* comme un prospectif emphatique.

Le problème posé dans Kagemni est celui d'une construction « auxiliaire + prospectif ». Le cas de Ptahhotep ouvre un autre débat sur une construction « auxiliaire + forme emphatique » (que cette forme emphatique soit vue comme *mrr=f* imperfective ou *sḏm*(*w*)=*f* prospective). Mais ces deux exemples ont en commun de s'inscrire dans le processus de grammaticalisation qui aboutira au *jw* circonstanciel du néo-égyptien.

Henri Doranlo
doranlo@wanadoo.fr
Candidat au Diplôme de l'Ecole Pratique des Hautes Etudes, IVè Section.
Paris, Sorbonne.

[29] Pascal Vernus, *Sagesses...* , 75-76.

Ancient Egyptian Mathematics and Forerunners Some Hints from Work Sites

Beatrice Lumpkin

Abstract

Archaeological evidence from many parts of Africa, from Kenya, the Lakes Region, South Africa and Egypt indicates that mathematical thinking developed over a long period of time in Africa's prehistory. This paper examines African artefacts and construction lines on ancient Egyptian monuments. Clues from materials found at work sites suggest the use of relatively advanced mathematical concepts not known from the written record.

Early Mathematical Thinking in Africa: Evidence from Tools

Many think of mathematics as a relatively modern subject, with little worthy of mention before the time of Classical Greece. In recent years, however, there has been a re-evaluation of the intellectual ability of earlier *Homo sapiens sapiens*. This re-evaluation is based, in part, on new evidence of very early abstract thinking in Africa. The evidence includes sophisticated harpoon bits carved with a serrated edge full of barbs. These fishing tools have been excavated on the Semliki River near Lake Rutanzige (Edward) in the Democratic Republic of the Congo and have been dated to ca. 90,000 B.P.[1]

Even older examples of planning and geometrical thinking have been found in East Africa. Partially undercut outlines for tool blanks, dated to ca. 200,000 B.P., have been discovered on a cliff face in Kenya.[2] These early Africans visualized and retained abstract geometric images of tools they had developed. Then, they made plans to produce such tools. Although the above dates may undergo some changes as dating methods are refined, these examples indicate that mathematical thinking began before anatomically modern humans spread from Africa to other continents.[3]

Numerical records also go back far into prehistoric times. The oldest known, to date, was found in Border Cave in the Limpopo Mountains of southern Africa and has been dated to ca. 37,000 B.P. This record of 29 tally marks was inscribed on a baboon bone, now fossilized. Perhaps it was a record of days in the lunar cycle; similar records are still made by the San of southern Africa for calendrical purposes.[4] Alexander Marshack describes similar tally records from many parts of the world.[5] It appears likely that quantitative thinking was part of the culture that anatomically modern humans took with them when they left Africa.

A more complex numerical record was found by Jean de Heinzelin at Lake Rutanzige in Central Africa and recently redated to 20,000 B.P. to 25,000 B.P. Known as "the Ishango Bone" after people of the region, the fossil shows tallies on one edge: 3, 6, space; 4, 8, space; 10, space; 5, 5, space; 7.

---

[1] J. E. Yellen, A. S. Brooks, et al. 1995. "A Middle Stone Age Worked Bone Industry from Katanda, Upper Semliki Valley, Democratic Republic of the Congo" in *Science,* **268:** p. 553.

[2] J. Gutin. 1995. "Do Kenya Tools Root the Birth of Modern Thought in Africa?" in *Science,* **270:** pp. 1118-1119.

[3] C. S. Henshilwood, et al. 2002. "Emergence of Modern Human Behavior: Middle Stone Age Engravings from South Africa" in *Science*, **295**: 1278-1280. Two pieces of ochre, dated to ca. 77,000 B.P., were found recently in a South African cave. They are inscribed with a pattern of rhomboids, product of abstract geometrical thinking.

[4] J. Bogoshi, K. Naidoo, and J. Webb. 1987. "The oldest mathematical artefact" in *The Mathematics Gazette,* **71,** p. 294.

[5] A. Marshack. 1991. *The Roots of Civilization.* Mt. Kisco, NY: Mayer, Bell.

Turning the bone shows 11, 13, 17, 19 and on its other edge 11, 21, 19 and 9.[6] It may be that more was occurring here than a simple count.

Over 14,000 years passed between the time of the Ishango Bone and the Naqada I period in Egypt. Perhaps future archaeological work will allow us to trace mathematical development during these years. We do know that measurement systems developed before writing and embodied a whole corpus of mathematical knowledge. For example, W. M. Flinders Petrie reported a red limestone balance beam and sets of graduated weights from Southern Egypt, dated to Naqada I. The weights demonstrate that the concept of ratios was already well developed.[7] Early hieroglyphic numerals on a knife handle from before the time of Narmer include a notation for the numerical value for 3,000, indicating a substantial development of numerals before that early date.[8]

Ancient Egyptian Mathematics: Evidence from Construction Sites

Of necessity, our knowledge of the prehistory of mathematics depends on archaeological and fossil evidence. In contrast, for literate periods, the history of mathematics is primarily based on written records. However, even for the literate period in ancient Egypt, valuable evidence can be gained from archaeology, in particular from construction sites.[9] The following examples of non-literate mathematical thinking used in the construction of great monuments suggest a rich potential source for students of ancient Egyptian mathematics.

Long before any of the surviving Egyptian mathematical papyri were written, an architect at the Saqqara step-pyramid complex sketched a plan for a large structure on an ostracon. The plan showed a concave curve and five equally spaced vertical segments extending down from the curve. Batiscombe Gunn took these segments as giving the horizontal location of five points on the curve. He wrote:[10] "The very fact that the distance is not specified makes it most probable that it is to be understood as one cubit, an implied unit found elsewhere." Each of the vertical segments is labelled with a hieroglyphic numeral that Gunn interpreted as the height of a point on the curve, expressed in cubits, palms and fingers. Gunn used the given values, in equivalent fingers, as vertical coordinates for points with standard horizontal spacing of one cubit apart. The coordinates were (0, 98), (28, 95), (56, 84), (84, 68) and (112, 41). Gunn connected these points with a smooth curve and got a close match to the sketch on the ostracon. Moreover, the sketch matched the curvature of the dome of a nearby structure.[11] Somers Clarke and R. Engelbach, in their classic on ancient Egyptian construction, called this early use of coordinates "of great importance."[12]

It is interesting to note that the concept of locating points on a plane by using rectangular coordinates does not appear in any of the surviving mathematical texts. However, construction guidelines used at the pyramids and mastabas of the Old Kingdom may have provided the basis for

---

[6] C. Zaslavsky. 1999. *Africa Counts*. 3rd edition. Chicago: Lawrence Hill Books, pp. 17-20.

[7] W. M. F. Petrie. 1920. *Prehistoric Egypt*, London, p. 28; also W. M. F. Petrie. 1926. *Ancient Weights and Measures*. London, p. 18.

[8] B. Williams and T. J. Logan. 1987. "The Metropolitan Museum Knife Handle and Aspects of Pharaonic Imagery Before Narmer" in *Journal of Near Eastern Studies*, **46**, no. 4, pp. 245-248.

[9] Such sites can also provide indications of systems of measures. See for example the work of Elke Roik on the so-called *nbj* measuring system, deduced from Old Kingdom mastabas (Elke Roik, "Das Maßsystem von Natursteinmastabas der 5. und 6. Dyn., Saqqara" in *Göttinger Miszellen*, **185**, pp. 97-111, also for literature references). I thank Michael St. John for this reference.

[10] B. Gunn. "An Architect's Diagram of the Third Dynasty" in *Annales du Service des Antiquités de l'Égypte* (Cairo) **26,** 1926, p. 200, n. 1.

[11] Ibid., pp. 197-202. The limestone flake containing this diagram was listed in the Cairo Museum *Journal d'Entrée*, number 50036.

[12] S. Clarke and R. Engelbach. 1990, originally 1930. *Ancient Egyptian Construction and Architecture*. New York: Dover, pp. 52-53.

using pairs of horizontal and vertical coordinates to locate points on a wall. In an 1865 study, *The Ancient Egyptian Cubit and its Subdivision,* Richard Lepsius discussed systems of horizontal and vertical guidelines at a number of pyramids. He showed that the spacing between guidelines was often an integral number of cubits. In fact, his study of these guidelines helped him fix the length of the royal cubit at 0.525 m.[13]

Lepsius described horizontal guidelines that are labelled by the number of cubits of distance from a "foundation line." The foundation or reference line was sometimes at the top of a room and the count of cubits started at the top and went downward. In other cases, the reference line was at the pavement level, and the count of cubits went from the bottom up. Lepsius also gave an example of a pair of vertical lines whose distance apart was stated. At Khufu's pyramid, in what Lepsius called "the second room from the top (Lady Arbuthnot's chamber)," two vertical lines intersect a horizontal line. The vertical line to the left is marked "3," and it is 3 cubits from the vertical line to the right.[14] Dieter Arnold also mentions vertical construction lines and describes examples at the unfinished pyramid at Zawyet el-Aryan. Arnold believes the vertical lines were used "not only for horizontal measuring, but also for marking directions."[15] Since there were both horizontal and vertical reference lines, it is possible that rectangular coordinates were used to locate points of interest during construction.

George Reisner, in a 1931 report, also discussed levelling lines at the Mycerinus (Menkaure) pyramid at Giza, built c. 2600 B.C.E. What Lepsius called "foundation lines," Reisner described as "zero lines." These "zero lines" were labelled *nfrw.* Reisner listed the following ancient technical terms, the first four collected by Ludwig Borchardt at the Abusir temples and the fifth by Petrie at Meidum:[16]

"zero"	*nfrw*
"zero line"	*m tp n nfrw* (literally, "the head of zero")
"above zero"	*ḥr nfrw*
"below zero"	*mḏ ḥr nfrw, ẖr n nfrw*

Even as late as the New Kingdom, *nfrw* signs were used with construction lines on foundations excavated at Hatshepsut's temple at Deir el-Bahri.[17] It may be worthwhile to look for and inventory *nfrw*-signs among stone masonry markings on other ancient Egyptian buildings and blocks.

It is interesting to observe that levelling lines were labelled as so many cubits above, or so many cubits below the reference line, labelled *nfrw.*[18] These numerical labels stated both distance (a quantity) and direction, above or below the *nfrw* line. Today, we would use the terms "directed or signed numbers" and "zero reference line." We can also observe a further level of abstraction beyond locating a reference line at pavement level to locating reference lines elsewhere on the wall.

In modern usage, the word zero is employed in different applications. In the history of mathematics, there has been much interest in the development of a zero symbol as a placeholder in a positional-value

[13] R. Lepsius. *The Ancient Egyptian Cubit and its Subdivision.* English translation, 2000, by J. Degreef from the 1865 original, *Die Alt-Ägyptische Elle und Ihre Eintheilung.* Edited by M. St. John. London: Museum Bookshop Publications, pp. 11-13.

[14] Ibid., p. 13.

[15] D. Arnold. 1991. *Building in Egypt.* New York: Oxford University Press, p. 18.

[16] G. A. Reisner. 1931. *Mycerinus: the Temples of the Third Pyramid at Giza.* Cambridge: Harvard University Press, pp. 76-77. R. Hannig. 1995. *Die Sprache der Pharaonen. Großes Handwörterbuch Ägyptisch-Deutsch.* Verlag Philipp Von Zabern, Mainz, p. 411, lists *m tp n nfrw* as "auf Level 0."

[17] Z. E. Szafranski. 1995. "On the Foundations of the Hatshepsut Temple at Deir el-Bahari" in *Gedenkschrift für Winfried Barta.* Münchener Ägyptologische Untersuchungen, 4, pp. 371-373.

[18] A system of levelling lines at Mastaba 17 at Meidum was recorded by W. M. F. Petrie in 1892. A copy of his sketch of the lines can be viewed at the following URL: http://www.petrie.ucl.ac.uk/digital_egypt/meydum/mastaba17/lines.html.

system of numerals. Since the ancient Egyptian (and the Greek Ionian) numerals did not have positional value, the zero placeholder is not relevant to this discussion.[19] We also use zero today on a number line for the reference point that divides positive and negative values, much as ancient Egyptian architects referred to the number of cubits above *nfrw* and the number of cubits below *nfrw*.[20] A third use is as a number, to express a quantity and to be used in calculations.

Of the above usages of zero, there is general agreement that the Egyptians did not have a zero placeholder. There is less agreement on how to interpret the *nfr* symbol (Gardiner code: F35) used to express zero remainders on a balance sheet in a Late Middle Kingdom papyrus, P. Boulaq 18.[21] Alan Gardiner and Raymond Faulkner took it to mean "zero,"[22] based on an old particle of negation, *nfr*.[23] James Allen however suggested that the P. Boulaq 18 sign was an abbreviation for the word *nfrw*, "depletion."[24] Whatever the correct interpretation may be, it is interesting that apparently related words were employed for two different uses for which we would use the concept of zero today: *nfrw* for a "level zero" reference line and *nfr* or *nfrw* for a zero remainder.[25]

---

[19] Historians credit the ancient Egyptians for the invention of cipherization of numerals, the use of an abstract symbol to represent a quantity. Hieratic numerals were fully cipherized but hieroglyphic numerals still used tallies for 1 to 9 and repeated powers of ten as needed. Cipherization is one of the two principles that advanced numerals beyond tally systems; the other principle was positional value.

[20] The word thus far addressed, *nfrw* as "level 0"-line on buildings, is written as F35-D21-G43. In the more distant Egyptological past it was only given the meaning "base" or "foundation/ground-level." Cf. DZA no. 25108390 and no. 25108400 (DZA = *Digitalisierte Zettelarchiv des Wörterbuches der ägyptischen Sprache*, on-line at URL http://aaew.bbaw.de/dza/index.html) and R. O. Faulkner. 1962 (1996 reprint). *A Concise Dictionary of Middle Egyptian.* Griffith Institute, Oxford. p. 132. Hannig, op. cit., p. 411, incorrectly lists our "zero line" and the five terms collected by Reisner (see main text of the present article) under a different *nfrw*. This other *nfrw* is written as 3xF35-Z1-O1 (or F35-Z2-O1), so with house determinative, and refers to the part of a building or tomb that lies at the end of the central corridor, as far as possible removed from the entrance. It means literally "end portion" or "end room," namely the "back part" or the "innermost room" of a tomb. Cf. Faulkner, op. cit., p. 132, Hannig, op. cit., p. 411. This second *nfrw* occurs, e.g., in Papyrus Abbott 3, 2 (see DZA no. 25108410; there and in BAR IV §517 still wrongly rendered "base") and in the Turin tomb plan of Ramesses IV (H. Carter. 1917. "A tomb prepared for queen Hatshepsut and other recent discoveries at Thebes" in *JEA* **4**, p. 110 n. 1; H. Carter and A. H. Gardiner. 1917. "The tomb of Ramesses IV and the Turin plan of a royal tomb" in *JEA* **4**, p. 143; J. Capart et al. 1936. "New light on the Ramesside tomb-robberies", in *JEA* **22** (1936), p. 78 (2,8); D. Meeks. 1998. *Année lexicographique Égypte ancienne* (1977). 2nd ed. Paris: Éditions Cybèle. p. 195). I thank Aayko Eyma for helping me to collect and interpret the linguistic material for this chapter and associated footnotes.

[21] P. Boulaq 18 is discussed in more detail in B. Lumpkin. 2002. "Mathematics Used in Egyptian Construction and Bookkeeping" in the *Mathematical Intelligencer*, **24**, number 2. I thank Frank Yurco for his kind help with some aspects of that article and the current one.

[22] A. H. Gardiner. 1978. *Egyptian Grammar*. Oxford: Griffith Institute, p. 574. R. O. Faulkner, op. cit., p. 132.

[23] This *nfr* particle means "be at an end," "finished," "be zero"(?), and occurs in constructions like *nfr n* (+ verbal form) "not," *r-nfr n* "so that not," and *nfr pw* "there is not." J. P. Allen. 2000. *Middle Egyptian. An Introduction to the Language and Culture of Hieroglyphs.* CUP. p. 190. Hannig, op. cit., p. 411. Faulkner, op. cit., p. 132.

[24] J. P. Allen, op. cit., p. 97. This third word *nfrw*, written F35-G43-Z2-G37, so with evil bird determinative, occurs in Rekhmire's "Duties of the Vizier," in which it seems to have the meaning "deficiency," "lack" (in BAR II § 706 wrongly rendered "Great Beauty"). Hannig, op. cit. p. 411. Faulkner, op. cit. p. 132. DZA no. 25108510, 25108520, and 25108530.

[25] The nouns *nfrw* ("depletion," "deficiency"), *nfryt* ("end," "bottom"), *nfrw* ("endmost part of building," "bottommost room"), and *nfrw* ("base" and "level 0") all derive from the mentioned negation particle *nfr* ("be at an end," "not at all"). Cf. Allen, op. cit., p. 190, and W. Vycichl. 1983. *Dictionnaire étymologique de la langue Copte*. Peeters, Leuven-Paris. p. 150. Vycichl thinks the common word *nfr*, "beautiful," "good," is unrelated to the above words (and several Afroasiatic etymologies for it have been proposed), but Dr. James Allen (pers. comm.) suggests that it may be related, in the sense of it meaning "perfect," as an ethical or esthetical "ultimate" or final state – as all the previously mentioned *nfr* words seem to convey a sense of "ultimate," either of negation or of position. A. Loprieno. 1995. *Ancient Egyptian. A Linguistic Introduction*. CUP. p. 129, opts for a meaning of "to be complete" for the root *nfr*, which in a positive sense would be "to be good" and in a negative sense "to be finished"; he further points at the analogy with the root *tm*, which has the meaning of "to be complete" but is also used as a negative infinitive ("to not be," "to finish"). For *tm*, see also Vycichl, op. cit., pp. 4, 214.

Summary

Archaeological finds from many parts of Africa – Kenya, the Lakes Region, South Africa and Egypt – indicate a long history of development of mathematical thinking going back far into prehistory. The evidence from tools reveals the use of abstract thinking from the earliest days of anatomically modern humans. Even with the development of writing, both our understanding and the historical record remain incomplete. Hints from the construction sites of ancient Egypt suggest that the builders were using mathematical concepts that do not appear in surviving papyri. Further study of construction technology can be expected to provide additional insights into the mathematical achievements of ancient Egypt.

Beatrice Lumpkin
bealumpkin@aol.com
Associate Professor of Mathematics, retired
Chicago City Colleges

On a Topos in Egyptian Medical History

Hedvig Győry

Introduction

According to classical tradition the ancient Egyptians had an anatomical book as early as the 1st Dynasty. The Greek historical work of the Heliopolitan high priest Manetho says, concerning king Djer (whose birth name was *jttj*), "Athothis ... built the palace at Memphis and his anatomical works are extant, for he was a physician" (according to the version of Africanus)[1] and, "... he practiced medicine and wrote anatomical books" (according to the version of Eusebius).[2] Based on written evidence from the archaic period,[3] it is not possible that an anatomical work as understood by the Greeks of the 3rd c. B.C. existed in the time of Djer, but the existence of a collection of utterances similar to the Pyramid Texts or Old Kingdom wisdom literature may be supposed. In fact, there may even have been several different traditions. The extant contemporary texts, however, do not contain, in strict sense, any documents, quotations or commentary in this respect. At the moment, the written evidence, archaeological material and conclusions worked out by Egyptologists do not allow us to state with certainty what facts lie behind this Manethonian topos, but probably do permit us to trace some of the reasons that led to it.

The topos has till now been explained by refering to a medical section of Papyrus Ebers, Eb 468, which mentions a medicine for the growth of hair being prepaired for Shesh (*šš*), mother of king Teti.[4] This Teti (*ttj*) would then not be the well-known 6th Dynasty king Teti (*ttj*), but a miswriting of Athothis (*jttj*; Abydos List no. 3). However, the argument that we should read Athothis rather than Teti not only seems to be rather circular, it is also unlikely as the mother of king Teti of the 6th Dynasty probably was called Seshseshet (*sšššt*), a name that could well be hiding behind Sheshe (*šš<t>*). For not only did several of the women (daughters or granddaughters) in Teti's family bear that name, of which Seshe (*sšit*) is an attested diminutive form, we also have the tomb of Mehu at Saqqara in which, besides domains of Teti, a domain of a "king's mother Seshseshet" is listed.[5] So it is very likely that Papyrus Ebers had the 6th Dynasty king in mind. Even if one were to presume that Manetho or his sources wrongly read Athothis into Eb 468 and based the topos on this, it should still be observed that this Ebers section does not speak of a medical book, only of one cosmetic product, linked to the king's mother, not the king. Therefore, it seems worthwhile to look to another section of Papyrus Ebers instead, a section that is not only with certainty associated with the 1st Dynasty but also with a medical treatise.

---

[1] W. G. Waddell, *Manetho, Aigyptiaka*, Loeb Classical Library 350, London 1964, Africanus frg. 6,2.

[2] W. G. Waddell, *Manetho, Aigyptiaka*, Loeb Classical Library 350, London 1964, Eusebius frg. 7,2.

[3] P. Kaplony, *Inschriften der ägyptischen Frühzeit 1-3*, ÄA 8, Wiesbaden 1963; id., *Supplement*, ÄA 9, Wiesbaden 1964; id., *Kleine Beiträge zu den Inschriften der ägyptischen Frühzeit*, ÄA 15, Wiesbaden 1966.

[4] Already in 1897, Bonifac Platz wrote, refering to Eb 468, that the 1st Dynasty ruler "Teta" wrote medical books – *O-Egyiptom irodalma*, Budapest 1897, p. 133 (he gives as his source: Lauth, *Aegyptens Vorzeit*, Berlin 1881, p. 112). The idea has been repeated till modern times, cf. W. Westendorf, *Handbuch der altägyptischen Medizin*, Brill, 1999, p. 27, and D. Redford, *Pharaonic King-Lists, Annals and Day-Books*, SSEA Publication IV, Mississauga 1986, p. 210, note 22 ("Eb 46B [*sic*] (ascribed to Atotis, wrongly interpreted as Teti)"). Westendorf (*Handbuch der altägyptischen Medizin*, vol. II, Leiden-Boston-Köln, Brill, 1999, p. 630, note 121) thinks *šš* is probably the diminutive form of a name of the form *nfr-sšm* + name of god or king (Ranke, *PN* I. 330.3).

[5] See J. Yoyotte, "A propos de la parenté féminine du roi Téti (VIe dynastie)", *BIFAO* 57 (1958), pp. 91-98.

Medicine in the Early Dynastic Period and the Old Kingdom

The idea that contemporary medical knowledge was the reflection of a time not much later than the unification of the Two Lands was certainly developed before the 18th Dynasty. The Papyrus Ebers[6] (18th Dynasty, Amenhotep I) in its 2nd Anatomical Treatise (Eb 856 = 102,14-103,18) and the later Berlin Medical Papyrus (no. 3038)[7] (19th Dynasty) in its anatomical section (hereafter: the Berlin Treatise; Bln 163 = 15,1-17,1) say that a medical treatise, of which they present different versions, was found during the reign of king Dewen (1st Dynasty).[8] The Berlin Treatise even adds a second discovery during the reign of king Senedj (2nd Dynasty)[9] and does not call itself a simple book but a *dmḏt*, a "collection", i.e. a choice of utterances possibly taken from several books.

> "Beginning of the Book (*mḏ3t*) of the Wandering of Pathogenic Elements (*wḫdw*)[10] in all limbs of a man, as was found in writings under the two feet of Anubis in Khem (*Ḫm*). It was brought to the majesty of the king of Upper and Lower Egypt, Khasti,[11] the justified." (Eb 856a (102,14-15))

> "Beginning of the Collection (*dmḏt*) of the Wandering of Pathogenic Elements (*wḫdw*), which was found in ancient writings in a case for scrolls under the two feet of Anubis in Khem (*Ḫm*), during the reign of the majesty of the king of Upper and Lower Egypt, Khasti, the justified, after he became miserable (*ẖsw=f*). It was [also] taken to the majesty of the king of Upper and Lower Egypt, Senedj (*Snḏ*), the justified, because of its excellence. Behold, this writing (i.e. ritual) is about undoing the sealing of legs[12] by the scribe of the divine words and the chief of the excellent physicians who satisfy the god. Behold, the writing (i.e. ritual) of the follower-of-the-Sun-disc (*šmsw n Ỉtn*)[13] is done. He gives offerings of bread, beer and incense on fire for the name of Isis, the

---

[6] W. Wreszinski, *Der Papyrus Ebers, I. Teil, Umschrift*. Leipzig 1913; H. Grapow, *Grundriss der Medizin der alten Ägypter IV-V.*, Berlin 1958; U. Luft, "Zur Einleitung der Liebesgedichte auf Papyrus Chester Beatty I ro XVI 9ff.", *ZÄS* 99 (1973), p. 110.

[7] W. Wreszinski, *Der Grosse Medizinische Papyrus des Berliner Museums (Pap. Berl. 3038) in Facsimile und Umschrift*, Leipzig 1909; H. Grapow, *Grundriss der Medizin der alten Ägypter IV-V.*, Berlin 1958; U. Luft, "Zur Einleitung der Liebesgedichte auf Papyrus Chester Beatty I ro XVI 9ff", *ZÄS* 99, 1973, pp. 110-111.

[8] T. A. H. Wilkinson, *Early Dynastic Egypt*, London-New York 1999, p. 76.

[9] T. A. H. Wilkinson, *Early Dynastic Egypt*, London-New York 1999, p. 88; W. Helck, "Sened", *LÄ* V, col. 849.

[10] The word *wḫdw* has been much debated, and the present author hopes to write about it elsewhere. For now it may suffice to say that it refers to some kind of morbific substance of a general nature, causing fever, weakness and different kinds of diseases in different parts of the body. At times it can also refer to a demon causing suffering and illness (e.g., Eb 242, 246, 385).

[11] Khasti (*Ḫ3stj*) is the Two Ladies name of king Dewen (Den) and this form will be consistently used in this paper; Gardiner codes: 2x N25 + X1 (Egyptologists previously read it as Semti [*Smtj*]). After the Old Kingdom the name was miswritten as *Ḥsptj* (also read as *Sp3tj*), i.e. 2x N24 (e.g. in the Abydos Kings-list, entry no. 5), and subsequently (with the regular mistake of Aa8 for N24) as *Ḳnḳn*, i.e. 2x Aa8, or *Ḳntj*, i.e. Aa8-X1-Z4 (e.g. in pEbers under discussion). Many Egyptologists assume that the reading *Ḥsptj*/*Sp3tj* led to Manetho's king Usaphais, and/or that the reading *Ḳnḳn* led to Manetho's king Kenkenes. Cf. J. von Beckerath, *Handbuch der Ägyptischen Königsnamen*, Mainz 1984, p. 38; S. Schott, *Zur Krönungstitulatur der Pyramidenzeit*, Nachrichten der Akad. Der Wiss. in Göttingen, Ph.-Hist. Kl., 1956, p. 76; G. Godron, *Études sur l'Horus Den et Quelques problemes de l'Égypte Archaique*, Cahiers d'Orientalisme 19, Genève 1990, p. 180; D. Redford, *Pharaonic King-Lists, Annals and Day-Books*, SSEA Publication IV, Mississauga 1986, p. 235.

[12] Cf. H. Győry, "The Seal is your protection", *Revue Roumaine d'Égyptologie*, 2-3 (1998-1999), pp. 35-52. pEbers ends with the statement that the *mtw* of the two legs "start to die"; in pBerlin the same statement is followed by a long treatment in several steps (without any prognosis).

[13] Compare U. Luft, "Das Verhältnis zur Tradition in der frühen Ramessidenzeit. Ein Vergleich zwischen dem Gefässbuch im Papyrus Ebers und dem Berliner medizinischen Handbuch", *Forschungen und Berichte* 14 (1972), pp. 65-66. Luft's translation of this whole Ebers section differs slightly from the present one.

great, of Horus, who-is-presiding-over-*Ḫtjj*[14], and of Khonsu-Thot, the god-who-is-in-the-stomach." (Bln 163a (15,1-5))

Apart from the tradition mentioned above, we can be fairly certain that during the archaic period and even earlier the master medicine-men transferred anatomical knowledge to their pupils by oral tradition, similar to the Pyramid Texts, most of the material probably having been carefully systematized. The earliest known medical title belongs to Hesira during the reign of Djoser (3rd Dynasty).[15] In addition to his high ranking official titles, two medical titles are listed on one of his wooden panels: "chief of dentists and physicians (*wr jbḥ swnw*)".[16] Later Old Kingdom titles[17] also attest that medicine was a specialized science: physicians were titled according to the various parts of the body, and were ranked in a developed hierarchical structure. Because a complicated order could not arise from nothing, it must already have had a rather long history even at this time. There must have been enough precedents to be worked with, and the 3rd Dynasty must also have been a time of serious systematization.

The other conclusion suggested by the specialization of titles is that the theoretical system of medicine must also have been articulated according to the parts of the body – a system often attested in later medical books. That is, the anatomical systems were built up of **separated** (groups of) body-parts. A parallel phenomenon can be documented for the field of (liturgical and funeral) theology: there are long lists with parts of the human body, as units attached to, or identified with, specific gods for the reason of protection.[18] As the religious background was an intimate part of every detail of life, it must have been the same in medicine. The harmony between anatomical titles and contemporary theological systems strengthens the view that this anatomically segregated approach must have been innate in Egyptian medicine.

There is only indirect testimony of medicine in the early periods. The special field of spells against snake-bites survived in the Pyramid Texts, and the spells are very similar to those of much later periods. The lack of discussion of other medical problems in the Pyramid Texts may reflect an anticipated non-existence of other diseases in the Netherworld or a triviality of their treatment there.

In all probability, the style of the medical texts followed that of the funeral texts, or used the treasury of the wisdom literature. For example, sentences might have been similar to those employed at the funeral offerings: "O Re, if you dawn in the sky, you dawn for the King, the lord of all things; if all things belong to yourself, then all things will belong to the double of the King, all things will belong to himself – the lifting up before him of a sanctified offering." (Sp. 50 = Pyr 37 b-e). Or: "O King, take the Eye of Horus, rescued for you; it will never escape from you – beer, an iron *ḥnt*-bowl. O King, take the Eye of Horus, provide yourself with it – beer, a *ḥnt*-bowl of *ḥtm*-material. Bring the two Eyes of Horus – an *jwnt*-bow." (Sp. 56-57 = Pyr 40-40+1).[19] The register aspect of the 2nd Anatomical Treatise

---

[14] A typical epithet of Horus at this time. Cf. B. Begelsbecher-Fischer, *Untersuchungen zur Götterwelt des Alten Reiches im Spiegel der Privatgräber der IV und V. Dynastie*, OBO 37, Freiburg-Göttingen 1981, pp. 231-232; P. Kaplony, *Die Inschriften der Agyptischen Frühzeit*, ÄA 8, 1963, pp. 109-120.

[15] Frans Jonckheere, *Les Médicins de l'Égypte Pharaonique. Essai de prosopographie*, Bruxelles 1958, no. 63. P. Ghalioungui, *The Physicians of Pharaonic Egypt*, Cairo 1983, no. 40.

[16] Gardiner's codes: G36-F18-T11; Cairo Museum, JdE 28504, panel no. 1426; cf. J. E. Quibell, *Excavations at Saqqara 1911-1912, The Tomb of Hesy*. Cairo 1913. For the title, see: H. Junker, *ZÄS* 63 (1928), 69; Jonckheere, 1958, fig. 21.

[17] E.g., Frans Jonckheere, *Les Médicins de l'Égypte Pharaonique. Essai de prosopographie*, Bruxelles 1958, nos. 16, 33, 37, 69; P. Ghalioungui, *The Physicians of Pharaonic Egypt*, Cairo 1983, nos. 15, 21, 153, pp. 43-45.

[18] E.g., Pyr 1303-1315; BD 42, 172; Ch Beatty VII, vs. 2,5-5, VIII. ro 7,1-9,6; the Magical Papyrus in the Vatican, cf. A. Erman, "Der Zauberpapyrus des Vatikan", *ZÄS* 31 (1893), p. 119; the Turin Mag. Pap., cf. A. Erman, *ZÄS* 31 (1893), p. 123; the socle of Behague, cf. Ét. Drioton, "Religion et magie", *Rev. Ég. Anc.* 1 (1912), p. 133; G. Lefebvre, *Tableau des parties du corps humain mentionnées par les Égyptiens*, Supplément ASAE 17, Le Caire 1952.

[19] All translations of Pyramid Texts used here are those of Faulkner. R. O. Faulkner, *The Ancient Egyptian Pyramid Texts*, Oxford 1969.

shows the similarity of approach: it lists the various parts of the body, and discusses them, the comments being instructions and explanations.

Letopolis and Horus Khenty-(en)-irty

The text of the treatise of the "Wandering of Pathogenic Elements (*wḫdw*)" connects the book to Khem (*Ḫm*), i.e. Letopolis (modern Ausim).[20] Judging by the Pyramid Texts, this town on the West bank of the Nile, opposite to Heliopolis, was an important cultic centre from the beginning.[21] In an anatomical list the chin was identified with the Letopolitan ram-god Kherti (*Ḫrtj*) (Pyr 1308a). But the Letopolitan god mentioned most often is Horus Khenty-en-irty (*(m)ḫntj-n-irtj*), "the one without the two eyes", a heavenly falcon god who was depicted without eyes[22], symbolizing the state of the sun and moon when they have set in the West. Of course in due time the two heavenly bodies would rise again, thus becoming the two seeing eyes of the god. The blind aspect of the god (*(m)ḫntj-n-irtj*) was symbolized by the shrew-mouse (a practically blind animal that lives under the ground), the seeing aspect (*(m)ḫntj-irtj*) by the ichneumon (a sharp-eyed animal that even tackles snakes, just as the sun-god fought the chaos serpent Apophis). Because the god regained his sight, he became the protector against eye-diseases and blindness. When connected with the Heliopolitan myths of the damaged (or lost) Eye of Re, Horus becomes the Chief Physician of Re, "the good doctor of his creator",[23] healing Re's solar eye(s).

In the Pyramid Texts it is stressed that Horus of Letopolis lives the life, has not died the death (Pyr 810-811), he makes the name and the pyramid of the King and its construction durable (Pyr 1670), and even raises the King in the Netherworld and guarantees him offerings (Pyr 1723). Though he was protected by "spittle" (Pyr 419), this very word gives protection against himself, when he is appearing "with evil coming" as *ḫnt-jrty* (Pyr 1270).[24] At other times he dispels the evil being before him or removes it from behind him (Pyr 908). This last statement was written in connection with the speech of the King when he arrived in the Netherworld: "I am hale and also my flesh, it goes well with me and with my name; I live and also my double" (Pyr 908). Horus can also take part in the expulsion of the *wfj*-viper (Pyr 419), and he cuts off "the heads of the mottled snakes" (Pyr 1211a). This quality could have been a natural consequence of his hawk shape and the snake-hunting instinct of the bird.

The four Children of Horus are the sons of Horus of Letopolis (Pyr 2078).[25] They later would be depicted on top of the canopic jars and on the sides of the canopic chests and would provide protection for the internal organs, but according to the Pyramid Texts, they are the friends of the King. They help him on the East side of the sky to reach Kheprer by the ladder, when he is living again in the Netherworld (Pyr 2078-79).

---

[20] F. Gomaá, "Letopolis", *LÄ* III, cols. 1009-1011.

[21] Even in the Ptolemaic era, Letopolis was a cult center second only to Memphis, as is shown by the close marriage ties between the families of the High Priests of Letopolis and of Memphis. Cf. J. Quaegebeur & A. Rammant-Peeters, "Un relief memphite de l'époque ptolemaïque de l'héritage de William Fox Talbot", *GM* 148 (1995), pp. 71-90.

[22] H. Bonnett, *Reallexikon der ägyptischen Religionsgeschichte*, Berlin 1952, p. 133

[23] H. Junker, *Die Onurislegende*, Wien 1917, p. 29, pp. 149-151.

[24] Pyr 419: "Babi stands up, having met *ḫnt-ḫm* whom the spittle protects (?), whom this one protects (?), (even) he the very well beloved. May the *wfj*-snake be got rid of; cause me to be protected." Pyr 1270: "If *ḫnt-jrty* comes with this his evil coming, do not open your arms to him, but let there be said to him this his name of "Spittle". "Go to *ddnw*, you will be found in a trembling condition for them [sic]; go northwards, go to Khem!"

[25] Pyr 2078-2079: "These four gods, friends of the King, attend on this King, (namely) Imsety, Hapi, Duamutef, and Khebhsenuf, the children of Horus of Khem; they tie the rope-ladder for this King, they make firm the wooden ladder for this King, they cause the King to mount up to Khepme when he comes into being in the eastern side of the sky." – cf. Pyr 1670: "*ḫnt-jrty* at Khem."

In certain aspects Horus was very close to Anubis who was also worshiped in Letopolis. They had common priests in the Old Kingdom, using a common title as priests of both gods.[26] Much later, in the Ptolemaic era, the two gods were even equated.[27] The identification was helped by the fact that they both were thought to be born to Uadjet (by the Greeks equated with their goddess Leto – hence the Greek name of the city).[28] As already indicated above, Horus Khenty-en-irty was linked via analogical magic with healing and medicine, particularly ophthalmology, because of his lost and regained eyes. It is evident that, at least in later times, Horus was worshiped as a physician for the gods,[29] and most probably there was a medical school attached to his cult, in Letopolis probably concentrating on ophthalmology. The question however remains: when was it established? If such a medical school existed in the archaic era, then it is possible that not only Horus but also Anubis was connected to that school, since both gods had a common cult in the Old Kingdom. That would explain perfectly why the medical treatise was found in Letopolis and under the feet of Anubis. But even without such a Letopolitan school and/or a connection to Horus Khenty(-en)-irty, there are several grounds on which the treatise could have been connected with Anubis in his own right.

Anubis, lord of life and death

The cult of Anubis[30] originated in Middle Egypt – his animal appears in the sign for the 17th Upper Egyptian nome (cf. its Greek name, Kynopolites).[31] His cult reached Memphis very early and spread to the neighbourhood: *nb r3-st3w* (Anubis of Memphis) and *nb sp3* (Anubis of Tura) are among his first attested titles. As we have just seen, he became closely connected to Horus in Letopolis, as is probably already reflected in Pyr 1723, in which the king receives various things from the Letopolitan Horus, but at the end these items are said to be offerings from Anubis: "... He who presides over Khem (*Ḫm*) raises you and has given a *t-wr* loaf and this grape-juice; the *jm3*-trees serve you, the zizyphus-tree bows its head to you, a king's boon is what is given to you, being what Anubis made for you." The reverence of the trees for the resuscitated King originates from Anubis in both Pyr 808 and 1019.[32] In general Anubis provided food in the Netherworld,[33] helped the deceased ones to get among the souls,[34]

[26] B. Begelsbecher-Fischer, *Untersuchungen zur Götterwelt des Alten Reiches im Spiegel der Privatgräber der IV. und V. Dynastie*, OBO 37, Freiburg-Göttingen 1981, pp. 19 and 27. Nos. 105, 109, 282, 300, 406, 522, 564 – Saqqara, Giza, 5th dynasty.

[27] pJumilhac V.22-23; in the Osiris litany in the chapel on the roof of the Dendera temple, cf. Mariette, *Dendera IV*, p. 73; "Anubis is *nb ḥt-rdw* ... He is Horus. He has spread his two wings (*dm3tj*) around you.", cf. A. Mariette, *Dendérah, Description Générale du Grand Temple de cette Ville, IV*, Paris 1880, p. 41 (no. 18); M. Rochmonteix, *Le Temple d'Edfou I*, MMAF 10, Le Caire 1897, p. 342.

[28] The local Hathor of Letopolis, originally the mother of Horus of Letopolis (but later [also] his wife), was identified with the Horus Eye, and as such she became equated with the uraeus snake and with Uadjet of Buto. Cf. Junker 1917, p. 44, Bonnett 1952, p. 424. In the theology of nearby Memphis, Anubis was thought to be the son of Bastet, a goddess who in time became equated with the Solar Eye and Uadjet. Cf. Bonnett 1952, pp. 42, 81-82, and pJumilhac VI.2-4 (Isis-Uto as mother of Anubis). For the parents of Anubis, see also Jean Claude Grenier, *Anubis Alexandrin et Romain*, Leiden, 1977, pp. 17-21. In another tradition, Anubis was made a half-brother of Horus, as illegitimate son of Osiris and Nephthys.

[29] W. Spiegelberg, "Horus als Arzt", *ZÄS* 57 (1922), pp. 70-71.

[30] Br. Altenmüller, "Anubis", *LÄ* I, cols. 327-333.

[31] H. Kees, "Der Gau von Kynopolis und seine Gottheit", *MIO* 6 (1958), pp. 157-175; Kees, "Anubis 'Herr von Sepa' und der 18. oberägyptische Gau", *ZÄS* 58 (1923), pp. 79-101; Kees, "Kulttopographische und mythologische Beiträge", *ZÄS* 71 (1935), p. 155; B. Begelsbecher-Fischer, *Untersuchungen zur Götterwelt des Alten Reiches im Spiegel der Privatgräber der IV. und V. Dynastie*, OBO 37, Freiburg-Göttingen 1981; D. Wildung, *Die Rolle ägyptischer Könige im Bewußtsein ihrer Nachwelt I*, Berlin 1969, pp. 21ff.

[32] Pyr 807-808: "A boon which the King grants, a boon which Anubis grants,....; the *jm3*-trees serves you, the zizyphus-tree bends its head to you, being what Anubis has done for you"; Pyr 1019: "a boon which Anubis grants: the *jm3*-trees shall serve you, the zizyphus-tree shall bow its head to you, you shall circumambulate the sky like *Zwnṯw*".

[33] E.g., Pyr. 806, 807, 896, 1281, 1723, 1936, the stela of Rudj-ahau in the Budapest Museum of Fine Arts.

and helped to preserve their bodies[35] – a funerary mirror to the activities of an earthly physician. In the house of Anubis, the fact of the king's death was announced (Pyr 1335), but also that "this King is hale" (Pyr 2069).

It is possible that in the beginning there were some links between Anubis and the idea of the two aspects of the heart (*ḥ3tj* and *jb*). We know an encouragement from the Pyramid Texts (Pyr. 1286) in which the following is said about Anubis: "You have relieved Horus of his girdle, so that he may punish the followers of Seth. Seize them, remove their heads, cut off their limbs, disembowel them, cut off their *ḥ3tj*-hearts, drink of their blood, and claim their *jb*-hearts in this your name of 'Anubis Claimer of *jb*-hearts!'" Anubis is mentioned here as a helper of Horus and the lord of the dead.[36] In fact, his epithet, highlighted at the end of the speech, contains a word for heart, *jb* (an ancestral word, derived from an Afro-Asiatic root), which most probably designates the abstract psyche.[37] The same epithet appears in Pyr. 1523, though in later times the "Claimer of hearts" role becomes characteristic of Sakhmet who has a similar double nature. This early epithet of the god accentuates his ruling role as the power over death and life.[38] But massacre is not his usual way of acting. He has the ability to kill, but he must be asked to do so for a good reason, just as a good ruler is asked to help in need. He is dangerous to the trespassers, not to the sinless subjects. He even protects them by his judgement: in his function as the counter of the *jb*-hearts he was one of the magistrates of the Tribunal in the Netherworld,[39] and the deceased could ask his help by wearing his amulet. Unas wore it in Pyr.157: "O Thot, go and proclaim to the western gods and their spirits: 'This King comes indeed, an imperishable spirit, adorned with Anubis on the neck, who presides over the Western Height. He claims *jb*-hearts, he has power over *ḥ3tj*-hearts. Whom he wishes to live will live; whom he wishes to die will die.'" So Anubis was thought to dispose over human life with the help of the heart(s) during the judgement of the netherworld funeral ritual of *wḥm ꜥnḫ*. The most basic aim of the earliest medicine was most probably to save life and to avert death. By analogy the Netherworld judgement would be a mythologized reproduction of earthly medical treatments, which would satisfactorily explain the common roots of the phraseology of the funeral and medical texts.[40] It is therefore possible that in the earliest phases of Egyptian religion, and still at the time of the unification of Egypt, diseased individuals asked Anubis to let them live and make them healthy again.

Unfortunately, the bulk of the written evidence that we have deals with funeral-liturgical and official matters. Scientific works were most probably mainly transferred orally, and texts about everyday life were simply not written down. Medicine was said to be a "craft of a secret knowledge" (*ḥmt sšt3*),[41] meaning that medical practices would be written down even less frequently than other

---

[34] E.g., Pyr 796, 797, 1015, 1676, 1909, 2002. His daughter was thought to be the "celestial snake"; she helped the deceased to reach the sky as a companion of Thot (Pyr 468), and refreshed the heart for the netherworld life (Pyr 1180, 1996, 2103).

[35] E.g., Pyr 1257 (prevent from rotting), 1122 (washes the entrails), 1380 ("Anubis of the Baldachin raises him").

[36] Cf. Pyr 135b, 573a, 1235a, etc.

[37] J. H. Walker, *Studies in Ancient Egyptian Anatomical Terminology*, The Australian Centre for Egyptology Studies 4, Warminster 1996, 147-186 (chapter "Heart", part D/12). Although *jb* and *ḥ3tj* occupy an absolutely identical locus in the body, and at times are able to function as synonyms, they are in principle (certainly before the New Kingdom) designations for separate aspects of an entity: "heart" (*jb*) as the abstract psyche or self, and "heart" (*ḥ3tj*) as the physical heart, the start of the blood circulation.

[38] In Pyr 2178 it is Nut who gives the *ḥ3tj* -heart to the dead king, but in the presence of Anubis. In Pyr 145b-c, the cutting off and the claiming of the heart are mentioned to show the reigning power of Osiris, Geb, Horus and Seth, but these gods do not get the "Claimer of hearts" epithet.

[39] E.g., Pyr 1713, 1919.

[40] For similarities see, e.g., J. F. Quack, "Magier und Totenbuch – eine Fallstudie (pEbers 2,1-6)", *CdÉ* 147/74 (1999), pp. 5-17.

[41] For instance, one of the titles of *Ḥwy* was *3ꜥꜥ ḥm(w)t št3t,* cf. Ghalioungui, 1983, p. 22, no. 43.

subjects. Our one-sided sources therefore do not allow us to search for any detail of medical practice in the early periods of Egyptian history.

Just as the physician was the specialist of a "secret" (*sštꜣ n swnw*), so Anubis was "he who presides over a secret" (*ḥrj sštꜣ*). In late archaic times his priest was called "*ḥrj-sštꜣ* dessen, der im Friedhof wohnt (*ḫntj-tꜣ-ḏsr*, Anubis) im (?) Tempel" and "*sm*-Priester im Tempel des *ḥrj-sštꜣ* (=Anubis!) *ḫntj-tꜣ-ḏsr*".[42] The *ḥrj sštꜣ* epithet in the New Kingdom hints at the canopic chests on which Anubis was lying, and to their contents, hence to mummification. However, the earliest known canopic set was found in the 4th Dynasty tomb of Hetepheres, so before that, during the archaic period, the epithet must have indicated that Anubis presided as a guard over something else, e.g. over his sanctuary,[43] over the tomb, or maybe over some sort of special box. The god is also given a role of guard in the medical treatise, by the implication that the papyrus was in his care ("under his two feet") for a very long time. The frequent representations of Anubis around the archaic kings also hint at his function of guarding the ruler.[44] In this role he could take care of a box which might contain the personal belongings of the king. If the king was not only identified with Horus but also with Anubis, as the tight link between the cults of the two gods might suggest, then the box may have contained also "medical" matters, like spells on the hearts for the king's use in his 'role' of Anubis.[45]

The heart motif in the "Tale of Two Brothers" (Papyrus d'Orbiney) is also based on the life-death question, but from a different point of view. In this myth of the 17th Upper-Egyptian nome about Bata, the bull-god, and Anubis, the jackal-god,[46] Anubis is asked by Bata to find his heart after his death. And Anubis makes his brother live again for a second earthly life by resuscitating his heart – a good parallel to a dead man resuscitated for the *wḥm ꜥnḫ* by a favourable Netherworld judgement with the help of the heart, or for a medical act of healing a seriously ill, eventually immobile,[47] patient by treating his heart (and circulation).

In the "Tale of Two Brothers" the actions of Anubis concern things of this earthly life. The story is known from the New Kingdom, but is certainly an older conception. Though the date is not yet settled, it seems to be generally accepted that the motif of the tale was known during the Old Kingdom, as the god's life-giving ability is stressed in several Old Kingdom settlement names,[48] and in the use of the personal name *jnpw m ꜥnḫ*.[49] This life-guarding aspect of Anubis could be felt at birth as well, even much later – he is present in Deir el Bahari, when Hatshepsut is born,[50] in Edfu at Ihi's birth,[51] and in

---

[42] P. Kaplony, *Die Inschriften der Ägyptischen Spätzeit*, Wiesbaden 1963, Bd. II. p. 1058, note 1811. K. T. Rydström, "*ḥry-sštꜣ* – 'in charge of secrets'. The 3000-year evolution of a title", *DE* 28 (1994), pp. 56-61, deals with the history of the title *ḥry sštꜣ*, and how it was originally associated with the funerary cult, but later in the Old Kingdom with the intimate surroundings and activities of the king.

[43] The spelling of the name of Anubis by a jackal on a shrine emerged only under Teti (6th Dynasty), cf. A. O. Bolshakov, "Osiris in the Fourth Dynasty again? The false door of *Jntj*, MFA 31.781", in H. Győry (ed.), *Mélanges offerts à E. Varga, le lotus qui sort du terre*, Budapest 2002, p. 75 and note 88. Anubis' epithet *mnjwj* (cf. Pyr 1380b; *Wb* II. 75,15-16) seems to indicate that originally he was guarding the *mnjw*, an object or dwelling the exact nature of which is not known.

[44] T. A. H. Wilkinson, *Early Dynastic Egypt*, London-New York 1999, pp. 280-281: wooden label of Aha, the Palermo stone, seals, stelae, etc.

[45] For this possible role, see, e.g., Pyr. 135b, 153c-d (king as *wtj-Jnpw*).

[46] H. Kees, "Der Gau von Kynopolis und seine Gottheit", *MIO* 6 (1958), pp. 157-175 – cf. pJumilhac VI.17-18.

[47] Cf. note 12, "sealing of the legs"; i.e., losing control of the legs because of serious illness or approaching death, the patient is confined to the bed not unlike a deceased to the bier.

[48] B. Begelsbecher-Fischer, *Untersuchungen zur Götterwelt des Alten Reiches im Spiegel der Privatgräber der IV und V. Dynastie*, OBO 37, Freiburg-Göttingen 1981, p. 31.

[49] Begelsbecher-Fischer 1981, p. 28; cf. *PN* I. 37,8.

[50] É. Naville, *The Temple of Deir el-Bahari*, II, London 1895, pl. 55.

[51] É. Chassinat, *Le mammisi d'Edfou*, MIFAO 16, Le Caire 1939, pl. LXVIII.

Dendera at Harpocrates' birth.[52] Also in the medical papyrus pLondon, incantations 27, 29 and 30, Anubis is asked to avert a premature delivery.[53]

King Dewen in later traditions

The medical treatise claims to have been found during the reign of king Dewen (1st Dynasty). What could be behind this claim? Interestingly, two chapters of the Book of the Dead that are known from the 18th Dynasty are likewise connected to the reign of Dewen. Chapter 64[54], probably originating from Memphis, says that the spell was found "in the wall of him who is in the *Ḥnw*" (i.e. Sokaris) during the reign of Dewen or "under the two feet of this god in the time of king Menkaure by Djedefhor". It speaks about the forthcoming of the deceased to the sun, his transformation ability and the preservation of his soul, but its knowledge – as is said *expressis verbis* – not only means justification in the afterlife but also on earth. While the spell was being cast, a nephrite heart-scarab was prepared, then myrrh was spread over it, the "Opening of the Mouth" ceremony was applied, and the heart chapter 30B recited. Chapter 130[55], probably originating from Heliopolis, describes the ceremonies for the birthday of Osiris. According to its description, the spell was found at two different places at the same time, during the reign of Khasti (i.e. Dewen)[56], in one of the rooms of the palace and in a mountain cave. It is said to have been invented by Horus for his father. The aim of the spell can be grouped around three notions: the guarantee of continuous funeral offerings, eternal life of the soul, and justification in the Hall of the double Maat. Thus the content involves the presence of both Anubis and the heart.

It may not be mere chance that both chapters of the Book of the Dead connected to Dewen allude to Anubis and the heart. It seems as if during the reign of Dewen something happened that made it worthwhile to keep his memory alive for centuries as part of the popular philosophy of history. Possibly something striking happened in connection with the royal cult which brought him into close association with Anubis and the heart concept. If the notion of Helmut Brunner[57] is right, namely that the word *ỉnpw* hints at an ancestral function of Anubis, and could be connected to a word for "prince" and "child from the royal house",[58] then a unification of the cults of Anubis and the king would be completely natural. This royal character of Anubis is strongly supported by his emblem *jmjwt,* which is usually depicted beside the king,[59] and by the fact that several high ranking officials close to the king fulfilled the common priesthood of Horus and Anubis even during the 4-5th Dynasties. Their title was *ḥm-nṯr Ḥr, Jnpw ḫntj pr šmswt.* These priests were thus the leaders of the *šmswt* house, which may be

[52] Fr. Daumas, *Les mammisis de Dendera,* Le Caire 1959, pls. XLIA, LIXbis.

[53] Christian Leitz, *Magical and Medical Papyri of the New Kingdom,* London 1999, pp. 68-70.

[54] The papyrus of Nu, Nebseni and Iuau: E. A. W. Budge, *The Book of the Dead: the chapters of coming forth by day; the Egyptian text according to the Theban recension in hieroglyphic, edited from numerous papyri, with a translation, vocabulary, etc.*, I, London, 1898, pp. 189, 199, 200 – for Dewen and Nu and Iuau (both longer versions) pp. 188, 193 – for Menkaure; T.G. Allen, *The Book of the Dead or Going forth by Day,* SAOC 37, Chicago 1974, pp. 58-59: papyri Ce, Aa, Ea – Dyn. 18th. Both versions go back to the coffin of queen Mentuhotep, 13th Dyn. See U. Luft 1973, p. 111, with references.

[55] Budge 1898, I, pp. 278-285; the chapel of Tutankhamon; cf. CT 1065, 1099.

[56] See note 11. The writing form used here is *Ḳntj.*

[57] H. Brunner, *Geburt des Gottkönigs,* pp. 27-29.

[58] *Wb* I. 96,5; Br. Altenmüller, "Anubis", *LÄ* I, col. 327; cf. J. Vandier, *Le Papyrus Jumilhac,* Paris, 1962, pp. 102-103 – an ancient Egyptian etymology. Compare also W. Federn, "*ḥtp (r)dj(w) (n) ỉnpw*; Zur Verständnis der vor-osirianischen Opferformel", *MDAIK* 16 (1958), p. 130, note 1. Federn thinks that, for the earliest times, the reading of the sign of the lying jackal is not certain, and that this sign might have been connected to Anubis only later, in Abydos. He suggests that *ỉnpw* originally designated the son of Osiris (before Horus took that position). If, however, the inclusion of Anubis in the family of Osiris was relatively late (cf. Grenier 1977, pp. 17-18), then the old Cynopolite tradition that makes Anubis a son of Re and the Hesat cow (Grenier 1977, pp. 17-21; *LÄ* I. col. 327, down; cf. CT 908, Pyr 2080e) could perhaps alternatively explain his name of "royal child", and would equally link him with the king (cf. Pyr 1029).

[59] H. Kees, "Kulttopographische und mythologische Beiträge", *ZÄS* 71 (1935), p. 154.

the meeting place or residence for the royal attendance. The funeral meal is supposed to be the aim of the house,[60] but as the life giving character of Anubis is frequently highlighted in the royal propaganda, and Horus seems to belong to it as well, it could also be some sort of performance room for the appearance of the king in his function as Horus or Anubis.

Both spells are connected to the religious beliefs around Memphis which is near to Letopolis, mentioned in the above treatise. If we may deduce from the two spells some sort of special connection between king Dewen, Anubis and the heart, and if we may combine that with the connection between Dewen, Anubis and some kind of ailment ("after he became miserable") in the medical treatise, may we then presume that both sources in their own way transmitted the memory of a heart-disease of this king? A disease that was followed by recovery but that needed some sort of continued medical support for the king's heart? As well as special magical support, e.g. on the birthday of Osiris (BD 130), one of the traditional days of the *j3dt rnpt* when Sakhmet attacked the hearts? May his weak physical state have been the reason for Dewen's frequent appearances in public (*ḫꜥt nswt-bjtj, ḫꜥt bjtj, ḥb-sd, ḏt* festival, the running of the Apis bull, etc.)?[61] As we do not have any contemporary private information about the king, any hypothesis in this respect would be mere guesswork.

King Dewen had a funeral cult in Saqqara during the 4th Dynasty, and his name is written several times on late dynastic bronze statuettes.[62] It is very likely that he and his cult place were not forgotten during the interval between his death and the 4th Dynasty. The cult of Anubis might easily have been connected to his cult, it being a funeral one (at least in a part) in the cemetery. But we know almost nothing about this king's cult. Finds from his reign, however, attest that it was a very important turning point both historically and in the material cult.[63] It was characterized by stabilization of the state, re-organization of the forms of the royal cult,[64] and probably systematization of knowledge. As kings, in their divine aspects, were responsible for everything that happened in the country, all major developments of their time were attributed to them. Thus these changes might have given him a mythical aura in later times. New Kingdom people had a very different view of life compared to people of the archaic period. They wrote letters, reports, annals, etc. about their investigations, inventions and results. They must have believed that their ancestors did the same thing, i.e. that these ancestors thus also wrote down important facts, including medical knowledge. The treatise of the "Wandering of Pathogenic Elements" thus reflects the perceptions of the period in which it was written, i.e. that the Egyptian medicine was as standard as the Egyptian state, and that it was under the protection of powerful gods who always help the people. Even the god of mummification can help them in their recovery, if he wants to.

In the 2nd Anatomical Treatise, Dewen is merely mentioned as the possessor of the Book. But it is already a copy, and the Berlin Treatise – a much longer one – adds that the king was "miserable". This comment suggests that the book was "found" in a search against this royal misery, and probably hints at a royal disease which was successfully cured via the methods described in the Berlin Treatise (although in that case one would perhaps expect that the effectiveness would have been explicitly stressed; but the king is not mentioned later in the text). The questions thus remain: what was written in the original text, and when was that original text written. Because of the phraseology and some grammatical phenomena common in both copies, the possibility of an Old Kingdom origin has been

---

[60] Begelsbecher-Fischer 1981, p. 19.

[61] Wilkinson 1999, p. 211, 212, 300, 305.

[62] D. Wildung, *Die Rolle ägyptischer Könige im Bewußtsein ihrer Nachwelt*, I, MÄS 17, Berlin 1969, pl. V.1.

[63] Cf. P. Kaplony, "The Bet Yerah Jar Inscription and the Annals of King Dewen – Dewen as 'King Narmer Redivivus'", *Egypt and Levant* 29, 2002, pp. 464-486, and P. Kaplony, "The En Besor Seal Impression – Revised", *Egypt and Levant* 30, 2002, pp. 487-498.

[64] E.g., the introduction of the title *nswt-bjtj* to the royal titles (Wilkinson 1999, p. 206), and the occasions of the *ḫꜥt bjtj / nswt / nswt-bjtj* rituals – see the Palermo stone (Wilkinson p. 211) etc.

raised.[65] But did the scribes use the original texts or did they only have annotated version(s) from later time? Was it copied without change, or were there several generations of copies with corrections and (increased) comments? And is the claimed discovery in the time of Dewen authentic or an invention of the New Kingdom? We cannot know with certainty.

Conclusions

Fictive dating is not uncommon in ancient Egypt. It was a method employed by scribes to make texts more precious, to legitimize the content by providing an acceptable authority, or to express views that were not otherwise safe to express.[66] At the same time the scribes took great care of the works of their ancestors, protected them, read and used them, and even taught them in school. These two aspects meant that the past continuously merged with the present in ancient Egyptian culture. Thus the past was regularly re-interpreted, and in that way often became disconnected from historical reality. When texts were attributed to the reign of a certain king of the dim past, this happened within this tradition: it could be based on a genuine memory or on a fictive one, but either way it did not happen without a reason. The treatise of the "Wandering of *wẖdw*" claims to have been found during the reign of king Dewen, under a statue of Anubis in Letopolis. As we have seen, there may have been good reasons for the separate elements of this claim, such as the guardian nature of Anubis, his dual nature of lord of life and death, and his connection with medicine via his links with Horus Khenty(-en)-irty, the heart-concept, and embalming. The claim was also supported by the circumstance that other texts, namely two chapters of the Book of Dead, associate king Dewen with the systematization ("rediscovering") of knowledge, and with Anubis and the heart.

The tradition of attributing texts to early periods was still alive centuries later, as is evident from Manetho's work. Even here such an attribution did not happen without a good reason. In the case of the Manethonian topos under consideration, the argument would have run like this: if a medical book like the treatise claims to have been found during the reign of king Dewen, and was thought to be a "collection" of earlier utterances, then logic dictates that it and/or its sources must have been written still earlier. As writing before the time of Menes was used only for a few names, and as this pre-dynastic time was associated by the Egyptians with the god-kings, it is highly probable that they thus will have placed the origin of the treatise in the time between Menes and Dewen – the reign of Manetho's Athothis (Djer) being a good candidate. And as we have seen, this choice seems to be not improbable, within certain limits.

Dr. Hedvig Győry
gyory@szepmuveszti.hu
Museum of Fine Arts, Budapest

---

[65] Grapow, *Grundriss* II., Berlin 1955, pp. 100-101.

[66] E. Blumenthal, "Nachleben", *LÄ* IV, cols. 287-88, W. Westendorf, *Handbuch der altägyptischen Medizin*, Brill, 1999, p. 5.

Altered States: An inquiry into the possible use of narcotics or alcohol to induce dreams in Pharaonic Egypt

Kasia Szpakowska

I have often been asked whether the Egyptians used drugs to induce dreams. This paper aims to address that question primarily as it relates to dream reports recorded prior to Egypt's Late Period. As I have noted elsewhere, the nature of references to dream reports changed substantially after that time, and deserve to be investigated separately.[1] The current enquiry will explore the mystical nature of drunkenness, the relationship of the goddess Hathor to ecstatic dreams, and the known dream experiences.[2] The first step in this inquiry is to consider what drugs could have been known and used during that time for other than medicinal purposes.

Drugs

The two most obvious candidates, opium (*Papaver somniferum*, probably ancient Egyptian *špn*(*n*)[3]) and cannabis (*Cannabis sativa*, probably ancient Egyptian *šmšm.t*[4]), can be eliminated, as little or no physical evidence remains of their use.[5] Certain fibres in textiles or rope have been classified as hemp, but this identification has yet to be substantiated by rigorous analysis.[6] The presence of opium is highly disputed as well, and although it may have been referred to in medical texts,[7] there is no evidence of its

---

[1] K. Szpakowska, *The Perception of Dreams and Nightmares in Ancient Egypt: Old Kingdom to Third Intermediate Period* (Dissertation, University of California, Los Angeles, 2000), currently under revision for publication under the title *Behind Closed Eyes: Dreams and Nightmares in Ancient Egypt* (Classical Press of Wales, 2003). For a brief overview of dreams in Ancient Egypt, see also K. Szpakowska, "Through the Looking Glass: Dreams and Nightmares in Pharaonic Egypt," in *Dreams: A Reader on the Religious, Cultural, and Psychological Dimensions of Dreaming*, ed. K. Bulkeley (New York: Palgrave, 2001), 29-43.

[2] The ritual of the Opening of the Mouth, scenes 8-10, may have included ritualized dreaming, but there is no evidence for the imbibing of any substance prior to the ritual; see Szpakowska (2000), 244-263. For the ritual in general, see E. Otto, *Das ägyptische Mundöffnungsritual* (Wiesbaden: Otto Harrassowitz, 1960), and H.-W. Fischer-Elfert, *Die Vision von der Statue im Stein* (Heidelberg: Universitätsverlag C. Winter, 1998).

[3] R. Hannig, *Die Sprache der Pharaonen: Großes Handwörterbuch Ägyptisch - Deutsch (2800-950 v. Chr.)*, (Kulturgeschichte der antiken Welt 64 / Hannig-Lexica 1; Mainz: Verlag Philipp von Zabern, 1995), 814; A. Erman and H. Grapow, *Wörterbuch der Ägyptischen Sprache* (Berlin: Akademie-Verlag, 1971), IV, 444-445.

[4] Hannig (1995), 824; *Wb* IV, 488.

[5] For examples of Cypriotic juglets possibly used to import opium, see D. Polz, "Bericht über die erste Grabungskampagne in der Nekropole von Dra' Abu el-Naga/Theben-West," *MDAIK* 48 (1992), 109-130. For an earlier view on this topic, see R. S. Merrilees, "Opium trade in the Bronze Age Levant," *Antiquity* 36 (1962), 239-242; and for an opposing view, see N. G. Bisset et al., "Was opium known in 18th dynasty ancient Egypt? An examination of materials from the tomb of the chief royal architect Kha," *Journal of Ethnopharmacology* 41 (1994), 99-114. Further examples can be found in S. Gabra, "Papaver Species and Opium through the Ages," *Bulletin de l'Institut d'Égypte* XXXVII, no. 1: Session 1954-1955 (1956), 39-56, as well as a detailed discussion in M. D. Merlin, *On the Trail of the Ancient Opium Poppy* (Rutherford, Madison, Teaneck: Fairleigh Dickinson University Press, 1984). A recent review of this problem can be found in M. Serpico and R. White, "Oil, fat and wax," in *Ancient Egyptian Materials and Technology* (ed. P. T. Nicholson and I. Shaw; Cambridge: Cambridge University Press, 2000), 404-405.

[6] G. Vogelsang-Eastwood, "Textiles," in *Ancient Egyptian Materials and Technology* (ed. P. T. Nicholson and I. Shaw; Cambridge: Cambridge University Press, 2000), 269.

[7] J. F. Nunn, *Ancient Egyptian Medicine* (Norman: University of Oklahoma Press, 1996), 153-156.

use in a religious context. Recent articles by the botanist William Emboden[8] and W. Benson Harer M.D.[9] have identified the blue water lily (*Nymphaea caerulea*), a plant native to Egypt, as a potential candidate.[10] This plant has commonly been confused by lay people and scholars alike—excluding botanists, but including Egyptologists—with the "Indian lotus," which belongs to the genus *Nelumbo*, and was not introduced into Egypt until the 6th century B.C.[11] Like many other imprecise terms in Egyptology, the blue water lily continues to be referred to as a "lotus." Even when the taxonomic differentiation between these plants is acknowledged,[12] the significance of the distinction between the plants remains unrecognized.[13] The following is a synopsis of the blooming patterns of the blue water lily as described by Ossian based on his own observations and those of Perry Slocum, a professional grower of water lilies.[14] The blue water lily, along with its relative the white water lily, forms its buds under the water. As the stalks grow, they emerge from the water with the buds still tightly closed. After a few days, the buds finally open, blooming quite quickly (usually in less than one hour), only to close again after a certain amount of time. The blooms stay resting delicately just on top of the water surface, closing and reopening again for 1-4 days, before finally sinking permanently beneath the water. This emergence from the water, followed by a dramatic opening of the blossom towards the sky and the sun, was the likely impetus for associating the lily with the sun god.

Most significantly, the blue water lily is considered by some to have narcotic effects to such a degree that at one time it was recommended as an opium substitute. Harer states, "In small doses these drugs induce a feeling of well being, drowsiness, giddiness and double vision. In larger doses they induce hallucinations and/or stuporous sleep with vivid dreams."[15] Harer and Emboden's results have been criticized, however, on the grounds that the specific genus of blue water lily that can be produced today is not necessarily identical to the variety produced in ancient Egypt. Other scholars firmly deny that the plant even contains any narcotic alkaloids.[16] Egyptian iconography commonly depicts people holding the flower to their nose, but even if we assume that the lilies did contain hallucinogenic

---

[8] W. Emboden, "The Sacred Narcotic Lily of the Nile: Nymphaea Caerulea," *Economic Botany* 32 (1978), 395-407; W. Emboden, "The Sacred Journey in Dynastic Egypt: Shamanistic Trance in the Context of the Narcotic Water Lily and the Mandrake," *Journal of Psychoactive Drugs* 21, no. 1 (1989), 61-75.

[9] W. B. J. Harer, "Nymphaea: Sacred Narcotic Lotus of Ancient Egypt?," *JSSEA* XIV, no. 4 (1984), 100-102; W. B. J. Harer, "Pharmacological and Biological Properties of the Egyptian Lotus," *JARCE* XXII (1985), 49-54.

[10] See also A. Wolinski, "The Shaman and the Blue Water Lily of Ancient Egypt," *The Glyph: Newsletter of the Archaeological Institute of America, San Diego Society* 1, no. 11 (1997), 10.

[11] Clair Ossian, "The Most Beautiful of Flowers: Water lilies & lotuses in Ancient Egypt," *KMT* 10, no. 1 (1999), 48-59. This article presents a clear and well-balanced discussion on the subject of this plant.

[12] As for example W. J. Darby, P. Ghalioungui, and L. Grivetti, *Food: The Gift of Osiris*, 2 vols., vol. 2 (London, New York, San Francisco: Academic Press, 1977), 619-644. On page 620, the authors state that "the aquatic lotus in the sense used today and as applied only to Egypt, covers *Nymphaea lotus* and *Nymphaea coerulea.*" Yet they continue to confuse the botanical name and the plant it represents as when they warn, "Persons who are unfamiliar with aquatic plants easily fall into the error of confusing the lotus with a water-lily" (p. 631), then continue to describe the plants as "white lotus, *Nymphaea lotus* … pink lotus, *Nelumbium speciosum* … blue lotus, *Nymphaea caerulea*" (p. 633).

[13] So for example in *LÄ* III, 1091-1096, where the various species of plants are mentioned, but their differences are ignored. A notable exception is of course S. Weidner, *Lotos im alten Ägypten* (Pfaffenweiler: Centaurs-Verlagsgesellschaft, 1985), who throughout his seminal work carefully distinguishes between the species in question, but does not mention any physiological effects of ingesting *Nymphaea caerulea*.

[14] Ossian (1999), 50.

[15] Harer (1984), 100. A number of first-hand accounts of experimentation with the blue water lily can now be found on the Internet (for example "The Vaults of Erowid.", n. p., at http://www.erowid.org/plants/bluelotus/bluelotus.shtml [cited 16 February 2002], and "The Blue Water Lily" by Colin Byrne [cited 20 April, 2002], which is a report on an experiment conducted in the British TV series "Sacred Weeds," at http://leda.lycaeum.org/Trips/The_Blue_Water_Lily.5862.shtml). These experiments, however, were not conducted in a controlled environment, and in some cases the blue water lily was ingested after the individual had already consumed another mind-altering agent (usually cannabis). These reports must therefore remain suspect.

[16] J. A. Tyldesley, *The Private Lives of the Pharaohs* (London: Channel 4, 2000), 171-172.

properties, sniffing them would not produce any sort of physical euphoria. It is the blossom and rhizome that have been suggested as the source of the hallucinogenic and narcotic agents nupharine, nupharidine and nuciferine.[17] These constituents are soluble in alcohol and it may be that a flavorful brew such as wine or beer would also help mask the bitter taste of the alkaloids—a feature common to most psychotropics and poisons. A potent narcotic could be produced by steeping the blue water lily in wine for an extended period of time, while adding a little of the juice before serving would create a milder tipple.

The fact that the blue water lily is often depicted in conjunction with the mandrake (*Mandragora officinarum*, ancient Egyptian *rrm.t*)—well-known for its sedative effects—could suggest that the Egyptians were aware of the plant's narcotic properties, but it could also point to their use as erotic symbols.[18] These two plants are often depicted in tandem, intertwined, or even with the mandrake fruit within a water lily flower. Emboden also lists instances of the poppy being shown with these other two plants. Additional support for the use of lilies in liquid preparations can be found in Harer (particularly 1984), who cites scenes of women harvesting and pressing the lilies, and the use of pouring juglets in "party scenes," which are too small to hold wine but would be ideal for adding small amounts of the lily oil to the guests' beverages. Artifacts include liliform chalices (one type representing the white water lily, the other the blue water lily), "unguent" jars, and an interesting lead hollow tube, angled with a strainer at the end. The latter was found in Amarna, and can now be found in the British Museum in two pieces as #55148 and #55149. Harer describes this piece as having a strainer intact at the bottom, with large holes, too large for beer or wine sediment, but an ideal size for straining water lily flower residue. A stele, which is also thought to come from Amarna, depicts an Asiatic man using a similar tube to drink from a jar.[19] Other evidence for the use of these jars with straws can be found in Michaelidis' article on the roles of Bes[20] that includes two depictions of Bes drinking from a large jug with straws.[21] However, the contents of these jugs are unknown and cannot be definitively proved to be beverages combined with narcotics. In addition, the theory that blue water lily was mixed with wine or beer should remain tentative at best, as scientific analyses of the dregs of ancient wine have so far failed to show any evidence of contamination with narcotic alkaloids.[22]

The textual evidence used to corroborate the use of the blue water lily for its narcotic properties is found mainly in medical texts. The term used was *sšn*,[23] and this seems to have been used for both the blue and white species of water lily. Harer cites Renate Germer as having found water lily blossoms mentioned in over 20% of ancient Egyptian plant remedies—the single most prescribed ingredient.[24] Nunn points out that the medical papyri always used the term the *ḫ3.w*[25] of the water lily.[26] Faulkner[27]

[17] F. Anderson, "A Flower for Eternity," *Garden* 7, no. 1 (Jan-Feb.1983), 12, where he also points out that these properties were first published in 1912 by botanist H. Pobéguin.

[18] P. Derchain, "Le lotus, la mandragore, et le persea," *Chronique d'Égypte, Bruxelles* 50 (1975), 65-86; P. Derchain, "Symbols and Metaphors in Literature and Representations of Private Life," *Royal Anthropological Institute News* 15 (1976), 7-10.

[19] This stele, #14122, is currently in the Ägyptisches Museum und Papyrussammlung, Berlin. Photographs of both of these artifacts can be seen in R. E. Freed, Y. J. Markowitz, and S. H. D'Auria, eds. *Pharaohs of the Sun: Akhenaten, Nefertiti, Tutankhamen* (Boston, New York, London: Bulfinch Press/Little, Brown and Company, 1999), 161, 239.

[20] G. Michaeilidis, "Bès au divers aspects," *Bulletin de l'Institut d'Égypte*, Le Caire XLV (1963-1964), 53-93.

[21] Michaeilidis (1963-1964), 64, fig. 17 a, b.

[22] To be fair, there have not been many analyses performed as yet, at least in part due to the lack of samples available. On this topic, see M. A. Murray, N. Boulton, and C. Heron, "Viticulture and wine production," in *Ancient Egyptian Materials and Technology* (ed. P. T. Nicholson and I. Shaw; Cambridge: Cambridge University Press, 2000), 577-608.

[23] *Wb* III, 485-486, gives both "vgl. Kopt. *šôšen* 'Lilie'," as well as "Lotusblume." Hanning (1995), 766, recognizes both its meaning as *Nymphaea lotos* and as *Nymphaea coerulea*.

[24] Harer (1984), 101, citing R. Germer, *Untersuchung über Arzneimittelpflanzen im alten Ägypten*, Doctoral Thesis, Hamburg, 1979.

[25] Hannig (1995), 577, "Blatt (*bes. vom Lotus*)"; *Wb* III, 219, "Blätter (?) des Lotus"; Hannig (1995), 766, *ḫ3w nw sšn*, "Blätter des Lotus (*Schwimmblätter, *Rhizom)."

translates this as "flowers," which would contain the active alkaloids. In the most recent complete translation of medical texts by Westendorf, the term is translated in two remedies as "rhizome," another section of the plant that would contain the pertinent alkaloid.[28] The rhizomes are to be mixed with other ingredients including beer and wine, and then "left in the dew overnight, strained and drunk for four days."[29] Nunn points out that this process of soaking the herbs overnight in alcohol could help to extract the narcotic ingredients,[30] *if* any of the plants did indeed possess any narcotic alkaloids. Otherwise, this process was simply used to mix the ingredients together thoroughly.

The symbolic significance of the plant is not in any doubt. The religious import of the blue water lily was enhanced by its very nature as a flower that rises from the water depths to bloom for only a few days, staying open for a short time before tightly closing again, only to reopen the next day. This repetitive behavior probably lead to its symbolizing rebirth, and it is no surprise that the god Nefertum is seen emerging from this particular flower. One cosmogony describes how in the beginning there was only primeval Nun, on which floated what we might call the Ur-lily, out of which Re emerged.[31] The *Lexikon* lists numerous deities associated with the water lily at various times, such as Ra, Khepri, Harpokrates, Khonsu, and Amun-Re,[32] and the flower is depicted in numerous religious contexts. The god Nefertum is particularly associated with the blue water lily, and he is depicted as an anthropomorphic deity with a headdress of blue water lily blooms.[33] At times called "the water lily at the nose of Re" (Pyr. 266), Nefertum is also the god of perfumes—a combination that may confirm that Egyptians held the flowers under their noses for symbolic reasons, or for their pleasant scent, rather than for any hallucinogenic purpose. The popularity of representations of the lily can be explained convincingly simply on the basis of symbolism, in the same way as the ubiquitous *ankh* or papyrus.

Drunkenness

While the evidence for the use of narcotics remains tenuous and inconclusive, there is more evidence that the state of "drunkenness," when mentioned in a religious context, whether public or private, quite likely indicated, not a state of pure alcohol intoxication, but of euphoria—an altered state of consciousness if you will. The use of "helpers" such as intoxicants, music, and dance, to reach an ecstatic state is common in religions around the world.[34] This "holy intoxication" or "mystical drunkenness" should not be confused with everyday drunkenness, as the Egyptians themselves were well aware. For while didactic texts advise the proper Egyptian not to drink to excess, other texts speak of drink and drunkenness as a ritual prerogative.[35] Inebriation could be a "sign of real devotion"[36]

---

[26] Nunn (1996), 158.

[27] R. O. Faulkner, *A Concise Dictionary of Middle Egyptian* (Oxford: Griffith Institute, Ashmolean Museum, 1981), 183.

[28] W. Westendorf, *Handbuch der altägyptischen Medizin*, vol. 2 (Handbuch der Orientalistik, Ser. 1 36, Leiden, Boston, Köln: E. J. Brill, 1999), 586; 631.

[29] L. Manniche, *An Ancient Egyptian Herbal* (London: British Museum Publications Ltd., 1989), 126.

[30] Nunn (1996), 158.

[31] E.-C. Strauß, *Die Nunschale - Eine Gefäßgruppe des Neuen Reiches,* ed. H. W. Müller, vol. 30, Münchner Ägyptologische Studien (Berlin: Deutscher Kunstverlag, 1974), 74.

[32] *LÄ* III, 1091-109.

[33] For Nefertum in general, see S. Morenz, *Der Gott auf der Blume: eine ägyptische Kosmogonie und ihre weltweite Bildwirkung* (Ascona: Artibus Asiae, 1954).

[34] See C. J. Bleeker, "Rausch und Begeisterung," in *The Sacred Bridge: Researches into the Nature and Structure of Religion*, ed. C. J. Bleeker, Studies in the History of Religions (Supplements to Numen) (Leiden: E.J. Brill, 1963), 159-179, for a detailed discussion of this topic.

[35] M.-C. Poo, *Wine and Wine Offering in the Religion of Ancient Egypt*, ed. G. T. Martin, Studies in Egyptology (London and New York: Kegan Paul International, 1995), 27-37. See also H. Brunner, "Die theologische Bedeutung der Trunkenheit," *ZÄS* 79 (1954), 81-83, for an examination of the religious meaning of drunkenness.

during feasts, and one text reveals: "It is while we are drunk in <your> presence that we speak our request like the favored ones" (KRI, III, 437:4). While discussing the role of the "*mnw*" juglet in what he calls the "fête solonnele de l'ivresse," Daumas explains that drunkenness could be used as a way of abolishing the barriers between this world and that of the gods. He goes on to state that, in the Hathoric or Osirian context, worshippers are advised to get drunk.[37] Indeed, while discussing the role of Hathor as "mistress of inebriety," Bleeker emphasizes "… she [Hathor] in no way despises intoxicants … The qualification implies that she is also the goddess who rouses ecstasy and religious fervor. This sounds peculiar to one of the 20th century, but it should be remembered that in ancient times drunkenness and ecstasy went together. The intoxicant had a sacred significance, not so much because it provided pleasure, but particularly because it was the medium through which contact could be effectuated with the world of the gods."[38]

Drunkenness in a spiritual sense—not the everyday sort of drunkenness that is so distasteful in didactic texts—was a state that was, like the dream, a liminal zone between the world of the "living" and the world of the gods. Assmann describes how songs often reveal basic ideas of the "union of heaven and earth, and the idea of the 'coming', the 'advent' of a god. The image of a union between heaven and earth translates precisely what we described as the suspension of the boundaries between inner and outer, secrecy and publicity, sacred and profane. It is this boundary which marks the normal state of reality."[39] It seems possible that intoxicants could have helped this crossing over the boundary of the "normal" to the liminal zone of dreams and gods.

The persistence of depictions of Hathor in her avatar as a wild cow wandering around in or coming forth from a field of water lilies and papyrus can be interpreted as emphasizing her role as a psychopomp through the liminal zone.[40] I know of only two dream reports surviving from ancient Egypt, and both contain themes of religious ecstasy and Hathor. Both are New Kingdom (Ramesside) and will be examined here in some detail. The famous royal dreams (those of Amenhotep II, Thutmosis IV, and Merneptah) will not be considered here, for they consist of brief passages used as a political device with the agenda of emphasizing the intense personal relationship of the pharaoh with the divine.[41] In contrast, the dreams in the texts below are highlighted, and purport to recount actual events.

Personal Dreams

The first text is a New Kingdom hymn to Hathor by a man named Ipwy, who described on a stele his vision of the goddess (l. 5-8).[42]

ẖr iry hrw ptr=i nfr.wt
wrš ib=i m ḥb iry

---

[36] A. I. Sadek, *Popular Religion in Egypt during the New Kingdom*, ed. A. Eggebrecht, vol. 27, Hildesheimer Ägyptologische Bieträge (Hildesheim: Gerstenberg Verlag, 1987), 214.

[37] F. Daumas, "Les Objets Sacrés de la Déesse Hathor a Dendara," *RdÉ* 22 (1970), 65.

[38] C. J. Bleeker, *Hathor and Thoth: Two key figures of the ancient Egyptian religion*, vol. XXVI, Studies in the History of Religions (Supplements to Numen) (Leiden: E.J. Brill, 1973), 51.

[39] J. Assmann, "Ocular desire in a time of darkness. Urban festivals and divine visibility in Ancient Egypt," in *Ocular Desire*, ed. A. R. E. Agus and J. Assmann, Yearbook for Religious Anthropology (Berlin: Akademie Verlag GmbH, 1994), 20.

[40] Illustrations abound, but for a selection of interesting examples, see L. Keimer, "La vache et le cobra dans les marécages de papyrus de Thèbes," *Bulletin de l'Institut d'Égypte* XXXVII, no. 1: Session 1954-1955 (1956), 215-257. A more recent discussion can be found in G. Pinch, *Votive Offerings to Hathor* (Oxford: Griffith Institute, Ashmolean Museum, 1993).

[41] Szpakowska (2000), 77-93.

[42] The publication of stele Wien 8390 can be found in H. Satzinger, "Zwei Wiener Objekte mit bemerkenswerten Inschriften," in *Mélanges Gamal eddin Mokhtar* (Cairo: Institut français d'archéologie orientale du Caire, 1985), 249-254.

mꜣ=ꞽ nb.t tꜣ.wy m ḳd
    ẖr di=s rš.w m ꞽb=ꞽ
wn.ꞽn=ꞽ <ḥr> wꜣḏ m kꜣ.w=s
    nn ḏd<.t> n=f ḥnr-n=ꞽ <ḥnr-n>=n

"(It was) on the day that I saw <her> goodness
    my heart was spending the day in festival thereof
that I saw the Lady of the Two Lands in a dream
    and she placed joy in my heart.
Then I was revitalized with her food
    without that one would say 'Would that I had, would that we had!'"

That the "Lady of the Two Lands" refers to Hathor is stated explicitly at an earlier point in the text. What is of interest here is the fact that Ipwy did not see the goddess at night, but rather while his "heart was spending the day in festival." Satzinger suggests that the use of *ib* here indicates that Ipwy was not physically at the festival, or a shrine of Hathor, but that he was there in his imagination.[43] One interpretation is that he was feeling the after-effects of festivities or intoxication and that this should read "while my heart was still in festival." Alternatively, Ipwy was physically in the presence of the goddess, as was the protagonist in the other dream report (see below), but that he was not in full control of his faculties, but rather in a state of mystical drunkenness. A mood of ecstasy is punctuated at a later point when he states:

ꞽꜥꞽ.tw tẖ.tw n mꜣꜣ=s

"One is bathed and inebriated by the sight of her."

The very sight of the goddess arouses an ecstasy and religious fervor in the lucky dreamer, and she places this joy directly into his heart, filling his very soul (l. 6-7). Ipwy then reveals that he was "revitalized with her food." This *kꜣ.w* was perhaps metaphoric food; in other words, it was the presence of Hathor, and the joy that she placed literally in his "belly" (*ib*) that nourished the man.[44] Satzinger suggests that the term *kꜣ.w* is used here—as opposed to *ḥtp.w* or *wdn.w*—to refer to actual Hathoric offerings, such as wine,[45] or perhaps the red beer associated with the goddess in the myth of the Destruction of Mankind discussed below. This may indeed be the case as Ipwy is represented on the stele standing before a large bulbous flask with a water lily blossom on top, as well as a bunch of lettuce. Although this is a standard offering scene, it is tempting to read in this iconography more than just a purely symbolic reference to divine offerings or sexuality (lilies, lettuce, and Hathor all have sexual connotations). It is tempting to imagine that the depiction hints at the actual context within which this vision took place. Red beer, wine, and the blue water lily are all associated with Hathor, and could be symbolic of the ritual drunkenness discussed earlier. By entering this intermediary state, Ipwy would have been able to lower the barriers between the mundane world and that of the divine, and ultimately to interact with the goddess.

---

[43] Satzinger (1985), 252 (b).

[44] J. C. Darnell, "Hathor Returns to Medamûd," *SAK* 22 (1995), 50, points out that this pun on *ib* (meaning belly, heart, desire, or wish) and food was not uncommon in a Hathoric context.

[45] Satzinger (1985), 253 (d).

The other known biography of a person other than a king interacting with a deity in a dream is that of Djehutiemhab.[46] In this case, it is more likely that the author had his vision while he was actually in the physical presence of the goddess, or her statue, or at least her sanctuary, when he states (l.9):

rš.wy ḥtp r gs=ṯ
    pꜣ ʿḳ n šwyt=ṯ

"How joyful it is, when the one who enters your shadow
    rests by your side."

It may be that the specific reference to resting by the goddess' side and entering her shadow means that Djehutiemhab was napping or in a stuporous state in the shadow of her statue, lying right next to it. Assmann takes the shadow to be a metaphor for protection,[47] but perhaps it has a literal meaning as well, and, by lying close to the physical earthly representation, a petitioner could hope for direct access to the divine. Djehutiemhab's contact with the goddess was certainly a special event in his life, and it was remarkable, for it included a direct verbal communication from the goddess Hathor.[48] He in turn prays directly to the goddess:

mntṯ i.ḏd n=i rꜣ=ṯ ḏs=ṯ
    ink Ḥnrii nfr.t
      iw iri.w=i [...] n Mwt
    iri=i ii<.t> r mtr=k
    ptr st=k imḥ.tw im=s
      n ḫd n ḫnty
iw=i m ḳd
    iw tꜣ m sgr
      m nfrw grḥ

"You are one who has spoken to me yourself, with your own mouth:
    'I am the beautiful Hely,
      my shape being that of Mut
    I have come in order to instruct you:
    See your place! And fill yourself with it,
      without traveling north, without traveling south.'
while I was in a dream,
    while the earth was in silence,
      in the deep of the night."

Unlike Ipwy's dream, Djehutiemhab's does not take place in the day (l.11-13); instead, his theophany takes place in the deepest darkest time of the night, while he is seemingly alone.

The element of ecstasy in the depth of night (*nfrw grḥ*) is also found in pLeiden I 350, dating to the time of Ramesses III. In the 60th chapter (III,12-III,13)[49] we read:

---

[46] For the publication of this stele, see J. Assmann, "Eine Traumoffenbarung der Göttin Hathor," *RdÉ* 30 (1978), 22-50. A translation can be found in *ÄHG*[1], 372-374, no.172.

[47] Assmann (1978), 31 (o).

[48] For a discussion on the use of dreams as a means of contact with the divine, see K. Szpakowska, "The Road Less Traveled: Dreams and Divine Power in Pharaonic Egypt," in *Prayer, Magic, and the Stars in the Ancient and Late Antique World*, ed. S. Noegel, J. Walker, and B. Wheeler (forthcoming).

[49] A. H. Gardiner (1905), 27-28; *ÄHG*[1] 314-5, no. 134.

ꜥṯ.w tw n=f hrw nỉ ḥb
grḥ sḏr rs m nfr.w grḥ
rn=f pẖr.w ḥr-tp ḥw.t
sꜣy ḥbsy m hrḥ ỉw=f wḫꜣ

"One brews for him on the day of festival:
the night of laying watchful in the depth of the night.
His name circulates upon the temple tops.
Sated(?) is he who sings in the night when it is dark."

Gardiner notes that the term "sated" should be understood in the "metaphorical sense of satisfied." Could this sense of satisfaction be related to Ipwy's feeling of being "bathed and inebriated" by the sight of the goddess? Was the individual in the pLeiden text left satiated and full as the result of a divine vision? This is not dissimilar to Ipwy and Djehutiemhab describing themselves as being nourished by the goddess Hathor. The difference, however, seems to be that this pLeiden text apparently speaks about an experience while being awake. The references to brewing, singing, and lying awake in the depth of the night, all could point to an attempt at entering an altered state of consciousness to help access the divine. The use of the term *wḫꜣ* in the last line, which Bleeker has shown refers to darkness rather than night, may also serve to emphasize the need for darkness in this ritual.[50] It is in the deepest darkest time of night that the supplicant prays to his god hoping perhaps that in the quiet stillness, without the interference of the din of the day, his voice may stand out and reach the ear of his god. Isolation is often used to stimulate a mental state conducive to visions, revelations, and dreams. It can be seen in traditions as diverse as the dream quests of Native American Indians[51] and the dream of Queen Maya (the Buddha's mother).[52] Bleeker notes that even Mohammed's initial revelation occurs in the isolation of the mountain of Mirah.[53] Apparently, a divine voice is best heard when complete silence occurs. In pLeiden (Chapter 90 IV, 6-IV,7)[54] Amun himself emerges from the primeval silence as the great goose:

wpi=f mdw.t m-ẖnw n gr
wn=f ỉri.t nb.w di=f dgꜣy=sn
šꜣꜥ=f sbḥ.w ỉw tꜣ m sgꜣ

"He commenced to speak in the midst of silence.
He opened all eyes, and caused them to behold.
He began to cry aloud while the earth was dumbfounded."

The image here is of a world of utter, complete stillness. The Great Cackler, *Ngg-wr*, breaks the silence for the first time, causing the newly created deities to open their eyes and begin to see, while

50 For terms describing the divisions of the night, see O. Neugebauer and R. A. Parker, *Egyptian Astronomical Texts* (Providence: L. Humphries, 1960), vol. I, 35, and A. Spalinger, "Night into Day," *ZÄS* 119 (1992), 144-156. For a discussion of the meaning of *wḫꜣ* as "darkness," as opposed to "night," see E. Hornung, "Lexikalische Studien I," *ZÄS* 86 (1961), 106-108.

51 See, for example, L. Irwin, *The Dream Seekers: Native American Visionary Traditions of the Great Plains* (Norman and London: University of Oklahoma Press, 1994).

52 S. Young, "Buddhist Dream Experience: The role of interpretation, ritual, and gender," in *Dreams: A Reader on Religious, Cultural, and Psychological Dimensions of Dreaming*, ed. K. Bulkeley (New York: Palgrave, 2001), 9-28.

53 C. J. Bleeker (1963), 175-176.

54 A. H. Gardiner, "Hymns to Amon from a Leiden Papyrus," *ZÄS* 42 (1905), 31-32; *ÄHG*[1], 316, no. 136.

the very earth was in silent astonishment at the sound of his voice. Moreover, it is precisely at this moment of total quiet and calm that one was able to see the primeval world, or a great god.

In the Djehutiemhab text, the time in which the dream occurred ("while the earth was in silence, in the deep of the night") may reflect a deeper significance. The darkest time of night is not usually midnight, but rather that time zone after the moon has set, but before the morning star or the sun have begun their rise.[55] In practical terms, REM activity progressively increases before dawn, and those pre-dawn dreams are the ones easiest to remember. This phenomenon may have encouraged the person in the pLeiden text to choose this dark and lonely time of night to induce a liminal state not unlike the dream-state—but a dream it was not.[56]

Djehutiemhab, after describing the dream that he had while he was dreaming in the deep of the night, remembers next that he woke at dawn, in an ecstatic state (l. 13):

> *ḥḏ tꜣ ib=i ḥꜥ.w tw=i m ršrš*
>
> "At dawn, my heart was exhilarated, I was rejoicing!"

Ipwy too was filled with joy (*rš.w m ib=i*), and indeed this rapturous state, whether instigated by a personal audience, or inspired by the mere thought of her, is one of the hallmarks of the cult of Hathor. When reciting hymns to Hathor, Bleeker notes "the note of optimistic joy" which can be heard in songs describing the singer's experiences at a festival. Enthusiasm surges forth from the song:

> "There comes wine together with the Golden One,
> And fills thy house with joy,
> Live in intoxication day and night without end,
> Be happy and care-free,
> Whilst male and female singers rejoice and dance,
> To prepare for thee a beautiful day."[57]

It is not surprising, nor coincidental, that in the dream reports the named deity was Hathor. Hathor was a goddess who aroused great reverence and gratefulness for her benevolence and mercy. She was a goddess who listened, answered petitions, and discharged guilt, while loving gaiety and promoting festivals.[58] She was a deity who remained accessible to the general public, apparently in dreams as well as in life. This does not imply, however, that Djehutiemhab or Ipwy actively tried to initiate contact with Hathor through dreams. In both cases, the dreams appear to have been spontaneous and unexpected; they do not indicate that any ritual preparations were made to instigate the dreams. Although the timing of Djehutiemhab's dream was meaningful, nothing indicates that he sought isolation to induce it. It is not totally clear whether Ipwy sought ritual drunkenness, but if he did so then holy intoxication would have been the liminal state aimed for—cf. *KRI*, III, 437:4 cited above—and the dream would have been a chance byproduct.

---

[55] For the term *nfrw grḥ*, "deep of night," see Spalinger (1992), 151, and Hanning (1995), 903. In the Destruction of Mankind myth, addressed below, *nfrw grḥ* is the time when Re executes his plan to flood the fields with a red brew without his Eye noticing it, so that the Eye would drink from the liquid at dawn.

[56] The "laying watchful in the depth of the night" in the Leiden text perhaps refers to a night-watch or vigil connected with the mentioned festival. It cannot be fully excluded that it may be an allusion to the "wakeful state" of dreaming, seeing that the etymology of *rsw.t* "dream" was "awakening." But the context of brewing and singing makes it more likely that "he" was awake. For a detailed discussion of the etymology of *rsw.t*, see K. Szpakowska (2000), 23-37.

[57] C. J. Bleeker, *Hathor and Thoth: Two key figures of the ancient Egyptian religion*, vol. XXVI, Studies in the History of Religions (Supplements to Numen) (Leiden: E.J. Brill, 1973), 64.

[58] Bleeker (1973), 83.

Hathoric Festivals

The festival that combined holy intoxication, Hathor worship, and ecstatic joy was the Feast of Drunkenness (*tḫy*).[59] Its roots are in the story of the Destruction of Mankind, in which the Solar Eye, in the form of the goddess Hathor, was sent out by the aging sungod Ra to punish mankind.[60] When the sun-god relented, however, he had to find a way to stop the goddess who was in the throes of blood lust. The solution was to mix breadpaste with red ochre[61] and pour it out on the fields, just before dawn, near the area where she was expected to continue her rampage. The ruse worked, and in the morning the goddess slaked her thirst with what she thought was mangled and crushed human flesh, and she became inebriated and pacified.

A cultic hymn dating from the Third Intermediate Period or the New Kingdom, Pap. Berl. 3014+3053, describes the Feast thus:[62]

> "Her aggression is directed against the rebels,
> her blood-lust against the inhabitants of the desert,
> her blaze against the limbs of the evil ones,
> the enemies of the Lord of Everything.
> Behold, she has returned from […]
> in this her name of Ipethemtes.[63]
> Come and let us perform for her
> (the ritual of) Spending the Night (*sḏr*) in the field,
> let us make it wet for her with the dew of heaven
> when the ostriches come to her
> clapping their wings
> with their chickens behind them.
> O our Mistress, Lady of Buto and Hieraconpolis,
> let us not be pierced by your arrows,
> let us be spared for your […],
> O our Mistress, Lady of Iamu,
> Show your might among us in the field."[64]

---

[59] For an analysis of the calendrical context of the Feast of Drunkenness, see A. Spalinger, "A Chronological Analysis of the Feast of *tḫy*," *SÄK* 20 (1993), 289-303.

[60] The primary text edition of the story is E. Hornung, *Der ägyptische Mythos von der Himmelskuh*, OBO 46, 1982.

[61] See A. Spalinger, "The destruction of mankind: A transitional literary text," *SAK* 28 (2000), 257-282, which adds much to our understanding of the myth. For the material used to pacify the goddess, see also D. Kurth, "Ein Mythos des Neuen Reiches in einer ptolemäischen Ritualszene," in *L'Égyptologie en 1979. Axes prioritaires de recherches* I, 1982, 125-132.

[62] The hymn has been translated (into Dutch) for the first time by J. van Dijk, "Hymnen uit het dagelijks tempelritueel voor de Egyptische godin Mut," in *Schrijvend Verleden: Documenten uit het oude Nabije Oosten, vertaald en toegelicht*, ed. K. R. Veenhof (MEOL 24; Leiden, 1983), 231-246. It belongs to a group of manuscripts that palaeographically belong to the Third Intermediate Period (and pharaoh Takeloth is mentioned in one of them), but that are probably copies from the New Kingdom (Ramesses IX is mentioned in two of them). A fragmentary parallel text was found in the temple of Nekhbet in El Kab, dating to the time of Psammetichus I.

[63] This is a wordplay between the name of the goddess *ỉp.t-ḥm.t=s* ("Her Majesty Ipet") and "the allotted duty (*ip.t*) of the maidservants (*ḥm.w.t*)." For the Destruction of Mankind tale explains that the mixing of the pacifying red solution is to be done by maidservants, the high priest of Re in Heliopolis supplying the red ochre (*didi*) from Yebu (Elephantine). The episode ends by saying about the resulting festival of Hathor: "This is how the making of *sḏr.w.t* (intoxicating nightly beverages) came to be, as the allotted duty (*ip.t*) of the maidservants (*ḥm.w.t*)." See J. van Dijk (1983), p. 241.

[64] Or alternatively: "Sakhmet in the field." The hymn continues with describing Libyan and Nubian dances that are to be performed, for the Eye returns from Nubia or the Libyan desert, and enters Egypt along the Nile. The rest of the hymn deals

Further details of the Hathor Festival can be found at Medamud near Thebes, in a text dating to the Ptolemaic period:[65]

"When the royal children pacify you with what is desired,
the officials consecrate offerings to you.
When the lector exalts you in intoning a hymn,
the magician reads the rituals.
When the organizer praises you with his lotus blooms,
the percussionists take up the tambourine.
The virgins rejoice for you with garlands,
the women with the wreath-crown.
The drunken celebrants drum for you during the cool of the night,
with the result that those who awaken bless you."

The text continues by locating this event to an area just in front of the temple, and then describes in detail the ecstatic cavorting and dancing of humans and animals who accompanied the goddess on her journey back to Egypt.

An inscription in the temple of Mut in Karnak, also dating to the Ptolemaic era, says about the feast:[66]

"A drinking vessel is brought to her in the Aa-meret barge together with Opet, the Great (*'Ipt-wr,t*), on the water.
A cry of joy is made to the Golden One in her feast of *tḫy* in the temple of Mut (of) Karnak."
(...)
"One pours down to the ground water for her in order to quench her anger.
Her heart was cooled..."
(...)
"A regular feast is made for her by Re since the beginning in order to cause her majesty to come out in procession from the place/seat she loves.
When she goes forth in the night, then she is clothed in the *snḏy* garment, completely sanctified (?)."
(...)
"Descending from Nubia and dancing takes place in front of her, Nubians are carrying braziers of silver; Nun is adoring her and Shu worshipping her *ka*—a great feast in the whole land (?)."
(...)
"Beer tinted with *didi* is abundantly poured for her at these occasion(s) of the Valley Feast, it being more precious/sublime (?) than blood, being the work of the beer goddess (=beer) in order to appease her heart in her anger."

Intoxication, singing and dancing, are prominent elements of the festivals, and apparently so was some kind of nightly ritual, an outdoor vigil.[67] It is tempting to assume that the experience described in

with other festivals and rituals for the goddess. For the complex mythology around the "Return of the Solar Eye," see H. Junker, *Die Onurislegende* (Wien, 1917).

[65] J. C. Darnell, "Hathor Returns to Medamûd," *SAK* 22 (1995), 47-94.

[66] A. Spalinger, "A Religious Calendar Year in the Mut Temple at Karnak," *RdÉ* 44 (1993), 161-183. This Ptolemaic text, like the hymn Pap. Berl. 3014+3053, covers in a narrative way all the festivals of the goddess throughout the year. Strictly speaking only the first stanza quoted here is "timed" during the Feast of Drunkenness, but the other stanzas clearly also have the motifs of the return of the Solar Eye and of the pouring of fluid to appease her.

pLeiden I 350, Ch. 60, quoted above, took place during the Feast of Drunkenness ("One brews for him on the day of festival: the night of laying watchful in the depth of the night"). Ipwy's dream may have occurred in the daytime after this Feast ("my heart was spending the day in festival thereof"), but we cannot be certain. It is important to note that nowhere it is implied in the descriptions of the Feast, which are admittedly not overly detailed, that this festival, pre-eminently dealing with (nightly) intoxication, was concerned with inducing dreams (or even visions). The rituals had to do with the anxiety over the changing of the year, about whether the goddess could be "pacified" and her "anger" and her "arrows" averted, and with the ecstatic joy that resulted when she proved to be benevolent and brought nothing but fertility and an abundant inundation to Egypt.[68]

Conclusion

In our effort to understand the past, we must resist the temptation to attribute practices prevalent in modern societies, especially those that are uncommon or unsanctioned and therefore in a certain sense "exotic," onto ancient cultures without carefully sifting through the actual evidence. In an effort to put this into practice, the following summarizes the evidence relating to the use of narcotics or alcohol to induce dreams. The first issue, that of narcotics, is particularly problematic as the evidence for both their presence and conscious use remains controversial and tenuous. It rests largely on modern interpretations of the physical components of the blue water lily, a flower ubiquitous in Egyptian iconography and loaded with symbolism. Some claim that the plant has potent narcotic alkaloids that could have been easily extracted, others claim that it does not contain these alkaloids at all, or that they would have been too difficult to extract. The only connection between the water lily and dreams is its appearance as part of a typical offering scene on a stele that happens to describe a dream.

The second issue is that of alcohol. There seems little doubt that Hathor was linked with festivals of sacred intoxication. Texts such as pLeiden I 350 which mention the brewing of beer on a festival day and remaining awake through the night likely may be referring to one of the goddess's festivals, perhaps one described in the Medamud hymn. Several of these elements recur in the only two non-royal dream reports from the New Kingdom that we have. First is the dream of Ipwy, who experiences his dream of Hathor, perhaps after a festival, and mentions the food of the goddess, which could refer to beer or wine. The second was Djehutiemhab, who experienced a theophanous dream of Hathor in the still of the night. The setting and contexts of the dreams, however, are not described in enough detail to conclude that they occurred in similar circumstances, or indeed in any ritual context. If participation in a festival or drinking alcohol played a role in these dreams, the connection was likely accidental. There is no indication that these dreams were initiated, sought, or incubated in any way. Indeed the Egyptians did not practice dream incubation (the custom of inducing dreams by sleeping in a temple or other holy place) until the Ptolemaic period, and even then there is no evidence that the practitioners imbibed anything to initiate the dream.

The evidence suggests that neither narcotics nor alcohol played a significant role in inducing dreams in ancient Egypt. The evidence for non-medicinal drug use is slim at best. Alcohol, on the other hand, was a part of everyday life and likely was used to induce a state of holy intoxication during certain rituals, particularly those dedicated to Hathor. However, dreams are not mentioned as playing any role in these rituals. Hathor is featured in dream reports, but the dreamers seem to have been astonished and amazed by their divine visions. These dreams were awe-inspiring in part because they were

---

[67] The Pap. Berlin speaks of "*sḏr* in the field"; *sḏr* means "to spend the night," either in sleeping or in doing something else, cf. Hannig (1995), 795. The Medamud text implies that "during the cool of the night" drums were used to awaken all who had fallen asleep.

[68] The pouring of red fluid on the fields, of course, is intended to bring about the inundation, when the Nile water, red with fertile Ethiopian mud, will flood the fields. See J. van Dijk (1983), pp. 241-242.

spontaneous and unexpected. Thus, rather than having been initiated by any human ritual, they were thought to be initiated by the goddess herself within the milieu of a dream, itself a liminal state. She specifically selected these individuals to receive her divine communication—an honor that had much greater impact because it was not solicited by the dreamer, but was rather a gift from god.

Dr. Kasia Szpakowska
Department of Classics & Ancient History
University of Wales Swansea

Weltende

Leslie E. Bailey

Introduction

When studying the cosmogonies of ancient Egypt, a logical question to pose is whether the inhabitants of the Nile valley had a vision for how the world would end. Based on their creation mythologies, it seems inherent that a return to chaos was possible since the forces of destruction were never fully eradicated during the initial conception of the world. The world was created out of the primeval abyss known as *Nun*, which was neither a void nor nothingness, but a limitless entity of water representing disorder. Therefore, chaos preceded the created world since it was not part of the creation process. Moreover, it remained juxtaposed to the world, never to be destroyed, because what was not created could not be destroyed. Periodically, the forces of chaos, as manifested in such things as foreign invasions and natural catastrophes, could impinge upon the created world of Egypt. In order to maintain balance between existence and non-existence, *Maat* had to be upheld by the king to ensure order.

Creation and Chaos

The theory of creation with the potentiality of the world devolving back into chaos represented itself as a distinct possibility for the Egyptian mind. The extent of how probable it was that chaos threatened Egypt can be seen diachronically with the First Intermediate Period's decentralization, the Second Intermediate Period's foreign invasions, and then the Third Intermediate Period's combination of both pernicious events. Synchronically, at the death of each pharaoh and until the coronation of the succeeding pharaoh, the Egyptians believed that chaos enveloped the land until the new pharaoh could reinstate *Maat*.

Since creation not only allowed physical entities including the gods to come into existence, it also produced abstract properties such as time and space. With the primal waters driven out of Egypt, anything outside the perimeter of Egypt was considered chaotic, such as the desert, foreigners, and the underworld. Some gods were given the epithet *nb r ḏr*, idiomatically translated as "Lord of All." The word *ḏr* may also be used in the expression *r ḏr=f* meaning "entire" or literally "to its end or limit" since *ḏrw* is defined as "end," "limit," or "boundary." In view of the fact that some gods had a life span, "Lord of All" seems like a misnomer for "Lord to the End or Limit."[1] The word "all" encompasses everything that exists as well as what does not exist [*nt(y) ỉwt(y)t*], indicating a limitless amount (positive and negative infinity), whereas if *ḏr* is translated more closely to its root designation a more mathematically discrete function arises. Furthermore, Erik Hornung interprets an inscription *m ḏrw nṯrw* found on Greco-Roman temples, usually translated as "in the realm of the gods," as meaning "so long as the gods are there."[2]

If space was a created and finite entity, then by analogy time, as a created entity, would be finite as well. The ancient Egyptians conceptualized two notions of time: *ḏt* and *nḥḥ*. Many scholars view *ḏt* as referring to linear time, and *nḥḥ* as referring to cyclical time. Outside of the created world, the two

[1] Erik Hornung, *Conceptions of God in Ancient Egypt: The One and the Many*, translated by John Baines (Ithaca: Cornell University Press, 1996), 235.
[2] Hornung, *Conceptions of God in Ancient Egypt*, 163.

concepts did not apply.[3] If assessed as mutually exclusive concepts, *nḥḥ* as circular time might not permit a termination point as long as perpetuity could be sustained; however, *ḏt* as linear time would allow for a continuum with two endpoints. The theory of the periodization of history is attributed to Greco-Roman influences on Egypt.[4] However, the presence of king lists in ancient Egypt, such as the Palermo Stone, delineating explicit intervals of king's reigns and containing specific unique events, documents that the Egyptians had a comprehension of, and belief in, linear time and its periodization, even though the theory of kingship employed the cyclical usage of the Osiris-Horus myth. Since the ancient Egyptians were fond of dualisms, it is more probable that *nḥḥ* and *ḏt* coexisted and supplemented one another. The combination of both entities then would represent the "sum of all conceivable units of time," and thus be equal to the "entire supply of time available to the world."[5]

Erik Hornung sums it up like this: "Since the created world is bounded and ordered in time and space, it follows that it has an end and must disappear; it is an island or an 'episode'…'between nothingness and nothingness.' It has duration, but there is no such thing as eternal existence, which would be a contradiction in terms."[6]

The act of creation had separated the demiurge from the primal waters from which he emerged. Subsequently, the other gods were brought into existence from him along with the rest of the world. Thus, the essential quality of the creator god was that at first he was one and then he is many. By the New Kingdom, the primeval god attained the epithet "the one who made himself into millions." Under both conceptions of *nḥḥ* and *ḏt*, the one who became millions could theoretically become one again, reabsorbing the millions that he had created.

One myth has Re, the sun god, being born on an island known as *iw nsrsr*, or the "Island of Flames." Significantly, a flame could erupt from this island, incinerating the earth.[7] As part of the solar theology, Re died every evening at sunset and was reborn every morning at sunrise. However, his circuit was continually threatened by his nemesis Apopis the serpent, who symbolized the forces of chaos.

Apopis was not of the created world since he was an antecedent of creation who was defeated during the initial act of creation known as the First Time, and then conquered every dawn by Re. Apopis is therefore another example of something that was not created initially being indestructible. In the later periods of Egyptian history, an "Apopis Book" was fabricated to aid Re in his daily struggle with the evil serpent. The rituals in the Bremner-Rhind Papyrus would ensure that the sun would rise and creation would be preserved. The existence of this book indicates that the threat to the fragility of creation was important enough for the Egyptians that they chose to document it. With the peril of Apopis always looming on the horizon, the sun was perceived as being vulnerable. Conversely, the circumpolar stars, *iḫmw skw*, who never descended below the horizon, were the imperishable or indestructible ones.

The Notion of *Weltende* in the Dynastic Period

In the religious literature of the Old through New Kingdoms, the notion of the world ending occasionally appears. In the Pyramid Texts, Utterance 254, the dead king threatens:

---

[3] Erik Hornung, *Idea into Image: Essays on Ancient Egyptian Thought*, translated by Elizabeth Bredeck (Princeton: Timken Publishers, 1992), 65.

[4] John J. Collins, *The Apocalyptic Imagination: An Introduction to Jewish Apocalyptic Literature*, second edition (Grand Rapids, Michigan: William B. Eerdmans Publishing Company, 1998), 33.

[5] Hornung, *Idea into Image*, 65-69.

[6] Hornung, *Conceptions of God in Ancient Egypt*, 183.

[7] Brunner, Hellmut, "Weltende," in *Lexikon der Ägyptologie*, Vol. 6, edited by Wolfgang Helck and Eberhard Otto (Wiesbaden: Otto Harrassowitz, 1986), 1214.

> "O Lord of the horizon, make ready a place for me, for if you fail to make ready a place for me, I will lay a curse on my father Geb, and the earth will speak no more, Geb will be unable to protect himself, and whoever I find in my way, I will devour him piecemeal. The *ḥnt*-pelican will prophesy, the *psḏt*-pelican will go up, the Great One will arise, the Ennead will speak, the earth being entirely dammed up; the borders will be joined together, the river-banks will unite, the roads will be impassable to travellers, the slopes will be destroyed for those who would go up."[8]

From this utterance, one can see the forces of chaos demolishing the world with a natural catastrophe brought about by the will of the divinity. Concerning the pelicans, Dimitri Meeks interprets this passage as the pelican who "liv[ed] on the confines of the universe...knew what would one day come to pass: the extremities of the creation would be reunited, or folded back one upon the other, until the created space was abolished and no avenue of escape was left."[9] The reference to the solar pelicans is meaningful as well, in that these birds were considered protective symbols against snakes.

A recurring theme interspersed within Egyptian texts is that of the sky falling down and crashing to earth. Pyramid Text Utterance 255 introduces the notion of pillars supporting the sky accompanied by a threat from the deified dead king to end the world by their removal. The text reads: "I will smite away the arms of Shu which support the sky and I will thrust my shoulder into that rampart on which you lean."[10] By the Middle Kingdom, the Coffin Texts personify *nḥḥ* and *ḏt* as Shu and Tefnut respectively as upholding the pillars of heaven and ensuring the existence of the created world.[11] In the post-Amarna period, Shu and Tefnut, as living pillars of heaven, appear in the Book of the Celestial Cow. In addition, the imagery of the threat of world destruction remains as one reads in The Contendings of Horus and Seth in the Chester Beatty Papyrus 1:

> "Thereupon Neith, the mighty, the god's mother, sent a letter to the Ennead, saying: Give the office of Osiris to his son Horus, and do not do those great acts of wickedness which are not in their place, else I shall be angry, and the heaven shall crash to the ground."[12]

By the Middle Kingdom, the breakdown of central authority in the First Intermediate Period was reflected in many pessimistic writings. The writings of Neferti, Khakheperre-Sonb, and Ipuwer describe conditions of national distress that bordered on chaos, so much so that the people of that time probably felt that the world was ending. Ipuwer writes: "If only this were the end of man, no more conceiving, no more births! Then the land would cease to shout, tumult would be no more!"[13] As it was pharaoh's duty to maintain order, Ipuwer blames the king for Egypt's despair. Neferti declares: "What was made has been unmade, Re should begin to recreate!"[14] In both Ipuwer and Neferti's writings, a prediction or prophecy is made in which a king will come and restore Egypt to order.

---

[8] R. O. Faulkner, *The Ancient Egyptian Pyramid Texts* (Warminster, Aris and Phillips Ltd., 1969), 63.
[9] Dimitri and Christine Favard Meeks, *Daily Life of the Egyptian Gods*, translated by G. M. Goshgarian (Ithaca: Cornell University Press, 1996), 17-18.
[10] Faulkner, *The Ancient Egyptian Pyramid Texts*, 66.
[11] Hornung, *Idea into Image*, 69.
[12] Alan H. Gardiner, *The Library of A. Chester Beatty: Description of a Hieratic Papyrus with a Mythological Story, Love-Songs, and other Miscellaneous Texts* (London: Oxford University Press, 1931), 15.
[13] Miriam Lichtheim, *Ancient Egyptian Literature: Vol. 1 The Old and Middle Kingdom* (Berkeley: University of California Press, 1975), 154.
[14] Lichtheim, *Ancient Egyptian Literature: Vol. 1*, 141.

The Coffin Texts of the Middle Kingdom allude to an end of the world scenario as well. An excerpt from Spell 1130 has Atum speaking as follows:

> "I have passed myriads of years between myself and yonder Inert One, the son of Geb [i.e. Osiris]; I will sit with him in one place, and mounds will be towns and towns will be mounds; mansion will desolate mansion."[15]

It appears from this text that two gods now will survive an eschatological scenario. As the prior ideas of world destruction seemed more final, Osiris and his regenerative powers were now added to incorporate the notion of a rebirth of creation. Perhaps after Egypt experienced firsthand the disorder of the First Intermediate Period, hoping for a saviour king to come and restore things to their right order, this hope was extended into the religious literature allowing for a revival of creation. When Atum threatens the world with annihilation again in the New Kingdom's Book of the Dead, Chapter 175, Osiris is still included among the survivors:

> "You shall be for millions on millions of years, a lifetime of millions of years. I will dispatch the Elders and destroy all that I have made; the earth shall return to the Primordial Water, to the surging flood, as in its original state. But I will remain with Osiris, I will transform myself into something else, namely a serpent, without men knowing or gods seeing. How good is what I have done for Osiris, even more than for all the gods! I have given him the desert, and his son Horus is heir on his throne which is in the Island of Fire; I have made what appertains to his place in the Bark of Millions of Years, and Horus is firm on his throne in order to found his establishments."[16]

A similar text to Chapter 175 appears in the Opet Temple at Karnak during the Ptolemaic Period: "There is no god, there is no goddess, who will make himself/herself into another snake."[17] Therefore, only Atum and Osiris had the ability to transform into serpents. Snakes, like amphibians, were seen as mysterious, primordial, and chthonic creatures that came forth from the mud, and godheads who first appeared in (and rose from) the watery Nun were often given the form of a snake, like the female members of the Hermopolite Ogdoade with their snake heads (the male members were frog-headed), or the creator Amun in his form of the snake Kematef.[18]

Magical texts as well evoked an image and sometimes a threat of ending the world. Papyrus Salt 825 (BM 10.051) is as follows:

> "It did not become day…gods and goddesses put their hands on their heads. The earth…The sun did not rise. The moon failed to appear…The sky fell into darkness…The earth had turned upside down, the water had changed. It did not move downstream…all of the world lamented and cried, the souls…gods and goddesses, people, ghosts and the dead, livestock and herds."[19]

---

[15] R. O. Faulkner, *The Ancient Egyptian Coffin Texts: Vol. 3 Spells 788-1185 and Indexes* (Warminster, Aris and Phillips Ltd., 1978), 168.

[16] R. O. Faulkner, *The Egyptian Book of the Dead: The Book of Going Forth by Day* (ed. Eva von Dassow, with additional translations and a commentary by Ogden Goelet) (San Francisco: Chronicle Books, 1994).

[17] Hornung, *Conceptions of God in Ancient Egypt*, 163-164.

[18] Cf. Hans Bonnett, *Reallexikon der ägyptischen Religionsgeschichte* (Berlin/New York: Walter de Gruyter, 2000), 6, 72, 373-374, 682, 729.

[19] Jan Bergman, "Introductory Remarks on Apocalypticism in Egypt," in *Apocalypticism in the Mediterranean World and the Near East*, edited by David Hellholm, second edition (Tübingen: J. C. B. Mohr [Paul Siebeck], 1989), 56. My translation of "Es wurde nicht Tag…Götter und Göttinnen legten ihre Hände auf ihre Köpfe. Die Erde…Die Sonne ging

In Papyrus Harris, a magician also refers to the termination of the world:

> "If the one on the water (the crocodile) opens his mouth,
> If he shakes with his two arms,
> I shall cause the earth to go down into the primal water.
> And the south will be the north
> And the earth will turn round."[20]

Scholars view these sources as exaggerated, egotistical, and sometimes intended for aggressive purposes, particularly in love charms. An example is: "If you refuse to listen and do not swiftly do what I tell you, the sun will not set beneath the earth, neither will Hades nor the Cosmos continue to exist."[21] The thought of the end of the world could also appear to a person in dreams, which were considered as a medium for revelations. In the Archive of *Ḥor*, *Ḥor* dreams that "a great destruction" (*sḳ ꜥ3*) will occur with Thoth appraising that "No province shall remain among them in any way on earth."[22] However, in this dream, known as the Little Apocalypse, *Ḥor* was asked to select a town that would survive the holocaust, and he requested his hometown of Pi-Thoth.

Apocalyptic Thinking in the Ptolemaic and Roman Periods

Beginning with Alexander the Great's conquests, apocalyptic texts seem to have flourished throughout the Near East. They may be viewed as a natural reaction to times of despair and persecution resulting from a loss of native rule, which continued into the Roman era. "The sense of present alienation from the present order is fundamental to many apocalypses, especially the historical type."[23] Apocalypses may also be seen as a form of protest literature that condemns the conquerors and encourages a hope that the existing order will be overthrown reverting to the days of yore.[24] Harking back to ideas of a saviour king restoring order, as seen in the Middle Kingdom complaint literature of Ipuwer and Neferti, ancient Egypt employed the genre of *Königsnovelle*, in which a "central or frame narrative concerns a legendary pharaoh encountering and resolving some catastrophe....The *Königsnovelle* thus gave the archaic oracles of woe a revealed or apocalyptic aspect: predictions of future woes to past kings as occurring after their reigns would implicitly apply to the current situation."[25]

Moreover, a pattern develops in historical apocalyptic texts: "foreign conquest of the land, termination of true worship, defeat of the foreign rulers by a native king, and restoration of proper order."[26] In particular, four documents survive from Egypt that were written in this style: the Demotic Chronicle, the Prophesy of the Lamb, the Oracle of the Potter, and the Perfect Discourse.

---

nicht auf. Der Mond blieb aus...Der Himmel fiel im Finstern...Umgestürzt war die Erde, umgewandt das Wasser. Es zog nicht stromab...Alle Welt klagte und weinte, die Seelen...Götter und Göttinnen, Menschen, Geister und Tote, Vieh und Herden..."

[20] Hornung, *Conceptions of God in Ancient Egypt*, 165.

[21] Hornung, *Conceptions of God in Ancient Egypt*, 210, n. 50.

[22] J. Gwyn Griffiths, "Apocalyptic in the Hellenistic Era," in *Apocalypticism in the Mediterranean World and the Near East*, edited by Hellholm, second edition, 284.

[23] Collins, *The Apocalyptic Imagination*, 24.

[24] James C. Vanderkam, "Prophecy and Apocalyptics in the Ancient Near East," in *Civilizations of the Ancient Near East,* Vol. 3, edited by Jack Sasson (New York: Charles Scribner's Sons, 1995), 2087.

[25] David Frankfurter, "The Legacy of Jewish Apocalypses in Early Christianity: Regional Trajectories," in *The Jewish Apocalyptic Heritage in Early Christianity* (Minneapolis: Fortress Press, 1996), 147.

[26] Vanderkam, "Prophecy and Apocalyptics in the Ancient Near East," 2087.

The Demotic Chronicle is an eschatological prophecy pertaining to the history of Egypt in the Persian and Greek periods but written after the fact in the third century B.C.E.

> "It is Harasaphes that will give orders to the ruler after the foreigners and the Ionians. Be joyful, thou prophet of Harasaphes! The prophet of Harasaphes will proceed in joy (in the time) after the Ionians, for a ruler shall have arisen in Heracleopolis."[27]

In addition, it is believed by scholars that Alexander the Great is referred to in this text as the "Great Dog." This might be a symbolization of Seth, who was affiliated with the desert and foreign lands, and was considered a force of chaos during this time.

The Prophesy of the Lamb is another *ex post facto* text written in Ptolemaic times, and it refers to an incident believed to have transpired during the reign of Bocchoris of the Twenty-fourth Dynasty. A lamb prophesizes to the king that adverse events await Egypt's cities: "There is much malediction against [Egypt]. Heliopolis weeps in the East, for it is vanquished… Bubastis weeps…; one makes the streets of Sebennytus into a vineyard."[28] After the predicted 900 years of anarchy with the gods' shrines being taken away to Nineveh (a reference to the Assyrian conquest of Egypt in 671-664 B.C.E.), order will be restored to Egypt. The lamb is interpreted to be an aspect of the ram god Khnum, a demiurge who brought beings into existence by forming them on his potter's wheel.

Another text, the Apology of the Potter to King Amenhotep, was written in 130 B.C.E. but is said to have taken place during the Eighteenth Dynasty. Egypt's demise is again foretold. A potter moves to Helios-Re (*iw nsrsr*) to practice his craft; however, the residents view this as a sacrilegious act, and they destroy his pottery and take him to see the king. To defend himself, the potter reads the smashing of his pottery as a prophetic sign and prognosticates:

> "The Nile will flow not having sufficient water, but only a little, so that the earth is scorched, but contrary to nature. For in the time of the Typhonians they will say: Poor Egypt, thou art abused by the terrible abuses wrought against thee. And the sun will be darkened, not wishing to behold the evils in Egypt. The earth will not respond to the sowing of the seeds…But that will happen at the end of the evils, when the foreigners fall away from Egypt as the leaves fall from a tree. And the city of the girdle-wearers will be made a desert, in the same way as my kiln experiences, because of the lawless deeds, which they did to Egypt. The divine statues of Egypt which had been carried away there, come home to Egypt, and the city by the sea will become a place for fishermen to dry their nets because Agathos Daemon and Mephis proceed to Memphis so that people passing will say 'So this was the all-nurturing city in which every race of men lived.'"[29]

Clearly disgruntled by Ptolemaic rule with its girdle-wearing Greeks and administration from Alexandria, the potter's vision of the future, as with the lamb's, epitomizes the *Königsnovelle* genre. Significantly, after both manifestations of Khnum issued their auguries, they promptly died and were given honourable burials by their respective kings emphasizing the texts' propagandistic value.

Egypt's discontentment with Greek and then Roman rule and the consequential cultural oppression culminates in a text from the fourth century known as The Perfect Discourse or the Apocalypse of Aesclepius.

[27] Griffiths, "Apocalyptic in the Hellenistic Era," 280.
[28] Griffiths, "Apocalyptic in the Hellenistic Era," 285.
[29] Griffiths, "Apocalyptic in the Hellenistic Era," 288.

> "A time will come when it will be seen that in vain have the Egyptians served the deity with piety and assiduous service, and all their holy worship will be found fruitless and to no profit. For the deity will retire from earth to heaven, and Egypt will be forsaken; and the land which was the home of religion will be left desolated, bereft of the presence of the gods."[30]

In this passage, Hermes Trismegistus, the amalgamation of Thoth and Hermes, envisaged a "final and definitive *Götterdämmerung*"[31] for Egypt. The "twilight of the gods" view is elaborated more fully in succeeding excerpts, which delineate such atrocities as death becoming more profitable than life, as the pious being deemed insane, and as wickedness being perceived as good. Accordingly, *Weltende* will ensue when

> "the earth no longer stands unshaken, and the sea will no longer be navigable; heaven will not support the stars in their orbits, nor will the stars pursue their constant courses in heaven; the voice of the gods will of necessity be silenced and dumb; the fruits of the earth will rot; the soil will turn barren, and the very air will sicken in sullen stagnation."[32]

However, as with the other texts of this genre, a glimmer of hope appears because it is prophesied that the god will recreate.

Summary and Conclusions

In sum, it can be seen that the Egyptians had several beliefs of how the world might end. Egypt's extinction could be brought about by fire, by the sky crashing to earth, or by the Nile's banks converging; all of these natural calamities deriving from divine volition. Texts that consider the ending of the world appear irregularly in ancient Egypt. It seems natural that little emphasis should be placed upon thinking about the unthinkable. However, the texts in this paper indicate that the Egyptians did realize the possibility of their existence being terminated, enough at least to speculate about it to a limited degree. Furthermore, the notion of the gods deserting their created realm of Earth would have challenged the existing order. In a majority of later Egyptian texts, the hope of a new creation beginning again after *Weltende* supplanted the writings of an absolute finality of the world.

In comparison to contemporary societies in the ancient Near East, Egypt did not endure as many natural cataclysmic events or foreign encroachments as the civilizations in the Levant and Mesopotamia. Thus, one would expect more prophets of doom and eschatological utterances deriving from these areas than from Egypt. With this in mind, many scholars have considered Egypt to be relatively static and optimistic in comparison to the rest of the ancient Near East. Interestingly, the end of the world only pertained to Egypt since whatever was outside of Egypt was seen as chaotic and not part of the created or ordered world. Therefore, this Egyptocentric view did not envision a worldwide annihilation, unlike apocalyptic writings in other cultures.

Leslie E. Bailey
Pompton Lakes, NJ
U.S.A.

---

[30] Griffiths, "Apocalyptic in the Hellenistic Era," 291.
[31] Garth Fowden, *The Egyptian Hermes: A Historical Approach to the Late Pagan Mind* (New York: Cambridge University Press, 1986), 40.
[32] Fowden, *The Egyptian Hermes*, 41.

CPSIA information can be obtained
at www.ICGtesting.com
Printed in the USA
FSOW04n0517180716
22837FS

9 781581 125641